Fundamentals of Fluorescence Imaging

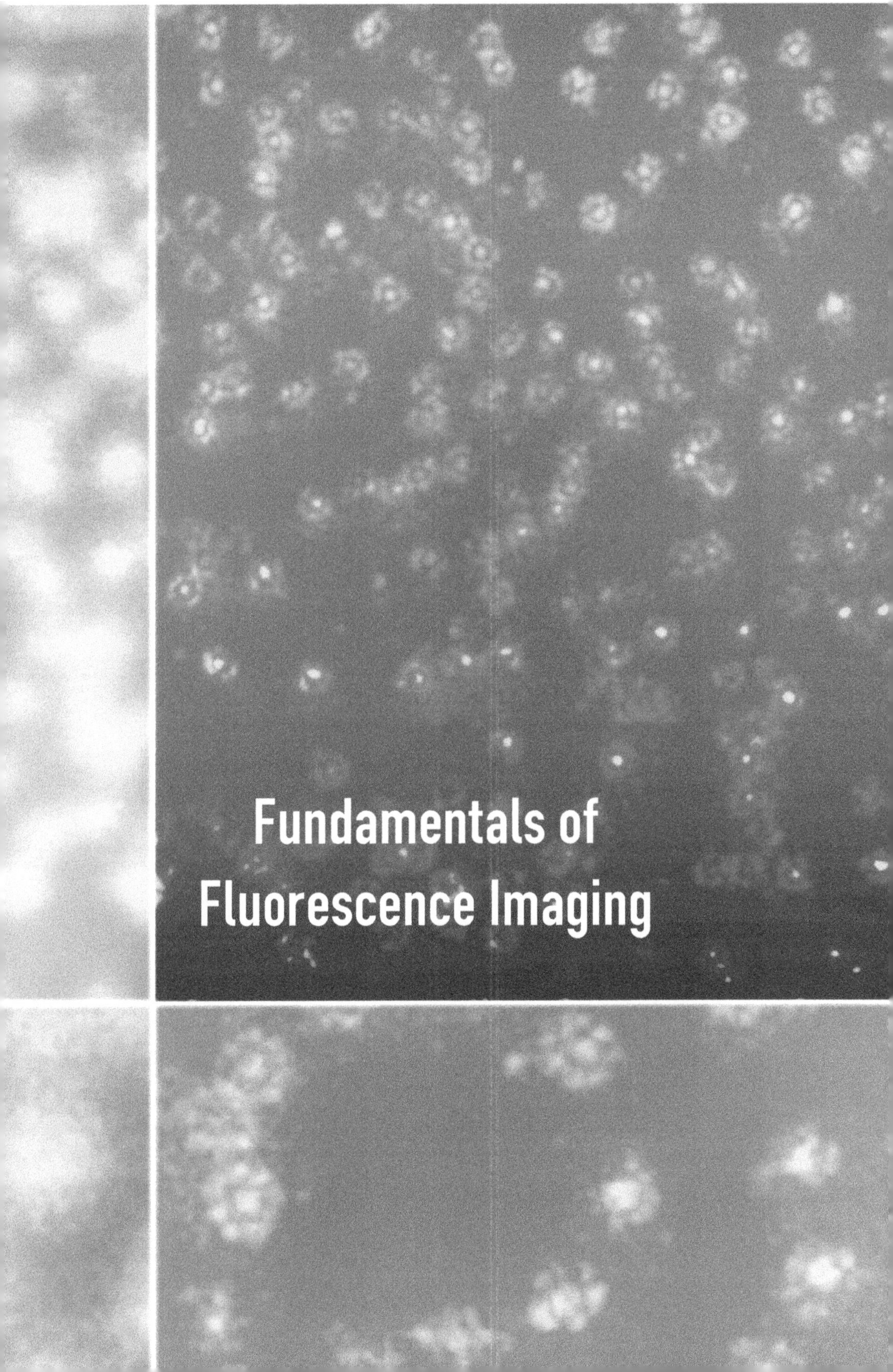

Fundamentals of Fluorescence Imaging

edited by

Guy Cox

JENNY STANFORD
PUBLISHING

Published by

Jenny Stanford Publishing Pte. Ltd.
Level 34, Centennial Tower
3 Temasek Avenue
Singapore 039190

Email: editorial@jennystanford.com
Web: www.jennystanford.com

British Library Cataloguing-in-Publication Data
A catalogue record for this book is available from the British Library.

Fundamentals of Fluorescence Imaging

ISBN 978-981-4774-85-7 (Hardcover)
ISBN 978-1-351-12940-4 (eBook)

In Memoriam

Professor Robert Clegg, brilliant researcher and inspiring teacher, was the senior author of Chapter 10. Tragically, he died during the preparation of this book. On his deathbed he gave instructions to his son as to where to find the files of his incomplete chapter and to send them to his coauthor, Chittanon Buranachai, for completion. This was duly done, and the result is a superb chapter. So I dedicate this book to Bob Clegg's memory.

G.C.

Contents

Preface xvii

1 Basic Principles of Fluorescence **1**
Scott Kable
1.1 Introduction 1
1.2 Basics of Fluorescence 4
1.2.1 Properties of Light 5
1.2.2 Energy States of Molecules and Jablonski Diagrams 7
1.2.3 Absorption of Light 10
1.2.4 Fluorescence and Phosphorescence 13
1.2.5 Photophysics and Photochemistry 17
1.2.6 Fluorescence Lifetimes, Quantum Yields 20
1.2.7 Quenching 21
1.2.8 Multiphoton Fluorescence 22
1.3 Linking to Microscopy 25
1.3.1 Autofluorescence 25
1.3.2 Fluorescence Labels 25
1.3.3 Photobleaching and FRAP 27
1.3.4 Laser Confocal Microscopy 28
1.3.5 Multiphoton Microscopy 29
1.3.6 Fluorescence Resonance Energy Transfer (FRET) 30
1.3.7 Fluorescence Lifetime Imaging (FLIM) 30
1.4 Summary and a Look to the Future 31

2 The Ploem Fluorescence Microscope **35**
Guy Cox
2.1 Introduction: Fluorescence Microscopes 35
2.2 The Ploem System 36

2.3 Filters and Mirrors 38
2.4 Light Sources 40
2.4.1 Arc Lamps 41
2.4.2 Metal Halide Lamps 43
2.4.3 LED Illuminators 44
2.5 The Complete Microscope 45

3 Confocal Microscopy **49**
Stephen H. Cody and Guy Cox
3.1 Introduction 49
3.2 The Confocal Principle 50
3.3 The Epi-fluorescence Confocal Microscope 52
3.4 Optimizing Confocal Performance 54
3.4.1 The Point Spread Function 54
3.4.2 Aberrations 56
3.4.2.1 Chromatic aberration 56
3.4.2.2 Spherical aberration 57
3.5 High-Speed Confocal Microscopy 59

4 Multiphoton Microscopy **63**
Mark B. Cannell and Guy Cox
4.1 Introduction 64
4.2 Multiphoton Excitation 66
4.2.1 Scattering Samples 69
4.3 Two-Photon Flash Photolysis 73
4.4 Practicalities and Advantages of MPM 75

5 Immunofluorescence **81**
Anthony F. Henwood
5.1 Antibodies as Histochemical Markers 81
5.1.1 Endogenous Fc Receptors 86
5.1.2 Antibody Validation 87
5.2 Fixation 89
5.3 Frozen Sections (Cryotomy) 90
5.4 Cell Cultures 91
5.5 Formalin-Fixed Paraffin-Embedded (FFPE) Sections 91
5.6 Immunofluorescence 92

5.6.1 The Direct Immunofluorescent Technique (DIF) 93
5.6.1.1 Renal immunofluorescence 94
5.6.1.2 Skin immunofluorescence 96
5.6.2 Indirect Immunofluorescence (IIF) 99
5.6.2.1 Autoimmune sera testing 100
5.6.3 Multi-antigen Immunofluorescence 103
5.6.4 Photobleaching 105
5.6.5 Autofluorescence 106
5.6.5.1 Autofluorescence in formalin-fixed sections 107
5.6.5.2 Formalin-induced fluorescence 108
Appendix A. Direct Immunofluorescence 114
Appendix B. Procedures to Reduce Autofluorescence 116

6 Fluorescent Proteins **117**
Anya Salih
6.1 Introduction 117
6.2 Historical Perspectives 119
6.2.1 Early Discoveries 119
6.2.2 FPs from Nonbioluminescent Organisms 122
6.2.3 The Green Glow of the Nobel Prize 124
6.3 General Properties of Fluorescent Proteins 125
6.3.1 General Structure 126
6.3.1.1 Blue and cyan FPs 127
6.3.1.2 Green FPs 128
6.3.1.3 Yellow FPs 129
6.3.1.4 Red FPs 130
6.3.2 Photoactivatable and Photoconvertible FPs 131

7 Dyes for Labeling Organelles and Compartments **139**
Iain Johnson
7.1 Labeling Protocols 141
7.2 Nucleus 144
7.2.1 Hoechst 33258, Hoechst 33342, and DAPI 144
7.2.2 Long-Wavelength Dyes 145
7.3 Cytoplasm 146
7.4 Cytoskeleton 146

7.5 Mitochondria 147
7.6 Intracellular Vesicles 149
7.6.1 Endosomes 149
7.6.2 Lysosomes 152
7.7 Endoplasmic Reticulum and Golgi 153
7.8 Plasma Membrane 155

8 Indicator and Reporter Dyes 163
Iain Johnson
8.1 Calcium and Other Cations 164
8.2 pH Indicators 169
8.3 Membrane Potential Sensors 171
8.4 Reactive Oxygen Species (ROS) 174
8.5 Enzyme Activity Indicators 175
8.6 Perspective 178

9 New Approaches to Cancer Therapy Using in vivo Fluorescent-Protein Imaging 185
Robert M. Hoffman
9.1 The in vivo GFP Revolution 186
9.1.1 Properties of GFP for in vivo Cellular Imaging 186
9.2 Fluorescence-Guided Surgery 190
9.3 Bacterial Therapy of Cancer 192
9.4 Tumor Grafts from Patients Made Imageable in Fluorescent-Protein Expressing Transgenic Nude Mice 193
9.5 Stroma Therapy 196
9.6 Fluorescence-Guided UV Therapy 197
9.7 Conclusion 197

10 Fluorescence Lifetime Imaging in Living Cells 209
Chittanon Buranachai, John P. Eichorst, Kei Wei Teng, and Robert M. Clegg
10.1 Introduction 210
10.2 Background 212
10.2.1 What Is the Meaning of the Fluorescence Lifetime? 212

10.2.2 The FLI Measurement: Basics of the Time Domain and Frequency Domain 213
10.2.3 Time Domain 215
10.2.4 Frequency Domain 217
10.2.5 The Frequency Domain Heterodyne/ Homodyne Measurement 219
10.3 Materials and Methods 221
10.3.1 FLI Instrumentation Setups and Data Acquisition 221
10.3.2 Light Source and Modulator 222
10.3.3 Microscope 224
10.3.4 Detectors 225
10.3.5 An Improvement in the Wide-Field Frequency Domain: The Implementation of the Spinning Disk Microscope to Provide Confocal FLI Images 227
10.4 Data Analysis 227
10.4.1 Time Domain 228
10.4.2 Frequency Domain 229
10.4.3 Rapid, Informative Analysis and Display Methods are Needed for FLI Data: The Polar Plot Analysis in the Frequency Domain 230
10.4.4 Phase Suppression 234
10.4.5 Spectral FLIM 235
10.5 FLI Studies on Biological Samples 236
10.5.1 Fluorescent Proteins in Transgenic Cells and FLI 236
10.5.2 FLI-Based FRET Imaging 237
10.5.3 FLI Applications with Environmental Probes and Sensors 238
10.5.4 Contrast Enhancement and Simultaneous Detection 239
10.5.5 The Advantage of Confocal Measurements 240
10.5.6 Phase Suppression of Biosensor for Locating Activity of MT1-MMP 240
10.5.7 Application of Polar Plot and FLIM with the Photosenstizer PPIX with ALA 242

11 Advanced Fluorescence Techniques: FRAP, iFRAP, FLIP, FLAP, Photoconvertible, Photoactivatable, and Photoswitchable Proteins **259**
Kelly Rogers and Sarah Ellis
11.1 Introduction 259
11.2 Fluorescence Recovery after Photobleaching (FRAP) 262
11.3 Fluorescence Loss in Photobleaching (FLIP) 267
11.4 Fluorescence Localization after Photobleaching (FLAP) 269
11.5 Potential Pitfalls in FRAP, FLIP, and FLAP Experiments 271
11.6 Photoactivable, Photoconvertible, and Photoswitchable Probes 272

12 Förster Resonance Energy Transfer Microscopy for Monitoring Molecular Dynamics in Living Cells **279**
Vinod Jyothikumar, Yuansheng Sun, and Ammasi Periasamy
12.1 Introductory Physics of FRET 280
12.1.1 Basics of FRET Microscopy 281
12.2 Overview on Microscope Setup and Fluorescent Imaging 283
12.2.1 Microscopy Setup 283
12.2.2 Fluorescence Illumination 283
12.3 Fluorescent Probes for FRET 285
12.3.1 Engineering of GFP-Like Proteins 288
12.3.1.1 Random mutagenesis 288
12.3.1.2 Site-directed mutagenesis 289
12.3.1.3 Multisite-directed mutagenesis 289
12.3.2 Cloning of Fluorescent Vectors 289
12.3.3 Functional Activity of Expressed Constructs 290
12.3.4 Expression and Overexpression 291
12.3.5 Guidelines for FRET Pairs 291
12.3.6 FRET-Based Sensors 293
12.4 Practical Considerations for All FRET Measurements 294
12.5 Application of FRET Microscopy 297
12.5.1 Wide-Field FRET Imaging 297

12.5.2 Confocal FRET Imaging 297
12.5.3 FLIM-FRET Microscopy 300
12.5.4 Three-Color FRET Microscopy 301
12.5.5 Other FRET Microscopy Application 301
12.6 Conclusions 304

13 Photodetectors for Fluorescence Microscopy 313
James Pawley
13.1 Historical Photodetectors 314
13.1.1 The Photomultiplier Tube 315
13.1.1.1 Photon counting 318
13.1.1.2 Hybrid PMTs 318
13.2 Solid-State Single-Channel Detectors 320
13.2.1 Multipixel Photon Counters 323
13.2.2 Single-Channel Photodetectors for Fluorescence-Lifetime Imaging (FLIM) 326
13.3 Imaging Photodetectors 327
13.3.1 The Charge-Coupled Device Detector 328
13.3.1.1 CCD readout schemes 332
13.3.1.2 Dark signal 335
13.3.2 Electron-Multiplier CCDs 335
13.3.3 sCMOS Imagers 337
13.3.4 Color Silicon Imagers 340
13.4 What Imager to Choose? 341

14 Practical Aspects of Localization Microscopy 347
Mark B. Cannell, Christian Soeller, and David Baddeley
14.1 Introduction 347
14.1.1 Anatomy of a Localization Microscope 349
14.1.1.1 Sensitivity 350
14.1.1.2 Stability 352
14.1.2 Multi-color and 3D Localization Microscopy 352
14.1.2.1 Multi-color imaging 352
14.1.2.2 3D localization 356
14.1.3 Sample Preparation 357
14.1.3.1 Switchable dyes/buffers 359
14.2 Practical Considerations 361

14.2.1 Illumination 361
14.2.2 Choice of 3D Method 363
14.2.2.1 How much experimental complexity can I live with? 363
14.2.2.2 How many molecules will be on simultaneously in a given frame? 364
14.2.2.3 How important is depth vs. lateral information? 365
14.2.2.4 How much background will I have? 365
14.2.2.5 How well behaved is the sample refractive index? 367
14.2.3 Analysis of Raw Data 370
14.2.4 Visualization and Postprocessing 372
14.3 Summary 376

15 Super-resolution Optical Microscopy with Structured Illumination 383
Liisa M. Hirvonen and Trevor A. Smith
15.1 Introduction 383
15.1.1 Spatially Modulated Illumination Microscopy 385
15.1.2 The Principle of SIM 386
15.2 Optical Sectioning with Structured Illumination 389
15.3 Resolution Improvement with Structured Illumination 390
15.4 Practical Implementation of SIM 392
15.5 Mathematical Background and Image Reconstruction 393
15.5.1 Image Formation 393
15.5.2 Image Reconstruction 395
15.5.3 Artefacts 396
15.6 Applications 399
15.7 Extensions of SIM 403
15.7.1 Three-Dimensional SIM (3D SIM) 403
15.7.2 Nonlinear Structured Illumination 403
15.7.3 Rapid Data Acquisition 405
15.7.4 Other SIM Approaches 405
15.8 Conclusion 407

16 STED Nanoscopy **415**
Christian Wurm, Andreas Schoenle, and Christian Eggeling
16.1 Introduction: The Diffraction Limit 416
16.2 STED Nanoscopy 417
16.2.1 The Basics 417
16.2.2 3D Resolution 422
16.2.3 Changing the Objective Lens 424
16.2.4 RESCue and New Dyes for Live Cell Imaging 425
16.2.5 Combination with Other Fluorescence Methods 433
16.3 Comparison with Other Nanoscopy Approaches 435
16.4 Conclusion 438

Index 451

Preface

The ancients must have had some awareness of fluorescence and phosphorescence, but scientific study of the phenomenon dates only from the late 19th century, with George Stokes's classic paper "Changes in the Refrangibility of Light," published in 1852. Soon after that came the first synthetic fluorescent dyes—fluorescein was first synthesized in 1871. The stage was set for fluorescence imaging, but it took a long time before it became an important scientific tool.

While the principles of fluorescence excitation and emission had been articulated, it was not until the arrival of quantum mechanics that it was really understood, and Chapter 1 explains how fluorescence works in terms of electron energy levels. Fluorescence microscopy was first invented in the early 20th century, but it was not until the second half of the century that it had real practical applications, initially in cytogenetics. Two factors led to its wider application: the development by the Dutch physicist (and pioneer computer artist) J. S. Ploem of a much better fluorescence microscope (Chapter 2) and the development of a wide range of methods for using fluorescent dyes in biology (Chapters 5, 6, 7, and 8).

It soon became apparent that fluorescence opened up new methods of imaging which were not possible with the stains used in conventional microscopy, enabling three-dimensional imaging (Chapters 3 and 4) and dynamic studies of cellular processes (Chapters 8, 10, 11, and 12). It could also be used at the macroscopic level (Chapter 9). Recording the images (often faint) needed to be quantitative to make full use of these capabilities, and this is covered in Chapter 13.

Most recently it has been realized that the unique features of fluorescence imaging can lead to methods of breaking the classical "half wavelength" resolution limit of optical microscopy. These techniques are covered in Chapters 14, 15, and 16.

As this book was going to press, news arrived that Jim Pawley, author of Chapter 13, died suddenly on March 7, 2019. He was a leading authority on confocal optical and scanning electron microscopy, a legendary teacher, and an all-round character. He will be missed.

Guy Cox
March 2019

Chapter 1

Basic Principles of Fluorescence

Scott Kable
School of Chemistry, University of New South Wales,
Sydney, NSW 2052, Australia
s.kable@unsw.edu.au

1.1 Introduction

It is very much easier to detect emission of light than absorption of light. A dark-adapted human eye can detect a few tens of photons per second. But that same eye struggles to distinguish a 1% difference in absorption, which is $\sim 10^{17}$ photons per second in room light. Instruments are the same. Photomultipliers can detect single photons, but in absorption the same detector can detect an absorbance of 10^{-6} ($\sim 10^{13}$ photons per second in room light). Likewise, fluorescence microscopy is much more sensitive and selective than transmission microscopy.

Fluorescence (or luminescence more generally) can be found everywhere in the universe, but is not all that commonplace. To explain, Fig. 1.1 shows a range of objects, with a length scale varying over 22 orders of magnitude, all of which we can see because of luminescence. The pink background in the Horsehead Nebula

Fundamentals of Fluorescence Imaging
Edited by Guy Cox

ISBN 978-981-4774-85-7 (Hardcover), 978-1-351-12940-4 (eBook)
www.jennystanford.com

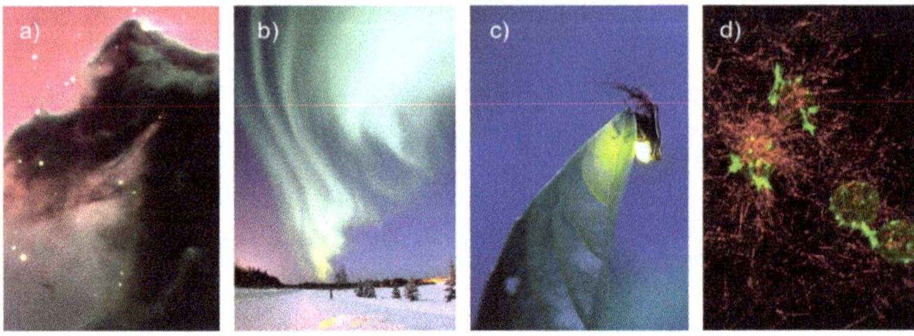

Figure 1.1 Four images of fluorescent objects: (a) the Horsehead Nebula [1], (b) Aurora Borealis [2], (c) firefly [3], and (d) fluorescence-stained adenocarcinoma cells [4]. The length scale of these objects is from 10^{16} to 10^{-6} m.

(size $\sim 10^{16}$ m) arises from hydrogen atom fluorescence, with the distinctive horsehead shape caused by absorption of the intervening dust cloud. The light from the aurora is due to fluorescence from a variety of species including N_2^+ and O_2^+. The light in a firefly tail is chemiluminescence from luciferins reacting with oxygen. The final image in Fig. 1.1 is a fluorescence microscope image of cultured breast adenocarcinoma cells (MTLn3) embedded in high-density collagen matrices. The length scale in this image is ~ 1 µm, which is 10^{22} times smaller than the nebula. Thankfully, luminescence is not very common or our eyes would be besieged by luminescent objects.

The two central images in Fig. 1.1 can be seen with our own eyes, but the largest and smallest objects require significant magnification. Single lenses were known at least as far back in history as the Assyrians (2700 years BP), where at least one artifact survives. They are frequently referred to in Greek and Roman writings, where they were used to start fires, and by craftsmen for magnifying fine work. The ability to resolve astronomical objects has been the subject of myths. For example, the binary stars Mizor and Alcor, in the constellation Ursa Major, were apocryphally used by the Roman army as an eye test.

The use of compound optics—the combining of more than a single optical element into a device—is controversially attributed to two Dutch spectacle makers, father-and-son team Zaccharias and Hans Janssen. They noticed that two lens placed inside a tube could make an object appear very much larger. Galileo learned about this finding and, famously, turned the tube to the skies. The invention of the "Galilean telescope" combined a convex and concave lens to magnify (Fig. 1.2a).

Robert Hooke published *Micrographia* in 1665, which contained a number of observations of the microscopic world using a primitive compound microscope (Fig. 1.2b). Hooke first coined the word *cell* when observing the boxlike structures in cork, reminding him of the cells of a monastery. Anton (or Antonie) van Leeuwenhoek was the first to describe single-cell organisms, although he used a simple, single-lens magnifier rather than a compound lens system, as he found he could get better resolution. He was the first person to see bacteria, red blood cells, and many other biological and physical objects in a long career.

Optical microscopes improved slowly throughout the centuries as better optical materials and better polishing methods were discovered. Conventional optical microscopy, however, remained limited by the fundamental resolution limit of the wavelength of light ($\sim$500 nm), and limited by the optical density of materials, in other words the penetrating depth of the light. Too little optical density resulted in insufficient contrast to measure an image, while too high an optical density meant that no light reached the detector (eye).

In the past couple of decades, there has been an explosion of new optical techniques. Super-resolution techniques, such as PALM, STORM, and GSD (see Chapters 14, 15, and 16) have allowed the fundamental resolution limit to be broken. Multiphoton techniques (Chapter 4) have allowed highly penetrating near-IR light to be used, while detecting visible light with high efficiency. Optical microscopy can be used on live samples (Chapters 7, 8, and 9) and in real time as a dynamical technique (Chapters 10, 11, and 12). Underpinning all of these new techniques is *fluorescence microscopy*. Fluorescence microscopy remains a valuable technique in its own right, but it is the fundamental properties of fluorescence that have enabled all of the more sophisticated techniques.

Many of these specialized fluorescence microscopy techniques are explained in much more detail in their own chapters. The objective of this chapter is to provide an understanding of the fundamentals of fluorescence, with a focus on concepts and techniques

Figure 1.2 Compound optical instruments were first developed in the late 16th and early 17th centuries and allowed scientists to observe objects far smaller and far greater than the naked eye could see. (a) Galilean telescope [5], (b) Hookes's compound microscope [6].

relevant to microscopy. The chapter is separated into two parts. The first describes the basic principles of fluorescence. The inherently quantum-mechanical concepts of "energy levels," "radiative and nonradiative processes," and "quantum yields" are developed from a nonmathematical framework, which is developed into a practical and useful understanding of fluorescence phenomena aimed at practicing microscopists. The second half of the chapter provides a very brief overview of how these fluorescence principles are used in several of the imaging techniques that are the subject of later chapters.

1.2 Basics of Fluorescence

To begin this section we will outline the basic features and properties of light, and energy states of molecules. Exploring the interaction of light with molecules then connects the two concepts. The key properties of absorption and fluorescence are discussed next, along with the properties that are bane of fluorescence microscopy—the so-called nonradiative mechanisms. After considering the isolated molecule, the interactions of molecules with their environment is brought in, which develops the concepts

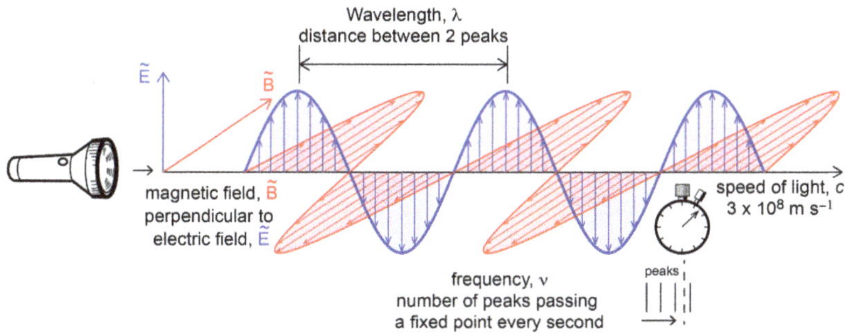

Figure 1.3 Schematic of an electromagnetic wave propagating left to right showing the relationship between wavelength, frequency, and speed.

underpinning photobleaching and quenching, but also the powerful techniques of FRET, FRAP, and FLIM (acronyms explained later). *En route* we will examine the excitation and emission spectra of typical dyes that are the staple of any dye catalogue (also Chapters 7 and 8), and biological structures that have their own fluorescence properties (also Chapter 6).

1.2.1 Properties of Light

Light is a form of electromagnetic (EM) radiation, consisting of an oscillating electric field with an oscillating magnetic field perpendicular to it (Fig. 1.3). These oscillating fields behave like waves travelling at the speed of light, c. The wavelength of the light, λ, is typically measured in nanometers (1 nm $= 10^{-9}$ m). Light can be distinguished by many other characteristics, including the frequency (ν), which is defined as the number of wavefronts passing a fixed point every second (Fig. 1.3). The frequency is therefore inversely proportional to the wavelength; at a fixed speed, more wavefronts of shorter wavelengths will pass a fixed point in a given time than will wavefronts of longer wavelengths.

The mathematical relationship that defines these three properties is

$$\nu = \frac{c}{\lambda} \tag{1.1}$$

The "optical" or visible wavelengths of light form a narrow range, extending from approximately 400 to 750 nm, defined purely anthropogenically as the wavelengths that the human eye can see. The wavelength range immediately to longer wavelength is called "infrared" (IR) while the adjacent region with shorter wavelengths is the "ultraviolet" (UV) (see Fig. 1.4). Historically, optical microscopy has dealt with the visible wavelength region because our eyes were the detectors. This was supplemented with emulsion film photography from the 19th century and charged coupled device (CCD) cameras since the 1990s. Furthermore, with the development of infrared and ultraviolet sensitive detectors, and with the advent of nonlinear optical techniques, these neighboring wavelengths regions have also become important.

About one hundred years ago, Planck postulated that the *energy* of an electromagnetic wave could not have any value, but that it was "quantized." This smallest unit of energy is proportional to its frequency and has a magnitude

$$E = h\nu \tag{1.2}$$

where h is now called the *Planck constant*. This smallest amount, or quantum, of energy is associated with one *photon* of light. The relevance for microscopy is that molecules usually absorb or emit one photon of energy at a time. The exception to this rule is

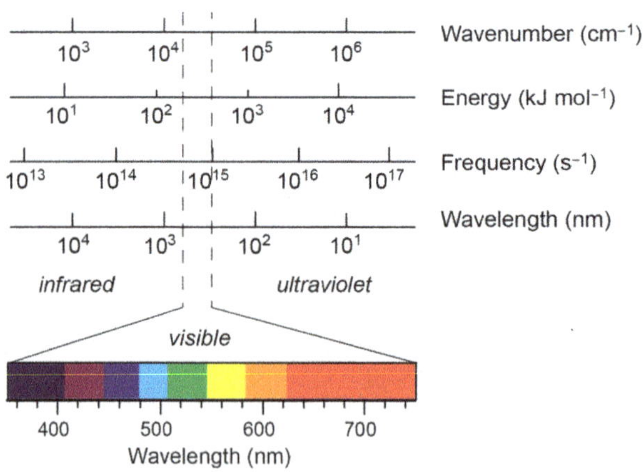

Figure 1.4 Characteristic values of wavelength, frequency, energy, and wavenumber for the visible, infrared, and ultraviolet regions of the electromagnetic spectrum.

under conditions of very intense radiation, such as a laser, where two or more photons can be absorbed simultaneously, leading to multiphoton absorption (see below).

The dichotomy of the two descriptions of light (the *wave* and *particle* descriptions) was the source of enormous debate for the first part of the 20th century and eventually united in quantum mechanics. For practical microscopy, both definitions are helpful in different situations. For example, reflection, refraction, diffraction, and polarization are best described using wave properties. Absorption and emission of light by molecules is more intuitively described using the particle (photon) description. In the discussion and chapters that follow, we will use the wave and particle description of light interchangeably, depending on the process.

1.2.2 Energy States of Molecules and Jablonski Diagrams

Energy can be stored in a molecule in many different ways. A quickly moving molecule has more kinetic energy than a molecule at rest; a rapidly rotating molecule has more rotational energy than a slowly rotating one; a molecule whose atoms are vibrating is "hot," while the atoms in a "cold" molecule are more stationary; excited electrons can emit visible, ultraviolet, or X-ray radiation; and excited nuclei can emit alpha, beta, or gamma rays. Each type of motion has its own characteristic energy—rotations are easy to excite, needing ~0.1 kJ/mol. Nuclear excitations can be 10^6 kJ/mol. The type of motion that is relevant to fluorescence is energy that is stored by electrons and by vibrating molecules. In typical molecules, the energy of these characteristic motions is ~5–40 kJ/mol to induce vibrational motion and ~200–500 kJ/mol to excite an electron (see Fig. 1.4).

The energy of a vibrating molecule is readily understood in terms of the motion of balls (atoms) connected by springs (bonds). Electronic energy does not have such a simple classical picture. *Molecular orbital theory* is often used to explain the energy of electrons. In this theory, there is a set of *molecular orbitals* (MOs), each of which can be occupied by up to two electrons. Figure 1.5 shows three MO energy level diagrams of a fictitious molecule. The lowest-energy configuration is when the electrons occupy the

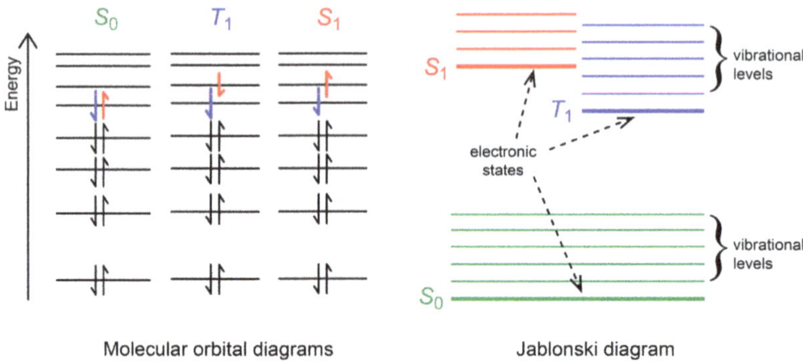

Figure 1.5 Showing the relationship between molecular orbital diagrams of the three lowest energy electron configurations of an arbitrary molecule, and a single Jablonski diagram, which only shows the total energy of each configuration.

orbitals with the lowest possible energy, given the constraint of no more than two electrons in any one orbital; this is called the *ground state* configuration. The combination of all orbitals gives rise an *electronic state* with an energy equal to the sum of the energies of the individual orbitals.

Spin is another intrinsic property of electrons and an electron can have two quantized spin states, assigned the values $m_s = \pm\frac{1}{2}$. The *Pauli exclusion principle* requires any two electrons in the same orbital to have opposite spin. The three MO diagrams in Fig. 1.5 all satisfy the Pauli exclusion principle. Most stable molecules have an even number of electrons. The ground state of a molecule, with electrons in orbitals of the lowest possible energy, must therefore have an equal number of electrons of each spin, as shown in the left-most MO diagram. The total, overall electron spin is therefore zero. An electronic state with zero overall spin is called a *singlet state* and given the letter S. The subscript "0" indicates that this is the ground state.

If an electron is promoted into an excited electronic state, then this produces two orbitals occupied by one electron each. There is no longer any requirement for the electron spins to be paired and so the electrons can have their spins opposite or aligned as the two excited state MOs in the figure show. If the spins are opposite, the total spin is still zero, and so the overall electron configuration is still singlet. If the electron was promoted to the next lowest energy orbital this configuration would be called S_1, meaning *first excited singlet state*. If the spins are aligned, the state has overall electron spin of 1, and the state is called a *triplet state*. The state shown in Fig. 1.5 is labelled T_1, meaning *first excited triplet state*. It is obvious in the MO diagrams that the electrons could have been excited to higher-lying orbitals. Each resultant electron configuration gives rise to a different excited state, each labelled by S or T and a given a subscript that simply ranks them in order of the energy of the overall electronic state.

Molecular orbital diagrams get complicated because a large molecule has so many electrons. All we really need to know is the overall energy of the electrons, not the individual orbitals, and even then, what is important is the energy relative to the ground state as a molecule cannot exist at lower energy. To simplify the diagram, we need only use a schematic representation of the energy levels of a fluorophore that is called a *Jablonski diagram*. A Jablonski diagram is like a cartoon version of a molecular energy level diagram that captures the essence of what is going on, without getting bogged down in the details. It starts off representing the energy of a molecule in its ground electronic state as a horizontal line as shown in Fig. 1.5. This state corresponds to the MO configuration labelled S_0 and is given the same label. Excited vibrational levels are represented by a set of horizontal lines lying above the ground state. As mentioned previously, the energy of an excited electronic state is usually much higher than the energy for a vibrational state, so an excited electronic state is shown above the set of vibrations. A molecule with an excited electron can still vibrate, so another set of vibrational levels are shown above the excited electronic state. Figure 1.5 shows a Jablonski diagram for the ground and first excited singlet and triplet states of a molecule, accompanied by sets of vibrational levels.

The ordinate (y axis) of the diagram is an energy scale. If shown, it typically does not have a scale because the diagram is indicative—there is no attempt to draw the energy levels to scale. The ground state, S_0, is defined as zero energy. Frequently, however, the scale is left off and the Jablonski diagram is just understood to have energy increasingly vertically. The idea of a Jablonski diagram is to represent a concept as simply as possible. If more states are needed, they are added to the diagram; if not, they are left off.

The concepts of MOs and electronic states are a simplification of real electronic motion. But they are powerful concepts and can be taken a long way towards understanding fluorescence properties of molecules, and all we really need for the concepts in this book.

1.2.3 Absorption of Light

When a molecule is placed in an oscillating electromagnetic field (i.e., light is shone on the molecule) the oscillating fields can push and pull the molecule around. In particular, charged parts of the molecule are affected by the oscillating *electric* field of light. If the charges on the molecule can oscillate in time with the oscillating field (this is called *resonance*) then the molecule can absorb energy from the field. As a physical analogy, consider a child on a swing. If you push the child randomly, then the swing never really gets going. But if you push *in resonance* with the oscillation frequency of the swing then the swing can absorb a lot of energy.

This concept of resonance pervades our daily life, from musical instruments to radio reception. One thing you should notice is that there is often one specific resonant frequency for a simple system. You can only push the child on the swing at one frequency. Middle "C" on a piano corresponds to the piano string vibrating 262 times per second. Molecules likewise have specific resonant frequencies. A typical C–H bond in a molecule vibrates about 10^{14} times per second (10^{14} Hz), while a C–C bond vibrates at about 3×10^{13} Hz. An oscillating electromagnetic field at 10^{14} Hz corresponds to light in the mid-infrared region of the spectrum (see Fig. 1.4). The corresponding range of infrared wavelengths is about to 3 to 30 μm. (see Eq. 1.1 and Fig. 1.4). A molecule exposed to mid-infrared radiation of the right frequency can therefore absorb energy from the field to become vibrationally excited.

Electrons oscillate at even higher frequencies (shorter wavelengths). A typical wavelength of light absorbed by the outer shell (valence) electrons is 200–1000 nm, corresponding mostly to the UV–visible range, but extending slightly into the near infrared

(Fig. 1.4). These outer (valence) electrons are the ones involved in the chemical bond. Exciting these electrons therefore changes bonds, which causes the molecules to change shape, or even to break the bond. A consequence of changing the shape of the molecule is that electronic transitions are usually accompanied by vibrational motion as the atoms move to their new positions.

On a Jablonski diagram, absorption is represented by an upward arrow, moving population from the ground state to a higher electronic state. The two upward arrows in Fig. 1.6 represent exciting a molecule to its first excited singlet or triplet state, accompanied by some vibrational energy change. The amount of vibrational energy involved in an electronic transition is inherent to the molecule itself, and even to the specific electronic state being excited. The rule of thumb is that the larger the change in geometry of the molecule, the more vibrational excitation accompanies the transition. This is a consequence of the *Franck–Condon* principle and interested readers are encouraged to explore textbooks on spectroscopy to explore this important principle further.

The innate strength with which a molecule absorbs light is characterized by the *absorption coefficient* (or *extinction coefficient*, ε). The amount of light absorbed (*absorbance, A*) is described by the *Beer–Lambert Law*, which relates this absorbance to the concentration of the compound (c) and the length of the path through which the light travels (l):

$$A = \log_{10}\left(\frac{I_0}{I}\right) = \varepsilon\, c\, l \qquad (1.3)$$

As Eq. 1.3 shows, the absorbance is defined in terms of the intensity of incident light, I_0 and the intensity of transmitted light, I. The unit of the pathlength is usually cm (indeed, the standard absorption cell is 1 cm in length). If the units of c are mol L^{-1}, then ε has units of L mol^{-1} cm^{-1} (which is the same as M^{-1} cm^{-1}).

An *absorption spectrum* is obtained by measuring the transmission of light as a function of wavelength. The combination of the resonance condition and the absorption strength gives rise to an absorption spectrum with a well-characterized intensity for every different wavelength. Figure 1.7 shows the absorption spectrum of a fluorescence reference standard, Rhodamine 6G (also called Rhodamine B, or just R6G), dissolved in ethanol. The spectrum

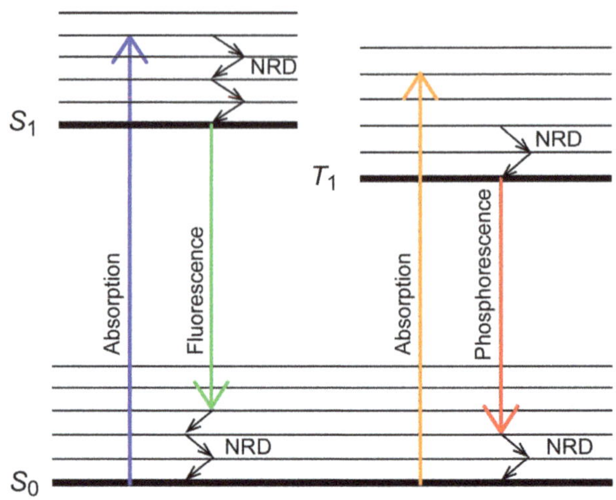

Figure 1.6 Simple Jablonski diagram. Absorption of light by a molecule is shown as an upward-pointing arrow indicating that the molecule has gained energy. Emission of light is a downward-pointing arrow. Nonradiative decay (NRD) is the rapid loss of energy to the environment.

shows a very strong absorption between 490 and 550 nm, with a maximum near 530 nm, where the molar extinction coefficient is 116,000 M^{-1} cm^{-1}. This transition corresponds to the R6G molecule (shown as an inset) being excited from its ground electronic state, S_0, to its first excited electronic state, S_1. The width of the transition is a reflection of the range of vibrational energy that can accompany the electronic excitation. The Jablonski diagram in Fig. 1.6 is a good representation of this transition in R6G. The spectrum also shows a number of other absorption features at shorter wavelength. These correspond to R6G being excited into higher and higher electronic states, but these transitions are not shown in Fig. 1.6.

Figure 1.6 shows absorption to a singlet or triplet state, but, in reality, the extinction coefficient for singlet–triplet absorption is typically several orders of magnitude weaker than the corresponding singlet–singlet transition. The reason is that an oscillating electric field is not effective at changing electron spin, therefore spin is approximately conserved for absorption and emission of light. These $S_0 \rightarrow T_1$ transitions are therefore very weak, and of no practical importance for fluorescence microscopy. However, the presence

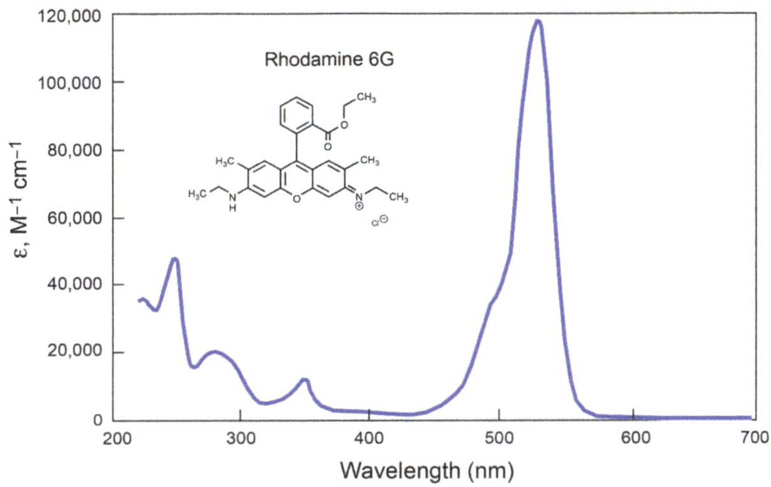

Figure 1.7 Absorption spectrum of Rhodamine 6G. The strong $S_0 \rightarrow S_1$ transition is at 530 nm with weaker transitions to other electronic states to shorter wavelength.

of triplet states plays a very important role in understanding photobleaching, as shall be seen later.

1.2.4 Fluorescence and Phosphorescence

A molecule in an excited state is unstable. It can relax (lose energy) in many different ways, including chemical reaction, exchanging energy with the environment, and by emitting light (*luminescence*). All of these processes are important for fluorescence microscopy.

Luminescence is the emission of light that occurs when an electron relaxes from a higher to lower electronic state. It can be any color of the spectrum as shown in part in Fig. 1.8, and can extend a little into the infrared, and a long way through the ultraviolet and into the X-ray region of the spectrum. X-ray fluorescence is a powerful imaging technique in its own right, however, it requires a synchrotron or neutron source to excite specific atoms and is not considered further in this book.

Fluorescence is defined as emission in which the spin state of the molecule is preserved, for example $S_1 \rightarrow S_0$. When the spin state

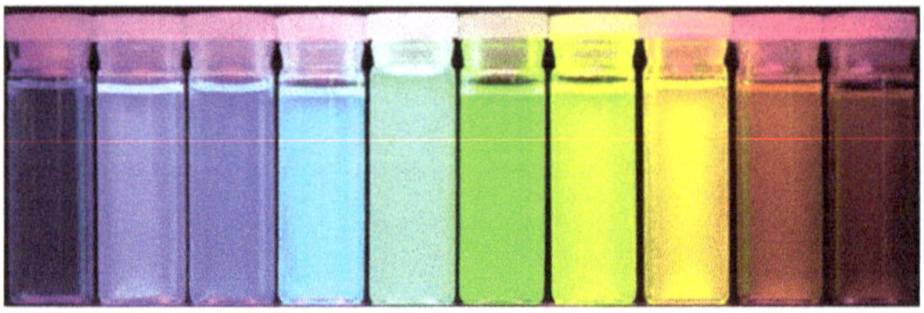

Figure 1.8 Cuvettes of fluorescent dyes demonstrating the wide range of fluorescence colors.

changes, e.g., $T_1 \rightarrow S_0$, the emission is called *phosphorescence*, as shown in Fig. 1.6. For exactly the same reason that $S_0 \rightarrow T_1$ absorption is very weak, so, too, is $T_1 \rightarrow S_0$ phosphorescence very weak.

One other process is also shown in this figure—*nonradiative decay* (NRD)—which is the loss of vibrational energy of a molecule through interaction with the surroundings. In a solution of a fluorescent dye, this interaction is invariably with the solvent. In the atmosphere it is caused by collisions with other gas molecules. In a fluorescence microscopy sample, the interaction is with whatever the fluorophore is bound to. No matter what the surroundings are, the fundamental principle is the same: following absorption of light, the molecule has excess electronic and vibrational energy. The vibrational energy can be quickly dissipated into the surroundings as the moving atoms collide with atoms of the surroundings. The natural lifetime of an excited electron in a singlet state is typically measured in nanoseconds (10^{-9} s), while it is more like microseconds (10^{-6} s) for a triplet excited state. In solution or solid samples, NRD typically occurs on a picosecond (10^{-12} s) timescale. Therefore, a molecule will generally lose all its vibrational energy before it loses electronic energy by emitting fluorescence or phosphorescence.

There are several important implications for NRD:

- A molecule emits from the lowest vibrational level in S_1, irrespective of where it was excited to initially.
- Fluorescence is always lower energy (longer wavelength) than absorption.
- If a molecule naturally incurs a lot of vibrational energy in absorption, it will also incur a lot of vibrational energy in emission.

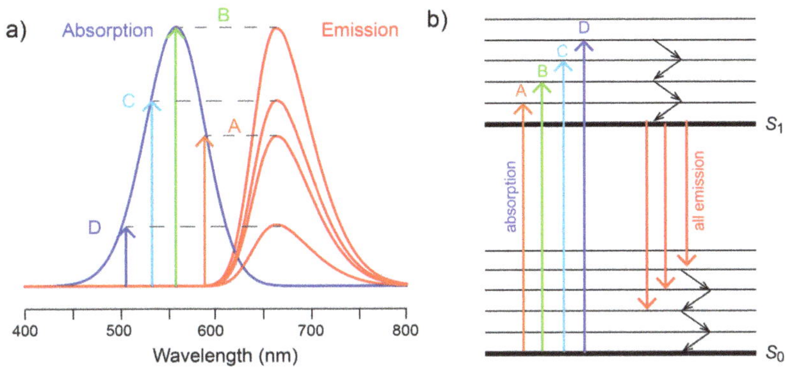

Figure 1.9 Simplified absorption and emission spectra with accompanying Jablonski diagram.

- The difference between the maximum intensity in absorption and emission spectra is called the *Stokes shift*.
- The larger the Stokes shift, the more energy is dissipated into the surroundings as heat.
- There is a *mirror symmetry* apparent between the absorption and emission spectra.

To explain these features further, consider the following example. Figure 1.9 shows some simplified absorption and emission spectra, accompanied by a Jablonski diagram. The blue spectrum to the left is an absorption spectrum with some points labelled A–D, which represent several wavelengths at which the molecule might be excited. The transitions that correspond to these excitation wavelengths are also represented on the Jablonski diagram. The highest intensity absorption is at position B. The excitations A, C, and D are part of the same electronic transition, but involve different amounts of vibrational energy.

After excitation, the molecule rapidly loses vibrational energy to the environment until there is no more vibrational excitation as shown by the NRD arrows in the figure. The molecule can then fluoresce at a variety of wavelengths, also shown in the Jablonski diagram. The range of vibrational levels produced after emission produces a broadened fluorescence emission spectrum, shown in red. The reason that many vibrational levels will be prepared in fluorescence is exactly the same reason that many vibrational states

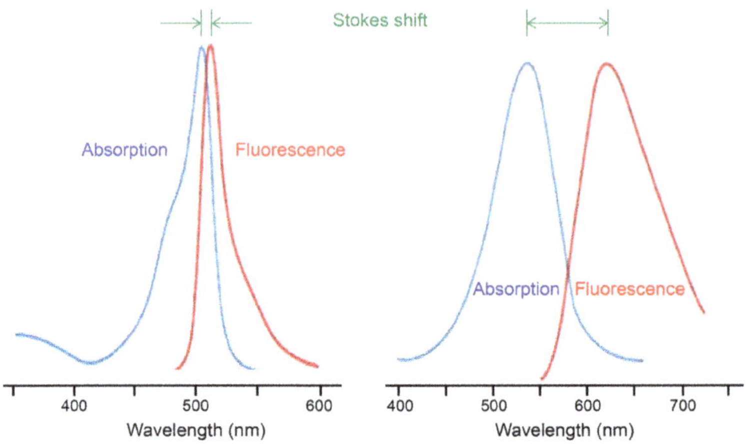

Figure 1.10 Absorption and emission spectra of two common fluorophores showing different Stokes shift between the peaks of the absorption and fluorescence spectra.

can be prepared in absorption. Vibrational excitation is produced by a change in shape of the molecule when the electron is excited, caused by changing the structure in the molecule. After fluorescence, the electron returns to its ground state and the molecule has to change back to its original shape. This requires just as much vibrational energy, in reverse. So, if the absorption transition is accompanied by a wide range of possible vibrational excitations, then the fluorescence emission is accompanied by a similar range of vibrational excitations. This gives rise to the well-known *mirror symmetry* between the absorption and emission, which is shown schematically in Fig. 1.9, and for real spectra of fluorescence dyes in Fig. 1.10.

Another impact of rapid NRD is also shown in Fig. 1.9. Notice in the Jablonski diagram that the different excitation energies populate the molecule in different vibrational levels. But no matter at what level is excited, NRD quickly removes all the excess vibrational energy. As a consequence, the emission spectrum is exactly the same, *no matter what the excitation wavelength is.* The *intensity* of the emission, however, will change. As shown in the figure, exiting a molecule away from its absorption peak, e.g., A, C, D, will reduce the overall emission intensity.

The difference between the peak of the absorption spectrum and the peak of the emission spectrum is called the *Stokes shift*. The larger the Stokes shift, the easier it is to separate the excitation wavelength from the emission wavelength using filters. The different fluorophores in Fig. 1.10 have quite different Stokes shifts. However, there is an accompanying trade-off that is often overlooked. The Jablonski diagram shows that the size of the Stokes shift is the amount of energy lost to NRD. This vibrational energy is dissipated into the environment in the form of heat. Therefore the larger the Stokes shift, the larger the local heating of the sample, which may cause problems of its own in specific circumstances.

1.2.5 Photophysics and Photochemistry

Fluorescence is a natural phenomenon and occurs throughout nature as discussed in the introduction to this chapter. If fluorescence was commonplace, then fluorescence microscopy would be hindered by all the natural fluorescence drowning out the features you are interested in. We have seen above that molecules lose vibrational energy to their surroundings very efficiently. To understand why fluorescence is not so common, we need to consider other pathways that molecules use to lose excess electronic energy.

The underlying principles of electronic relaxation are embedded in the quantum description of the electronic states. It is fair to say that, even at the present time, the nature of interactions between electronic states is not completely understood, and largely beyond our ability to calculate. However, we don't really need to know the inner details in any case—just that it happens and under what circumstances.

It is convenient to consider the loss of electronic energy in two realms: as an intrinsic property of the molecule, or its electronic state (*radiationless transitions*), and as an extrinsic property, induced by its environment (*quenching*). We will consider extrinsic loss below under quenching and deal with the intrinsic property here.

The energy levels of a molecule are traditionally discussed as arising from a single, local motion. For example, rotational energy levels are associated with motion of the whole molecule; vibrational

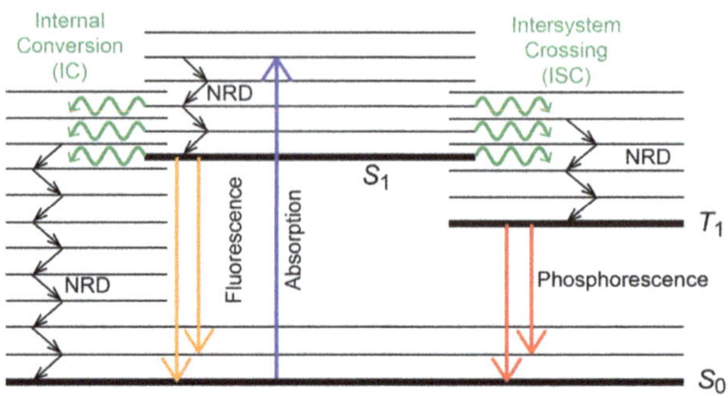

Figure 1.11 Jablonski diagram showing the radiationless processes, internal conversion and intersystem crossing.

energy is associated with movement of the atoms, and electronic energy with the motion of the electrons. However, this is an approximation, called the *crude Born–Oppenheimer approximation*. A vibrating molecule does indeed rotate with a slightly different energy to one in its vibrational ground state and Coriolis forces in a rapidly spinning molecule can force the molecule to vibrate differently. A vibrating molecule can also affect the energy of electron orbitals. The interaction between electronic and vibrational energy underpins the mechanism by which an electronically excited molecule loses energy.

The Jablonski diagram in Fig. 1.11 shows the general idea. In this diagram, the molecule absorbs a photon and is excited to an excited electronic state. It sheds vibrational energy through NRD. Before it emits, the molecule converts the electronic energy into vibrational energy of a lower electronic state in a *radiationless transition*. When the spin state is preserved, this electronic relaxation is called *internal conversion* (IC) and when the spin state changes it is called *intersystem crossing* (ISC). As this transition is not accompanied by a photon to remove the energy, the transition must conserve energy. Therefore, radiationless transitions are drawn as horizontal arrows (same energy) on a Jablonski diagram. The ensuing vibrational energy can be removed by NRD as before. The highly vibrationally excited S_0 molecule will return to the ground state, without ever emitting light. Following ISC, the T_1 molecule can either phosphoresce, or undergo ISC again, converting T_1 electronic energy into S_0 vibrational energy. Again, the molecule sheds the excess vibrational energy, returning to the ground state.

The two processes that deplete the triplet state population—phosphorescence and ISC—are spin-forbidden. If the deactivation process is weak, then the triplet state will survive for a considerable time. A molecule in its triplet state is considered to be *metastable*, which means that it is apparently stable on a short timescale, although, energetically, it must eventually dispose of the excess energy. All the processes considered so far are termed *photophysical*; no change in the chemical state of the molecule results. The nett effect of absorption of light is that it is converted to heat through sequential NRD processes. However, a metastable, excited molecule is also prone to chemical reaction. Metastable triplet molecules are particularly sensitive to other molecules in triplet states because they also have unfilled orbitals. As nature would have it, the ground state of molecular oxygen, O_2, has triplet character. Any O_2 in the sample is therefore prone to reaction with the metastable triplet fluorophore in a photochemical reaction. The sequence of non-radiative photophysical events reduces the brightness of a fluorescence image, but the fluorophore is left unchanged and able to re-absorb another photon. However, the photochemical processes actually destroy the fluorophore permanently (see *photobleaching*, below).

Internal conversion and intersystem crossing are even more important for higher excited electronic states. You will notice in the molecular orbital energy levels in Fig. 1.5 that the electronic orbitals are more closely spaced with increased energy. This is generally true. As a result, the energy of successive electronic states is also closer together with increasing energy as shown in the Jablonski diagram in Fig. 1.12. When electronic states are closer in energy, IC and ISC are more efficient. As a result, higher electronic states undergo radiationless transitions more efficiently than does the first excited electronic state because the next lower state, the ground state, has the biggest energy gap.

The consequence of the rapid radiationless relaxation of higher states leads to a general principle of fluorescence, called *Kasha's law*, which is that emission is only observed from the lowest excited state of each spin state (i.e., S_1 or T_1). Figure 1.12 shows an absorption

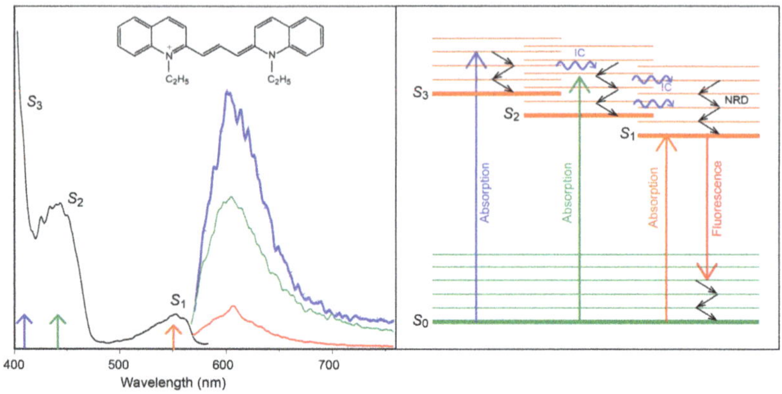

Figure 1.12 (a) Absorption and emission spectra of a fluorescent dye. (b) Jablonski diagram showing why the fluorescence spectrum is the same, irrespective of excitation wavelength. This is an illustration of *Kasha's law*.

spectrum of a dye molecule. Following excitation of the dye at the three wavelengths shown, the three color-coded fluorescence spectra were measured. Clearly the spectrum is essentially the same with the only difference being the overall intensity of the emission. On the right side of Fig. 1.12 is a Jablonski diagram explaining the observed spectra. Three different excitations are shown, to the three different electronic states, S_1, S_2, and S_3. After excitation, the molecule undergoes a series of rapid IC and NRD steps. These are usually so quick that it is hard to tell them apart. The only important feature is that the bottleneck in the relaxation pathway is when the molecule reaches the bottom of S_1, where it can fluoresce, irrespective of which state it was excited to.

1.2.6 Fluorescence Lifetimes, Quantum Yields

All the emission, photophysical and photochemical processes described above occur with their own inherent efficiency, or rate constant. The more efficient the process, the faster the rate constant. The ideal fluorophore will have fluorescence as the dominant process, and other radiationless processes much less efficient. In other words, a very strong fluorescence transition will have a very fast radiative rate constant, with the rate constants for IC and ISC much slower. One measure of the importance of each process is the *quantum yield*. Mathematically, the quantum yield ϕ_i for process, i, is

$$\phi_i = \frac{k_i}{\sum k} \tag{1.4}$$

where k_i is the rate of process, i, divided by the sum of all rates. Conceptually, the quantum yield is simply the fraction of times that a molecules undergoes a particular process. For example, a quantum yield for fluorescence, $\phi_f = 0.90$ means that after absorption of a photon, 90% of the molecules will re-emit the energy as fluorescence. Clearly, the higher the quantum yield for fluorescence, the brighter a fluorophore will appear (more photons out for fewer photons in). In terms of photobleaching of the fluorophore, the quantum yield for intersystem crossing, ϕ_{ISC}, should be as small as possible.

All the photophysical processes described above (other than NRD) are intrinsic to the molecule. They do not require external influence; they even occur in an isolated molecule in vacuum. The rates of the processes are therefore characterised by *first-order kinetics* and the population of the excited molecule will decay exponentially, with a natural lifetime, $\tau = 1/k$. The lifetime is closely related to the half-life: $t_{1/2} = \tau \times \ln(2)$. In first-order kinetics, the observed lifetime is independent of how much fluorophore is present. [Radioactive decay is another example of first order kinetics and the radioactive half-life is also independent of the amount of radioactive isotope.] The constancy of lifetime is an important contribution to the effectiveness of *fluorescence lifetime imaging* (FLIM, see Chapter 10).

1.2.7 Quenching

In addition to the intrinsic (first-order) properties above, interaction with a *quencher* is another way in which a molecule can lose its excited state energy. There are two main types of fluorescence quenching, both of which are important in fluorescence microscopy.

Static quenching occurs when the quencher molecule forms a complex with the fluorophore. For example, many fluorescence markers have a large, flat structure with aromatic rings. These flat molecules can align themselves in the hydrophobic region of a lipid bilayer, or intercalate into DNA or proteins. The quenching interaction is often very specific and the change in fluorescence can provide information about whether the fluorescence label is bound (nonfluorescent, or *quenched*), or free (unquenched).

Dynamic quenching occurs when another molecule deactivates the excited electronic state of the fluorophore in a collisional process. Collisional quenchers tend to be small molecules or ions that are highly mobile. Common examples include oxygen (O_2) and heavy metals such as mercury (Hg^{2+}). Oxygen as a quencher is particularly harmful because, as discussed above, the ground state of oxygen is a triplet, and so quenching by O_2 involves the triplet state of the fluorophore, and often leads to chemical reaction with oxygen, thereby permanently removing the fluorophore.

Static quenching is an equilibrium process, while dynamic quenching is a kinetic process. Nonetheless, the efficiency of both depends approximately linearly with the concentration of quencher. If quenching is an undesirable outcome (e.g., O_2 quenching), then lowering the concentration of the quencher will reduce the quenching. If quenching is beneficial then the amount of quenching is proportional to the amount of quencher in the sample.

Photobleaching is usually an irreversible process, leading to lowered fluorescence emission intensity as a function of time of illumination of the fluorophore. For example, the images in Fig. 1.13 show a fluorescence image of fluorescein-stained HeLa cells taken after 0, 10, and 20 seconds of illumination. The image has almost completely faded after 20 seconds. Practical solutions to this problem include the use of anti-fade reagents (containing antioxidants and commonly used for fixed-specimen fluorescence microscopy, but incompatible with live cell investigations), development of fluorophores less susceptible to photobleaching (lower rate of ISC), minimization of illumination intensity or duration, and use of multiphoton excitation (see below).

1.2.8 Multiphoton Fluorescence

The discussion above on absorption of light and excitation of fluorescence above could be termed *one-photon* excitation, i.e., one photon is absorbed, leading to one photon being emitted. The newer

Figure 1.13 Photobleaching of fluorescein dye attached to actin in HeLa cells, after 0, 10, and 20 seconds of illumination [7].

suite of fluorescence microscopy techniques now routinely includes *two-photon* or more generally *multiphoton* fluorescence. These techniques have several advantages and are becoming increasingly common. Multiphoton microscopy uses pulsed laser beams and optics to focus an intense beam of laser light to a precise spot. Using such intense radiation allows the possibility of more than one photon to simultaneously excite a fluorophore.

A simple understanding of two-photon excitation can be achieved using the particle description of light. Consider firstly a photon (particle) with exactly half the energy required for a specific electronic transition. If two photons impinge on the molecule within the timescale it takes for an electronic transition ($\sim 10^{-18}$ s), then the molecule can absorb both two photons together to impart twice the amount of energy into the molecule as would be expected. The likelihood of two photons impinging on the molecule at the same time is proportional to the number of photons squared. Therefore, the number of two-photon absorption events scales with the square of the laser power.

The total energy of the two photons must still correspond to a real energy level of the molecule, as shown in the Jablonski diagram in Fig. 1.14. This diagram also shows that the resultant fluorescence, following two-photon absorption, is still the normal one-photon emission. Two-photon spontaneous fluorescence has never been observed because the intensity of the fluorescence is much too weak. (Compare the fluorescence intensity with that of an intense laser field.)

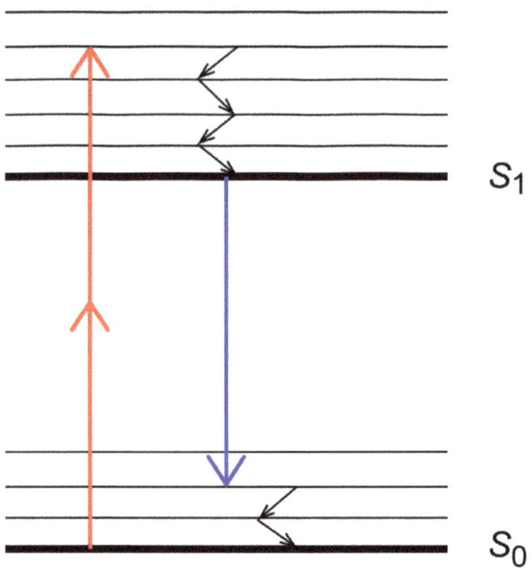

Figure 1.14 Jablonski diagram showing two-photon absorption followed by normal fluorescence. Fluorescence is blue-shifted compared to the absorption wavelength.

Multiphoton excitation provides some interesting and useful variations from one-photon fluorescence microscopy. Consider two-photon excitation, where each photon carries half the energy required for one-photon excitation. This produces the same fluorescence output, but this time the fluorescence is *blue-shifted* (shorter wavelength) than the incident light. A practical consequence of having longer wavelength excitation is that the excitation source avoids the high energy UV region in favor of the lower energy red to near infrared region of the spectrum. This provides a number of very important advantages:

- The penetration of longer-wavelength radiation into dense media such as biofilms of microorganisms and animal or plant tissues is much greater, allowing analysis of structures at much greater depth into the tissue.
- Excitation only occurs at the focal point of the excitation laser. Fluorescence therefore only arise from this focus, and not

throughout the whole depth of the sample, providing depth resolution equivalent to confocal techniques.

- Photobleaching is much reduced because only molecules at the focal point are excited. Those out of the focal volume cannot be bleached.
- UV dyes can be used, without exposing the sample to UV light.

Multiphoton microscopy is dealt with in much more detail in Chapter 4.

1.3 Linking to Microscopy

1.3.1 Autofluorescence

Autofluorescence, or intrinsic fluorescence, occurs when there are naturally occurring fluorophores in the sample under investigation. This can be advantageous if the fluorescent species is the structure of interest because, obviously, no further fluorescence staining is required. The autofluorescence from daunorubycin later in this chapter (Fig. 1.18) is an example.

However autofluorescence is frequently detrimental—causing a broad featureless background in some instances, thereby reducing the contrast in the structure of interest. For example, many fixatives fluoresce under blue or near ultraviolet excitation.

Strongly autofluorescencing structures can also drown out weaker fluorescence from the stained regions of interest. Chlorophyll, for example, is a strongly fluorescent molecule, with broad emission, as shown in Fig. 1.15. Despite the beautiful structure of spirogyra apparent from the chlorophyll fluorescence, any staining of other organelles is swamped by the strong chlorophyll fluorescence.

1.3.2 Fluorescence Labels

The variety of fluorescence labels these days is bewildering—four chapters of this book are devoted to fluorescence markers and labels. In this short section we relate the principles of fluorescence to the choice of labels, without considering the issues of biology

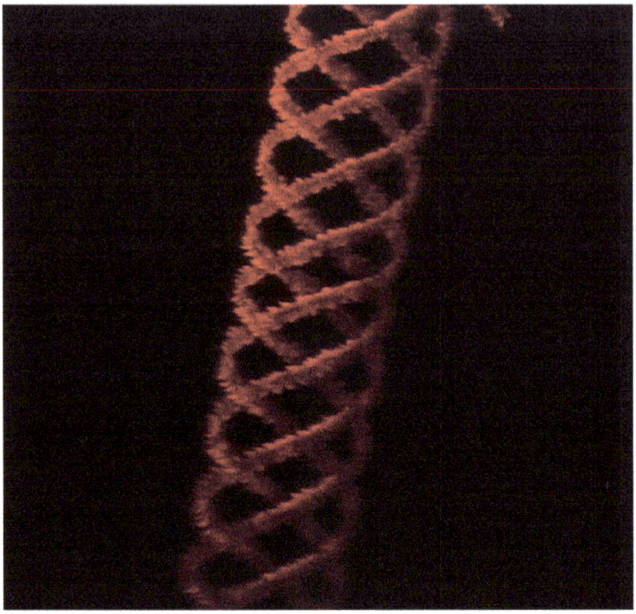

Figure 1.15 Autofluorescence of chlorophyll in *Spirogyra* [8].

or chemistry that are often even more important in the correct choice of fluorophore. From a spectroscopic perspective the crucial things to consider are the excitation wavelengths available in your microscope and the set of filters available for viewing the fluorescence while discriminating against scattered excitation light. If you intend to use multiple labels, then the restrictions become more stringent and there are two options: (i) having labels that absorb at different wavelengths so that each can be exited in turn, or (ii) having labels with different emission spectra that can be successfully discriminated using different filter sets.

The image below (Fig. 1.16) shows a multiple-labelled fluorescence image of human postmortem brain tissue showing nuclei (Hoechst33342, blue dye), microglia cells (Iba-Alexa555, green), and tau filaments (Ttau-Alexa647, red). In this case the emission spectra can be separated with different filter blocks and this image is a composite of the three individual images measured at different emission wavelengths.

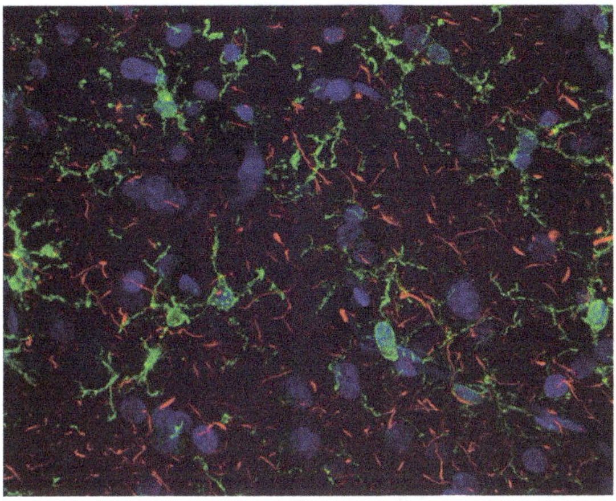

Figure 1.16 Triple-labeled fluorescence images of brain tissue (see text) [9].

1.3.3 Photobleaching and FRAP

Photobleaching is usually a problem in fluorescence microscopy. However, in some cases, photobleaching has been used to advantage in experimental investigations of dynamics. *Fluorescence recovery after photobleaching* (FRAP) yields important dynamical information and is described in detail in Chapter 11. In short, an intense laser is used to intentionally bleach the fluorophore from a local area in a specimen. Observation of the bleached regions over subsequent time can elucidate whether fluorescently labelled molecules are able to move into the bleached area, the magnitude of this movement and the diffusion coefficients of the labelled molecules.

The images in Fig. 1.17 show a region of a lamellipodia of a HT1080 fibrosarcoma cell transfected with either GFP or GFP-β-actin, which was bleached and allowed to recover. The GFP alone recovered evenly and quickly across the bleached region, indicative of diffusion. In contrast, the GFP-β-actin was incorporated into branched actin filaments at the leading edge and as the filaments age, they move further into the lamellipodia. The time scale for reappearance of fluorescence provides information about the diffusion of the fluorophore, and hence about the medium it is diffusing through.

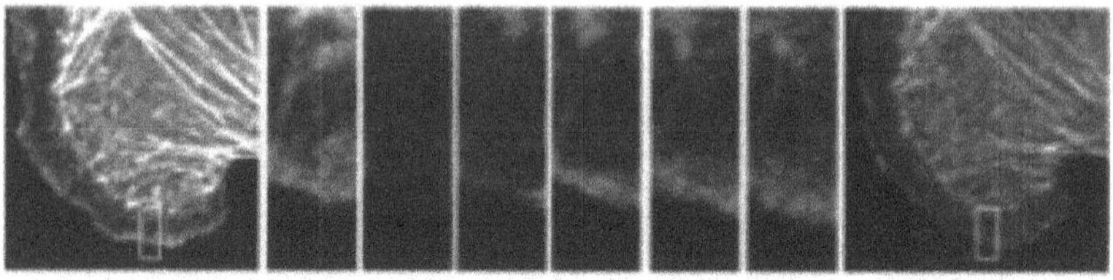

Figure 1.17 Fluorescence recovery after photobleaching. The fluorescence is bleached by a laser at zero time and recovers full intensity after about 1 min [10].

1.3.4 Laser Confocal Microscopy

Although patented in 1957, confocal microscopy revolutionized optical microscopy in the late 1980s with the inclusion of lasers into commercial microscopes. Confocal microscopy allowed, for the first time, optical sectioning of a sample. This means that the three dimensional (3D) structure of a sample can be reconstructed by measuring different 2D slices at different depths using purely optical techniques. Until this time, 3D images had to be reconstructed from different physical sections, e.g., by successive microtomes.

Confocal microscopy achieves the ability to measure an image at a specific depth by clever imaging. A laser can be focused to a point that is limited, fundamentally, by the wavelength of light. In the z, or depth, direction, the focal depth is determined by the focal length of the lens with shorter focal lengths giving better z resolution. In practice, typical x, y resolution of about 0.5 μm can be achieved, but z resolution of about 1 μm. The same very short focal length lens is usually used to image the fluorescence onto the detector. Light from the focal point is imaged perfectly (optical aberrations notwithstanding), while light emanating from regions illuminated before and after the focal point are blurred. By focussing the image and passing through a pinhole, the blurred light is strongly discriminated against and the signal very strongly dominated by emission from only the focal region.

The image in Fig. 1.18 shows a 3D rendered confocal image of HeLa cells. The green stain is green fluorescent protein (GFP)

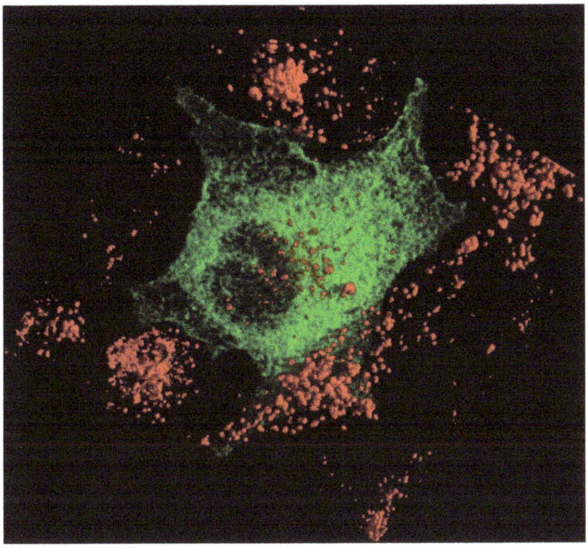

Figure 1.18 Double-labelled confocal image of Hela cells [11].

attached to the p-glycoprotein and the fine structure is the endoplasmic reticulum. The red stain is autofluorescence from daunorubycin in neighboring cells Confocal microscopy is described in much more detail in Chapter 3.

1.3.5 Multiphoton Microscopy

Multiphoton microscopy combines all the advantages of confocal microscopy without the need for pinhole imaging. The other advantages, including greater penetration, using blue dyes, less photobleaching, etc., were described above.

The fluorophores available for normal fluorescence microscopy are generally suitable for two-photon excitation. The electronic states don't change for one- versus two-photon excitation. However, the inherent two-photon absorption strength is determined by different physical and chemical properties than for one-photon. It is very difficult to measure a two-photon absorption spectrum. Where such spectra are available, the curves do not seem to change much. However, dyes that are the strongest one-photon fluorescent species may or may not be the strongest for two-photon fluorescence. Multiphoton microscopy is described in more detail in Chapter 4.

1.3.6 Fluorescence Resonance Energy Transfer (FRET)

Fluorescence resonance energy transfer is a technique that takes advantage of efficient quenching. Above, it was explained that radiationless transitions (IC and ISC) are more efficient if the electronic states are close in energy. Similarly, quenching can be enhanced if the quencher has an electronic state that lies at slightly lower energy to the fluorophore. In this case the electronic energy can be very efficiently transferred from the fluorophore (the "sensitiser" or "donor") to the quencher ("acceptor" or "emitter"). If the quencher is fluorescent, then the emission can be detected from a different molecule to that which was excited. The distance dependence of FRET is very steep, with FRET efficiency dropping off with $1/r^6$, where r is the distance between the donor and acceptor. Consequently, FRET emission is only observed when the two species are extremely close, usually taken to be interacting with each other. Therefore, FRET provides information on specific molecular interactions, which occur on a length scale much smaller than optical microscopy usually allows. FRET is explained in much more detail in Chapter 12.

1.3.7 Fluorescence Lifetime Imaging (FLIM)

Fluorescence lifetime imaging microscopy utilizes the property that different fluorophores can have a wide range of difference fluorescent lifetimes. As explained above, the fluorescence lifetime is an inherent property of a fluorophore. In the same way that different excitation or emission wavelengths can be used to discriminate different fluorophores, so can the fluorescence lifetime. In FLIM, the detector can be triggered to measure fluorescence in a fixed time window after the excitation laser pulse. If the trigger occurs immediately after excitation, fluorophores with short lifetimes will be enhanced, whereas a late trigger will favor fluorophores with long lifetimes. At the most basic level, this provides another degree of separation for different fluorophores and can be used to

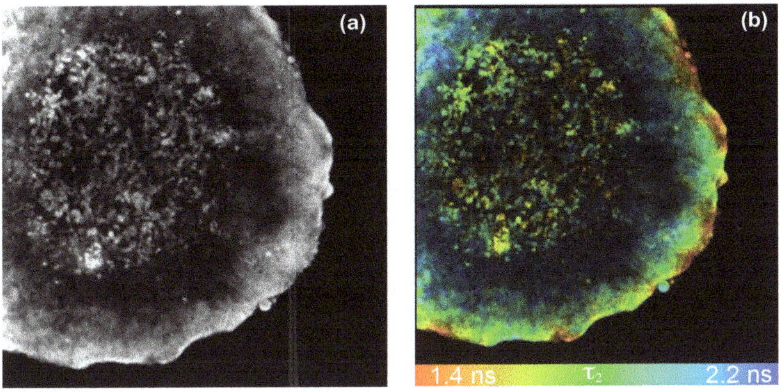

Figure 1.19 (a) Confocal fluorescence microscopy images of spheroid treated with 20 mM of curcumin. (b) Lifetime maps of the treated spheroid using the same compound (scale bar = 100 μm) [12].

discriminate against unwanted autofluorescence, for example. But at a more sophisticated level, the environment of a fluorophore also affects its lifetime and so FLIM can be used to probe for information about the different physical or chemical environment of the fluorophore, for example, free in the intracellular medium, or bound to a cellular structure.

The image in Fig. 1.19a shows a confocal microscope image of curcumin in a spheroid showing that the compound is observed only at the outer domains of the spheroid. Fluorescent lifetime imaging was used to show the release of curcumin from a hypoxia activated cobalt chaperone in DLD-1 colorectal cancer spheroids. This work takes advantage of the shorter fluorescence lifetime of curcumin compared to the parent compound allowing visualization of ligand release [12]. FLIM is discussed in more detail in Chapter 10.

1.4 Summary and a Look to the Future

Optical microscopy has been one of the most valuable tools in the history of science. After a long period of relatively incremental improvements, the past two decades have provided revolutionary advances in sensitivity, three-dimensional imaging and resolution. The fundamental properties of fluorescence needed to understand these techniques have been explained in this chapter.

All of the techniques alluded to in this chapter are limited by the lower limit of resolution, which is the wavelength of light. Some clever optical approaches and tricks have lowered this marginally, to maybe ~200 nm. But this is nowhere near the resolution afforded by techniques such as X-ray or neutron diffraction, or electron microscopy. These techniques utilize either a much shorter wavelength region of electromagnetic radiation (X-ray), or the de Broglie wavelength of light particles (electrons and neutrons). Both provide commensurately better resolution, to ~0.1 nm because the wavelengths involved are much shorter. But these techniques have their own limitations in terms of sample preparation and/or the need for crystalline samples. The ideal technique would be an optical microscopy technique with the resolution of electron methods

In the past few years, the fundamental resolution limit has been defeated with a new suite of optical microscopy techniques: photoactivated localization microscopy (PALM), stochastical optical reconstruction microscopy (STORM), and ground state depletion (GSD). These techniques are known collectively as *super-resolution microscopy* techniques, and all, so far, utilize fluorescence as the detection medium. Resolution down to ~20 nm can be obtained with commercial instruments and the challenge to improve super-resolution techniques further is the subject of competitive research programs. This book finishes with three chapters on super-resolution techniques, and, maybe, presages an upcoming revolution in optical and fluorescence microscopy.

Acknowledgments

The author wishes to thank the many people who gave permission to reproduce their images in this chapter, especially Nicole Bryce, Arthur Chien, Danielle Davies, Fu Dong, Eleanor Kable, Radim Schreiber, and Ying-Ying Su.

Image Credits

1. Cavadore, C., ESO. Source: http://www.eso.org/public/images/eso0202a, available under Creative Commons.
2. Strang, J., U.S. Air Force, ID 050118-F-3488S-003, available under Creative Commons.
3. Schreiber, R., http://www.nwf.org/wildlife/wildlife-library/invertebrates/firefly.aspx, used with permission.
4. Chien, A., Macquarie University, used with permission.
5. Engraving from Smith, A. (1848). *Smith's Illustrated Astronomy* (Cady & Burgess, NY).
6. Hookes, R. (1665). *Micrographia* (Martyn & Allestry, London).
7. Kable, E. P. W., Australian Centre for Microscopy and Microanalysis, University of Sydney, used with permission.
8. Su, Y. Y., Australian Centre for Microscopy and Microanalysis, University of Sydney, used with permission.
9. Davies, D., Brain and Mind Research Institute, University of Sydney, used with permission.
10. Bryce, N. S., and Weaver, A. M., Department of Cancer Biology, Vanderbilt University Medical Center, Nashville, TN, USA, used with permission.
11. Kable, E. P. W., and Dong, F., Department of Pharmacy, University of Sydney, used with permission.
12. Bryce, N. S., University of New South Wales, used with permission.

Chapter 2

The Ploem Fluorescence Microscope

Guy Cox
Australian Centre for Microscopy & Microanalysis,
University of Sydney, NSW 2006, Australia
guy.cox@sydney.edu.au

2.1 Introduction: Fluorescence Microscopes

Fluorescence microscopy was first introduced at the very beginning of the 20th century, and August Köhler—he of Köhler illumination—was one of the pioneers. These early fluorescence microscopes used a transmission light-path and ultraviolet illumination. The images tended to be very dim, since the blocking filter needed to be very dense to avoid blinding the operator. With the primitive filters of the time this meant that a lot of the fluorescence was lost as well. These microscopes were dangerous to use—accidental removal of the barrier filter could, and did, blind the operator (the author knew a victim of this—fortunately, it was a monocular microscope).

Despite their problems, these microscopes became widely used in one specialized field. The early cytogeneticists found that fluorescent dyes would label characteristic "bands" on chromosomes and

Fundamentals of Fluorescence Imaging
Edited by Guy Cox

ISBN 978-981-4774-85-7 (Hardcover), 978-1-351-12940-4 (eBook)
www.jennystanford.com

this enabled the geneticists to localize genes to specific regions of specific chromosomes.

A later refinement was to use a dark-field condenser to illuminate the specimen, so that no direct UV light entered the objective. This both improved safety and made the task of the barrier filter much easier. Image brightness could be improved. However, the risk remained—it is easy to misadjust a dark-field condenser. Fluorescence microscopy remained a highly specialized craft.

2.2 The Ploem System

In the late 1960s a Dutch professor, J. S. Ploem, changed fluorescence microscopy completely [1]. Ploem's revolution was to illuminate the sample from above, through the objective lens (Fig. 2.1). Illuminating light which does not excite fluorescence mostly passes harmlessly through the slide and cannot reach the eye. Since the illumination comes through a high numerical aperture objective, light-gathering is optimized and illumination is efficient. Optimal adjustment is much simpler than with any sort of transmission system.

To make this work required a new type of beam-splitter, a *dichroic mirror* which reflects short wavelengths and transmits longer ones. Multilayer *interference filters* (see below) were already being developed, although they were not yet in general use, so the required technology was available. Nevertheless, it took several years of collaboration between Ploem and the Schott company to develop usable designs. A Russian group had in fact already developed such mirrors for fluorescence microscopy [2], but this was the height of the Cold War and the groups were not aware of each other's work.

Ploem's other breakthrough was to realize that fluorescence did not necessarily require UV excitation. Dyes such as fluorescein and rhodamine (widely used at the time for applications such as water tracing) were bright, and excited by blue and green light respectively. These wavelengths were less damaging both to living cells and to human eyes. Just as important was the ability to distinguish different dyes by exciting at different wavelengths. Ploem's designs therefore,

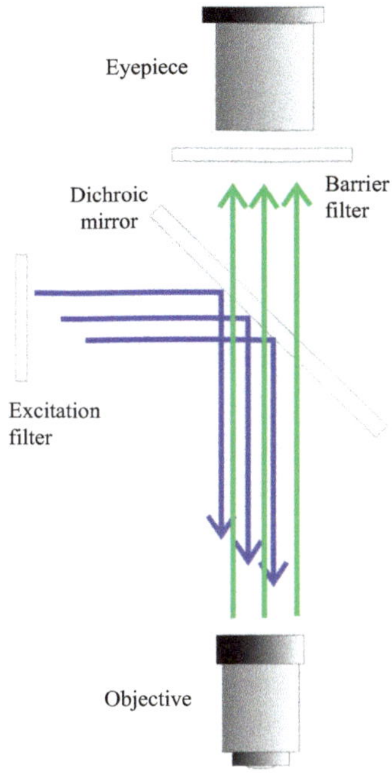

Figure 2.1 The Ploem system. Light enters through an excitation filter (here selecting blue light) and is reflected by the dichroic mirror, which reflects all wavelengths below 500 nm. Green fluorescence from the sample passes back through the dichroic mirror to the eyepiece(s). A barrier filter prevents any stray blue light from entering the eyepiece and can also exclude unwanted (e.g., red) fluorescence.

from his earliest prototypes, incorporated several dichroic mirrors on a slider.

A logical development from this, used by almost all manufacturers, was to combine the excitation filter, dichroic, mirror and barrier filter (Fig. 2.1) into a single filter cube (Fig. 2.2). This meant that a naïve user could not select an unworkable filter combination, and was another important step in taking fluorescence microscopy from an arcane technique practiced by specialists to a technique which any graduate student could use in his or her research.

Figure 2.2 A filter cube from a modern Nikon fluorescence microscope (see Fig. 2.8). This is designed for UV illumination so we see that the barrier filter (top) blocks any wavelengths shorter that 420 nm while the excitation filter (lower right) passes 330–380 nm. The dichroic (reflecting below 400 nm) is inside the block.

2.3 Filters and Mirrors

From the beginning of the twentieth century, great technological improvements were made in the fundamentally ancient technique of making glass filters. (Colored glass goes back to the Sumerians, around 2000 BC.) The driving forces were partly scientific, but the rapidly expanding popularity of photography also played a major part. These filters are made from glass with metal oxides added, and absorb light that is not transmitted. They still have a part to play in modern fluorescence microscopy, but have relatively gentle transitions between transmission and blocking (Fig. 2.3).

Interference filters were first developed in the 1940s [3]. They work on the principle of the Fabry–Pérot interferometer, which dates back to 1880. An F-P interferometer consists of two parallel partial mirrors, separated by a precise distance d (Fig. 2.4a). Light entering through the first mirror will bounce back and forth between the two. For a wavelength where $d = \lambda/2$ (or an exact

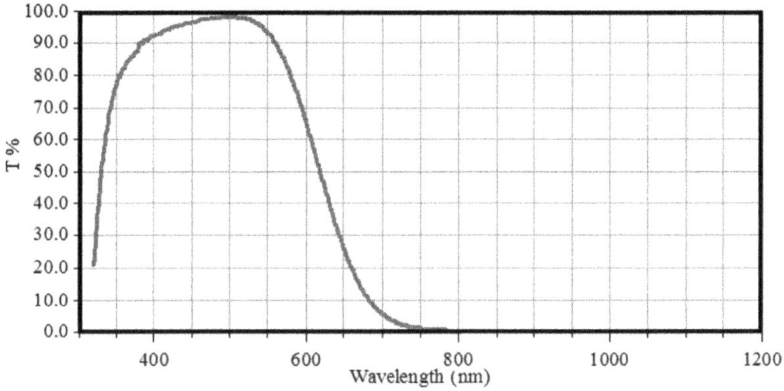

Figure 2.3 Spectrum of a BG38 glass filter. This is a typically used as an infrared blocker. The transition from transmission to blocking takes place gradually from 550 nm to 750 nm. Courtesy of Chroma Corporation.

multiple n of this) constructive interference will take place at each reflection and the light will propagate. Other wavelengths will not propagate in the cavity and will be reflected back.

Early applications of this principle [3] used evaporated metal layers as the mirrors on either side of a clear dielectric layer. This had rather poor transmission (~40%) although blocking was very efficient. Modern versions use a multilayer stack as the mirror as well (Fig. 2.4a). This consists of alternating layers with low and high refractive indices, each $\frac{1}{4}\lambda$ in optical thickness (Fig. 2.5). Reflected light will interfere constructively, transmitted light destructively. Since the layers have low absorbance (at least until one gets into the violet–UV region) the transmission of the complete filter will be high. This structure—the single F-P interferometer—is called the basic cavity. A complete filter will typically have multiple cavities (Fig. 2.4b).

One thing that should be obvious from this simple explanation is that a filter passing a given wavelength will generally also pass half that wavelength. In the case of a complex filter other wavelengths for which the numbers just happen to work out may also be passed. Therefore, additional blocking may be needed—often just a colored glass filter (Fig. 2.4b).

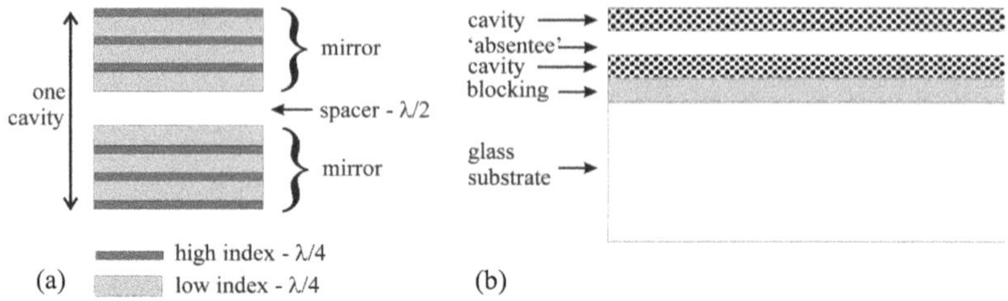

Figure 2.4 (a) A Fabry–Pérot interferometer, which forms the "basic cavity" of an interference filter. (b) A simple filter, with two basic cavities and a colored glass blocking layer.

Dichroic mirrors typically have to do without the blocking layer, since both transmitted and reflected light are used. This means that they are particularly prone to reflecting and transmitting unexpected wavelengths (Fig. 2.5). In general, a dichroic mirror is designed to work with specific excitation and barrier filters and the combination should be kept together. Since the angle at which rays pass through a cavity affects the path length, a dichroic is normally computed for 45° incidence and use at any other angle will change the wavelength characteristics.

Interference filters typically have a very sharp transition from transmitting to blocking (Fig. 2.5), and this is very valuable in fluorescence microscopy where many fluorochromes have a rather small Stokes shift (Chapter 1).

2.4 Light Sources

The ideal light source for fluorescence microscopy has (probably) not yet been developed. In principle one needs a wide spread of wavelengths running from at least 350 nm in the ultraviolet to about 600 nm in the red, since dyes excited at all these wavelengths are now in common use (Chapters 6, 7, and 8). It also needs to be a small source, since it must be focused on to a small area of the specimen. Carbon arcs were used in early, experimental work and meet most of these criteria, but are unstable, difficult to control, and hazardous.

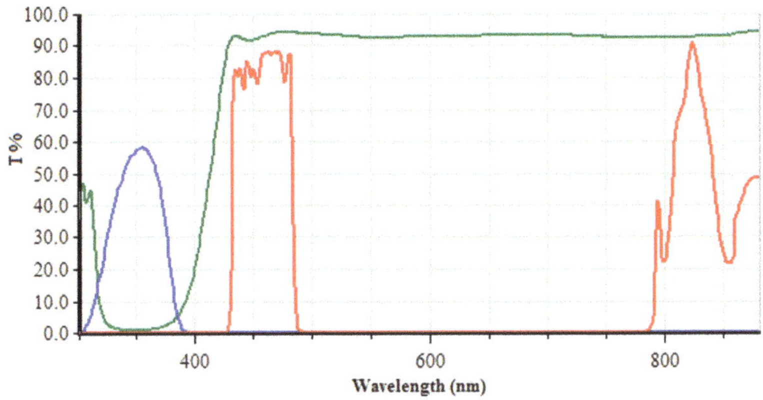

Figure 2.5 A filter set for imaging the DNA stain DAPI (Chapter 7). It illustrates both the advantages and quirks of interference filters. The 350 nm excitation filter (blue) has a rather low overall transmission because the materials used to make the evaporated layers do not transmit well in the UV. This is a not usually a problem since light sources at this wavelength are typically quite powerful. The dichroic mirror (green) changes from reflecting to transmitting over a small wavelength range (400–425 nm) *but* also transmits at 300 nm, so it must be used with the correct excitation filter. The emission filter (red) transmits a very precise narrow band, 430–480 nm, excluding other fluorescence that might be excited by 350 nm excitation, but it also transmits in the near-IR, well beyond visible and camera detection. *Take-home message*: filter sets are designed to work together; mix and match at your own risk! Courtesy of Chroma Corporation.

2.4.1 Arc Lamps

From Ploem's time up until the turn of the century the common light sources used in fluorescence microscopy were sealed high-pressure arc lamps. These use fixed, inert-metal electrodes in an atmosphere of xenon or mercury vapor, contained in a quartz envelope. An arc has the interesting property that as the current increases the resistance decreases, which would lead to a runaway situation without a suitable control system. This can be just a large metal resistor, since that has the opposite property—increased current increases the resistance. (The author, in his school days, ran an ancient carbon arc lamp from a cinema projector via a domestic radiant electric heater.) For microscopy something more sophisticated is required, so these light sources are not cheap.

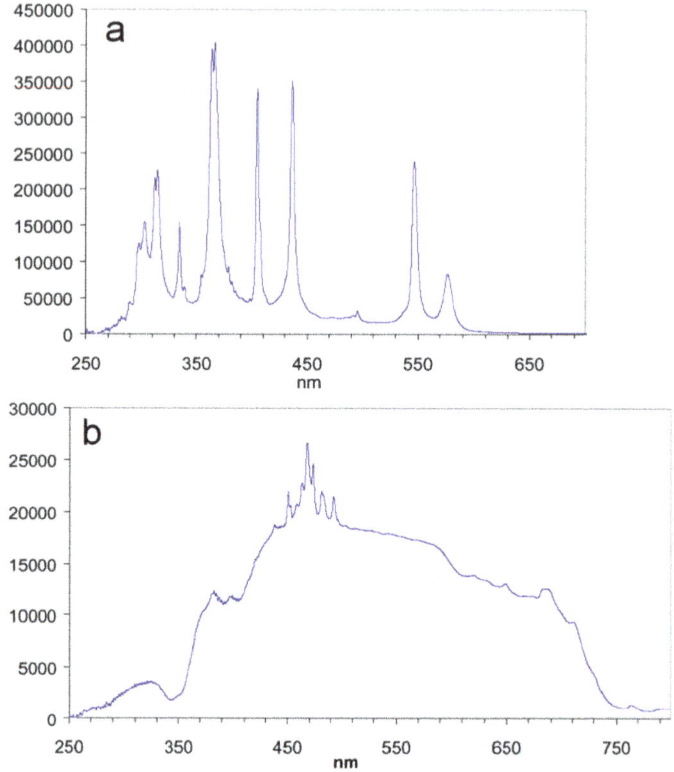

Figure 2.6 (a) Spectrum of a short-arc mercury lamp (HBO 50). There are prominent peaks in the UV, violet, and green, but blue is somewhat deficient. (b) Spectrum of a short-arc xenon lamp. The continuum is more important than the peaks, providing illumination throughout the visible spectrum. Excitation in the UV is weak.

Mercury arcs also need a system for vaporizing the mercury at start-up.

Mercury lamps have a spectrum with strong peaks in the green, violet, and ultraviolet (Fig. 2.6a), which favors some dyes over others, though there is a strong background continuum so that dyes such as fluorescein which do not match a peak are still excited. However no excitation is available in the red and yellow part of the spectrum. Xenon lamps have a more uniform spectrum, with only minor peaks, but are not strong in the ultraviolet

(Fig. 2.6b). Both run at very high temperature and pressure, and so can potentially explode in use, though when cold they are extremely robust. Lifetimes of the lamps are short, particularly for the mercury versions, and mercury lamps have the added requirement that once turned off they must be cooled completely before restarting, or the arc may become unstable and flicker. All these disadvantages mean that arc lamps are now decreasing in popularity, and account for only about 5% of new sales of fluorescent illuminators.

2.4.2 Metal Halide Lamps

The most popular illuminators at the time of writing are based on metal halide lamps. Their most familiar use in home or lab will be as the lamp used in video/data projectors, but they are also widely used in stadium and other public illumination. These are also high-pressure arcs but contain a complex mix of mercury, an inert gas (argon or xenon) and metal iodides or bromides. The inert gas helps the arc to start and the metal salts add different components to the spectrum. For example, sodium will add yellow, indium will add blue, and lithium will add red—all components lacking in the mercury spectrum. This gives tremendous versatility and just about any spectrum can be designed. They have a much longer life than mercury arc lamps, in the thousands of hours, but are very expensive to replace.

The lamps run at a high temperature and this must be kept constant—otherwise the spectrum may vary since the different salts have different boiling points. In microscope (and data projector) use this means a cooling fan is needed, and to avoid vibration this means that the lamp housing must be remote from the microscope. The connection is usually made by a liquid light guide since this has better performance in the ultraviolet than a fiber optic bundle. Liquid light guides contain an oil of high refractive index, and this must be enclosed in a tube made of (or coated with) a solid of lower refractive index for total internal reflection to take place. This solid is typically a member of the polytetrafluoroethylene (PTFE, Teflon) family of plastics. The liquid light guide is the Achilles heel of metal halide illuminators, since it is delicate (kinking it will destroy it) and

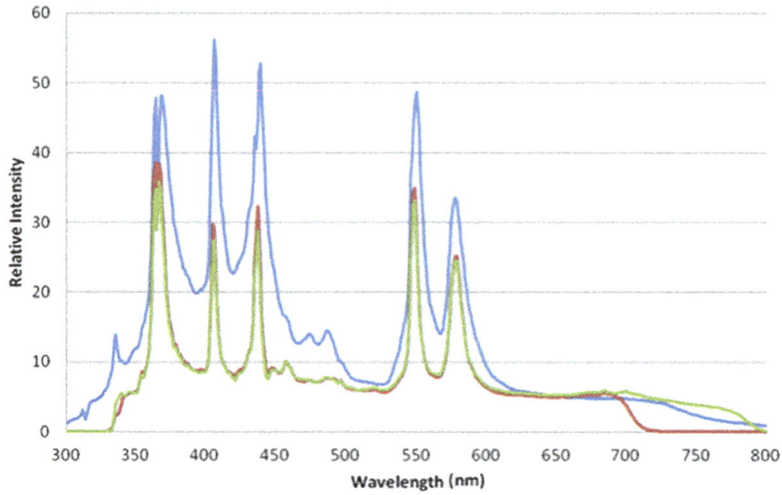

Figure 2.7 Spectra of metal halide fluorescence illuminators, measured at the end of the liquid light guide. Three models of X-cite illuminators: blue is the everyday 120, green is the 200 DC which runs on direct current for greater stability and has lower peaks, and red is the similar Exacte model designed for extreme stability and with an IR blocking filter. Courtesy of Excelitas Corp, and special thanks to Kavita Aswani for her help.

it has a finite life of just a few years before bubbles form and degrade performance.

2.4.3 LED Illuminators

The recent development of high-intensity and short-wavelength light-emitting diodes (LEDs) has provided another alternative for fluorescence illuminators. These run cool and are much more controllable than other light sources: their brightness can be varied and they can be switched rapidly. The capability for rapid switching makes them ideal for applications such as lifetime (Chapter 10) and ratiometric (Chapter 8) imaging, where otherwise expensive high-speed shutters would be needed. However, diodes have a limited spectrum, so several will be needed for advanced applications (Fig. 2.8). (White light diodes in consumer applications are nothing of the sort; low-end ones are near-UV diodes with fluorescent pigments, and high-end ones are three-diode chips in one package).

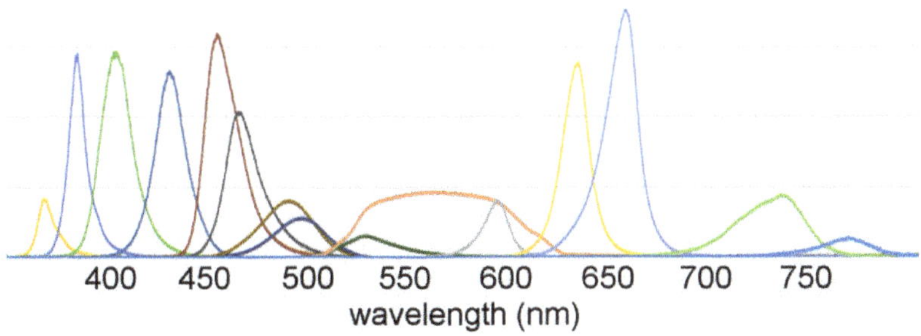

Figure 2.8 Spectra of a commercial multiple LED illuminator. The broad band in the yellow region (red line) comes from an LED with added fluorescent materials. Courtesy of Coolled Ltd., with thanks to Peter Call for his help.

Of course, fluorescent pigment technology can also be applied to LED fluorescence illuminators, to fill gaps in the spectrum (Fig. 2.8). Using multiple diodes also has the advantage that the user can select which wavelengths are turned on. All other illuminators have all wavelengths on at once.

LEDs have long lifetimes, and at the time of writing seem likely to take over as the preferred light source for fluorescence. Their initial high cost is dropping rapidly. Will something better come along? Maybe white light lasers? Who knows.

2.5 The Complete Microscope

Research-level fluorescence microscopes these days are often computer controlled, which tends to be great for the experienced user who can navigate the subtleties of the software but limiting to the naïve user who only selects the default options. Here we illustrate a Nikon Eclipse E800 of about 10 years ago—a fully featured research microscope in which all functions are manual—to show what the user can control.

The illumination source, as usual at that date, is a mercury arc, and since the brightness of this cannot be varied, neutral-density filters, in sliders, are used to reduce the intensity when required. Multiple filter blocks are housed in a slider so that

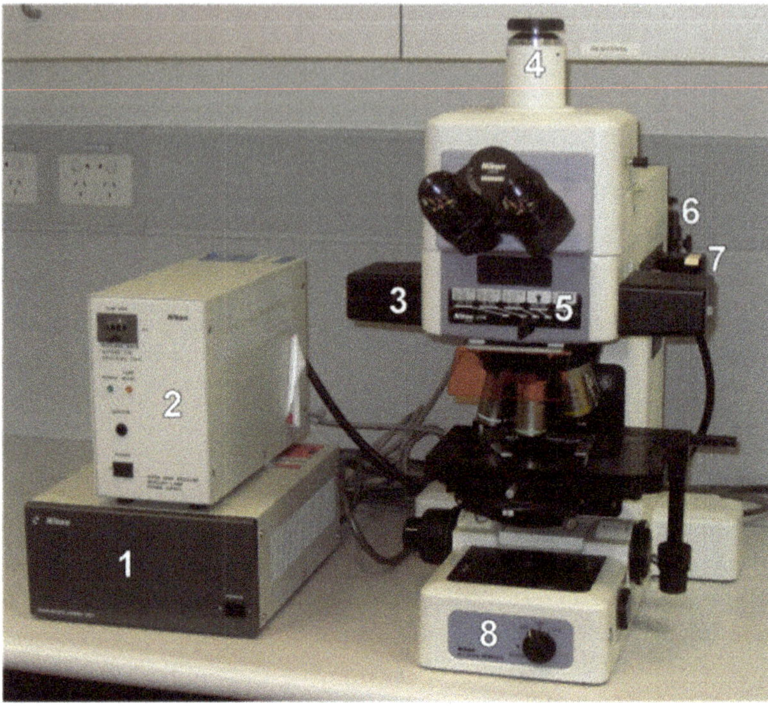

Figure 2.9 Nikon E800 fluorescence microscope. 1. Power supply for transmitted light illumination. 2. Power supply for fluorescence (HBO 50 mercury lamp) illumination. 3. Housing for sliders carrying filter blocks (Fig. 2.2). 4. Camera adaptor (no camera fitted). 5. Slider for selecting filter cubes. 6. HBO 50 mercury arc illuminator (at rear). 7. Neutral density filters. 8. Transmitted light control.

several different combinations can be selected as required. It must be evident that computer control can give much greater flexibility, particularly with the wavelength selection available from an LED illuminator. However, in the older microscope the individual components are much easier to identify. Computer control *is* the future, no doubt about it, but to get the best out of a microscope one must understand what is actually happening in the hardware.

References

1. Ploem, J. S. (1967). The use of a vertical illuminator with interchangeable dichroic mirrors for fluorescence microscopy with incident light. *Z. Wiss. Mikrosk.* **68**, 129–142.
2. Brumberg, E. M., and Krylova, T. N. (1953). O fluoreschentnykh mikroskopopak. *Zh. Obshch. Biol.* **14**, 461 (in Russian).
3. Hadley, L. N., and Dennison, D. M. (1947). Reflection and transmission interference filters. Part I. Theory. *JOSA* **37**, 451–465.

Chapter 3

Confocal Microscopy

Stephen H. Cody and Guy Cox
[a]*Microscopy Centre, Monash University, Monash, Vic. Australia*
[b]*Australian Centre for Microscopy & Microanalysis, University of Sydney, NSW 2006, Australia*
guy.cox@sydney.edu.au

3.1 Introduction

Commercially available confocal microscopes introduced in the late 1980s have revolutionized research in the bio-medical sciences. Coupled with the development of a vast array of fluorescent dyes and proteins, these technologies together represent one of the most enabling technological advances in biomedical research [1]. To be able to take full advantage of the capabilities that confocal microscopy offers, scientists must have a good working knowledge of the principle of confocal microscopy, the aberrations that degrade the performance of a confocal microscope, and the guiding principles to minimize these deleterious aberrations. This requires a rudimentary understanding of some principles of optics. It is the purpose of this chapter to explain, in easy-to-understand language, the principles of confocal microscopy, some important aberrations, and methods to optimize confocal microscopy.

Fundamentals of Fluorescence Imaging
Edited by Guy Cox

ISBN 978-981-4774-85-7 (Hardcover), 978-1-351-12940-4 (eBook)
www.jennystanford.com

3.2 The Confocal Principle

Confocal microscopy has been "invented" several times in history. This is largely because Marvin Minsky, who made and used the first confocal microscope, never published it in the scientific literature, though he did take out a patent [2, 3]. The confocal microscope has also been called a "dual focusing" microscope because, as Fig. 3.1 shows, a point source of light is focused on the specimen by an objective lens, and that illuminated spot is then imaged by an objective lens on to a pinhole in front of a detector.

The two objectives must not only be focused perfectly on the specimen, but must be perfectly aligned with each other in X, Y,

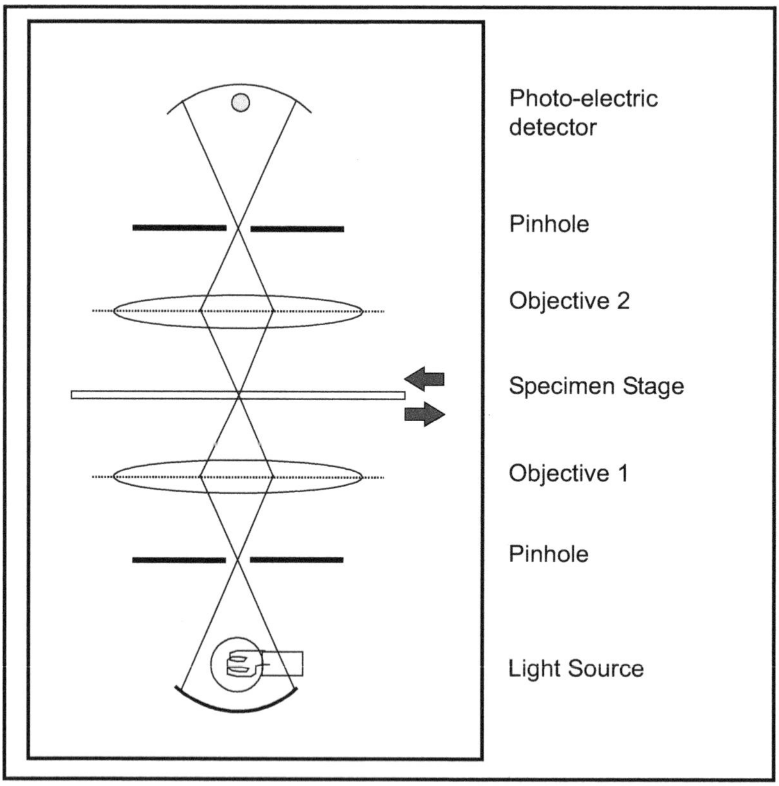

Figure 3.1 Diagram (modified from Minsky [3]) of Minsky's early transmission confocal microscope.

and Z dimensions (i.e., "confocal"). The difficulty in alignment of this optical system was a major drawback to this design. In order to build up a two-dimensional image, Minsky scanned the stage in X and Y, so that over time an image of the specimen was built up.

Minsky was not the first to use a confocal optical system, but he was the first to make a confocal microscope which formed an image. Confocal optics had been used some years earlier by Hiroto Naora [4, 5] in a microspectrophotometer, which produced a spectrum from a single point on a microscope slide.

Various groups subsequently developed confocal microscopes and we cannot deal with all of them here, but modern development stems from the work of Sheppard and Wilson at Oxford [6] (whose work led to the first production confocal microscope) and Brakenhoff in the Netherlands [7]. Their microscopes all simplified the design by using the same objective for both illumination and imaging, working either in reflection mode or fluorescence.

This all might seem like overkill, since either the illumination system or the imaging system is itself sufficient to form an image. Why do we need both? Confocal imaging has three important benefits:

(1) It eliminates scattered light from the image. The spot that is imaged is the only spot that is illuminated, so the image cannot be degraded by scattered light from elsewhere in the sample.
(2) It is focal-plane selective—that is, only the plane that is in focus is imaged. This is described in detail below.
(3) It can in principle provide a modest increase in resolution (a factor of $\sqrt{2}$) [8]. This is not really useful in cell biology since it requires a very small pinhole size which does not pass enough light for fluorescence microscopy [9].

Each of these features has been the main focus of different investigators. Naora was only interested in the exclusion of scattered light, since he wanted his spectrum to come from the point of interest. His samples were thin so focal plane selectivity was of no interest. This was also the case for the much more recent development of a confocal scanning electron microscope by Nestor Zaluzec [10]. Electron microscopes have very small numerical apertures, so other benefits are irrelevant, but the exclusion of scattered signal is very significant.

Minsky was aware of the potential resolution benefits, as his patent shows, but for his own research—looking at thick brain slices—focal plane selectivity was all important. And this is where, now, the significance of confocal imaging in cell biology lies.

Sheppard and Wilson were primarily interested in boosting resolution, and thought that signal attenuation would stymie any substantial effort at focal plane selection at depth. They therefore concentrated on material science imaging in reflection, where the signal was bright enough to permit the use of a small pinhole (Colin Sheppard, personal communication).

3.3 The Epi-fluorescence Confocal Microscope

The development of epi-illumination fluorescence microscopy (Chapter 2) provided a natural model for confocal microscopes to be built with a Ploem-style light path (Fig. 3.2). The epi-illumination light path greatly simplifies the alignment, as with this design the same objective lens acts as both the condenser and objective, and thus is well matched and always in alignment. This provides an ideal platform for a fluorescent confocal microscope, and Brad Amos at the Medical Research Centre, Cambridge, realized this [11]. Amos's design was the basis for the Bio-Rad MRC series of confocal microscopes, the first successful confocal microscope aimed to the cell biology market.

In epi-illumination, the light source (which is usually a laser to maximize the light intensity in the focused spot) is reflected from the surface of a chromatic beam splitter. This may be a simple dichroic mirror, as described in Chapter 2, a polarizing beam splitter (taking advantage of the fact that laser light is polarized and fluorescence generally is not) or a more complex acousto-optic device. Each has its own advantages. Dichroic mirrors are available in a wide range of single, dual, or triple wavelength configurations, but in a confocal system exchanging them requires very precise alignment—or sometimes realignment. Polarizing beam splitters avoid this problem but are less efficient, so additional blocking filters may be

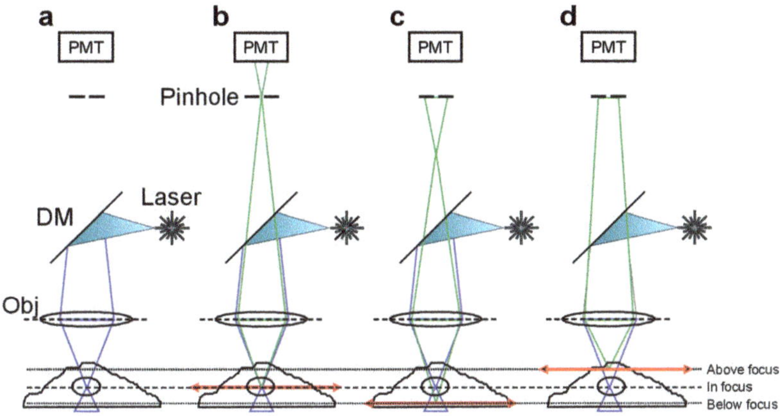

Figure 3.2 The principle of confocal fluorescence microscopy. (a) A light source (usually a laser) is reflected from a beam splitter (here a dichroic mirror, DM). The laser is focused through an objective lens (Obj). The excitation beam (in blue) is shown coming to focus within the nucleus of a cell. This is the "In focus" plane (dashed line). (b) Fluorescence is emitted from the point of focus in all directions. Only that cone of light that shines on the objective is drawn (green). All light from the in focus plane can pass through the confocal pinhole (Pinhole) to the detector, usually a photo multiplier tube (PMT). (c) This diagram shows fluorescence emitted from below the plane of focus. Some of this light enters the objective (green). However, as this signal is from an out of focus plane, it does not come to focus at the pinhole, and thus the great majority of this signal does not reach the PMT. (d) Similarly, the majority of fluorescent signal that is emitted from above the plane of focus, does not pass the pinhole.

required, and the detected signal may be reduced. Acousto-optic devices (based on tellurite crystals) are controlled electronically and can be extremely precise, switched very rapidly with no alignment issues, and have high transmission efficiencies. The downside is the cost, so they are typically found in high-end, expensive systems.

The excitation laser source is focused by the objective lens, which in this case is acting as a condenser lens. It is brought to focus within a specimen at the plane of focus. Fluorescent signal emitted from the plane of focus, and that is collected by the objective lens, will pass through the beam splitter and be brought to a focal point at the aperture of the confocal pinhole. This in focus light is thus able to pass through the confocal pinhole efficiently onto the detector

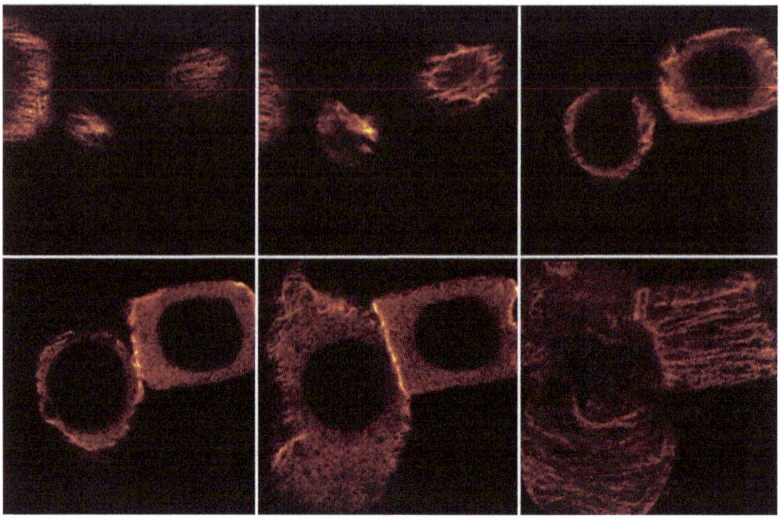

Figure 3.3 A confocal image stack. Six optical sections from a set of 74, showing microtubules in onion root tip cells.

(usually a photo multiplier tube, PMT). While the laser is most intense at the plane of focus, it also excites fluorescent molecules above and below the plane of focus. Light emitted from planes other than the plane of focus will have focal points either in front or behind the confocal pinhole, and the confocal pinhole will effectively block this out of focus signal from being detected. Hence, only the in focus signal is imaged, producing an optical section. By changing the focus of the microscope and collecting a series of images, a three-dimensional (3D) data set can be acquired very simply (Fig. 3.3).

3.4 Optimizing Confocal Performance

3.4.1 The Point Spread Function

The point spread function (PSF) is, in principle, a simple concept, but it is often difficult for biologists to understand since it seems counterintuitive. The PFS describes what happens to the light from every point within a specimen after it has passed through the optics of a microscope. Imagine observing something very small such as

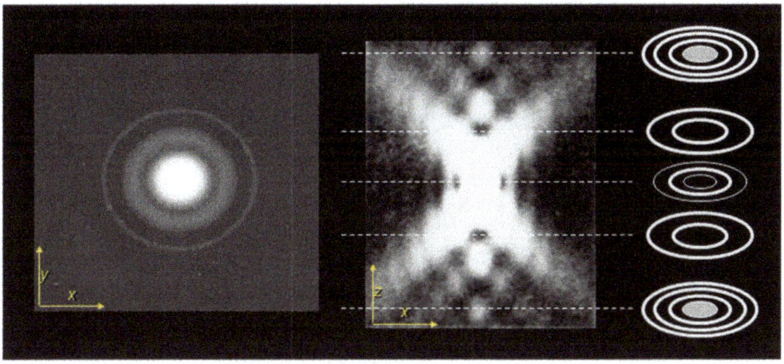

Figure 3.4 The point spread function (PSF) of an optical system. Left: The image of a point source at the plane of true focus. Right: A vertical section through the PSF, with XY slices at various distances from the plane of focus.

a single fluorescent molecule under the microscope. When it is in focus, what you observe down the microscope will not be a point, but a disk with, if you look carefully, a halo around it (Fig. 3.4). This is called the Airy disk, after the 19th-century scientist George Airy, who first described it. The diameter of the Airy disk depends on the numerical aperture (NA) of the objective, and not on any property of the sample. So far is pretty familiar to most microscopists. However, as you defocus, you expect the image of the point to become blurry, but it is not as simple as that. When you look closely at this blurred image it has structure, there are concentric rings that change as you focus up and down. This can often lead the user to think that there is something wrong, but in fact it shows that everything is OK.

The confocal pinhole should be set to just pass the Airy disk, and so the out of focus parts of the PSF will be excluded. In fluorescent microscopy the potential gain in resolution from a smaller pinhole is not practically achievable [9] and so any smaller setting will just lose signal. Equally anything larger will just pass out of focus light, degrading the image. Most microscopes will indicate the correct "1 Airy" setting. Figure 3.4 also explains why the depth resolution of a confocal microscope is always worse than the lateral resolution—the central spot extends further vertically than laterally. This is inescapable (except with the super-resolution techniques described in Chapter 14) but the difference is least at the highest numerical apertures. For this reason most manufacturers supply high-NA, low-magnification lenses specifically for confocal microscopy.

3.4.2 Aberrations

Where we have to look more seriously at the PSF is when we consider aberrations. Simple lenses have many imperfections and these are typically very well corrected in our expensive microscope objectives. But there are limitations, and in advanced systems such as confocal microscopes we are often coming up against these limitations. Of the many aberrations lens designers have to cope with there are just two that impact on us.

3.4.2.1 Chromatic aberration

Simple lenses bring different colors to different focal points since glass naturally has different refractive indices to different wavelengths. Achromatic lenses have a straightforward correction which brings two wavelengths, typically red and blue, to the same focus. Other wavelengths will not be too far out but in confocal microscopy we need precision—excitation and fluorescence must come to the same focus.

Fluorite lenses—usually with "fluor" in their name—used to use the mineral fluorite (calcium fluoride) in their optics, and maybe some still do, but the term has become generic. They are technically achromats, but the remaining wavelengths come much closer to the same focus. They can often be a good choice since their design depends more on special glasses than multiple elements so they have a good transmission.

Apochromat lenses are designed to bring three wavelengths to the same focus. Traditionally these were red, green, and blue, but with the wide range of lasers now available manufacturers are making tailored lenses for specific wavelengths—a popular choice is "violet corrected" apochromats to cope with the deep violet (~405 nm) lasers now in common use, which are outside the range of conventional apochromats. Deep-red shifted versions are also now on the market. From the point of view of accurate wavelength correction they are the "gold standard," but they do contain a lot of glass which reduces their transmission efficiency. They are also very expensive.

3.4.2.2 Spherical aberration

Spherical aberration is caused by the outer parts of a simple lens being more powerful than the center. It can, of course, be corrected, but there is a catch—in fact a double catch. First of all, it becomes a much larger problem as the numerical aperture (NA) increases—in fact it behaves as the fourth power of the NA. But high NA is very important in confocal microscopy since the Z (depth) resolution increases as the square of the NA.

Secondly, it can be corrected—perfectly—but only for one particular object and image condition. Yet the whole benefit of confocal imaging is the ability to obtain a series of in-focus images at different depths! For this reason high-NA objectives have markings indicating what they are corrected for. One—largely of historical relevance—indicates the image position. On older microscopes this will be 160, 170, or 250, indicating the distance in mm behind the objective at which the image should be formed for perfect correction, but on modern microscopes it will be ∞, implying that the image will not be formed by the objective alone, but can be formed by a "tube lens" at whatever position suits the microscope designer. It is the other end that is more important to us.

The highest NA objectives will be designed for oil immersion and will have "oil" engraved on the lens barrel. This implies that the refractive index of the medium beyond the objective will be oil of refractive index 1.515 (approximately the same as that of the front lens of the objective). If everything from that point on (coverslip and mounting medium) is the same refractive index we will be able to focus in until the objective hits the coverslip without loss of quality. There are bound to be minor differences, and they will affect the image, but this is close to an optimal scenario—especially since the working distance of a high-NA oil-immersion objectives is typically small.

Dry objectives are problematic. They are designed for use with a coverslip of thickness 0.17 mm (which will be, once again, engraved

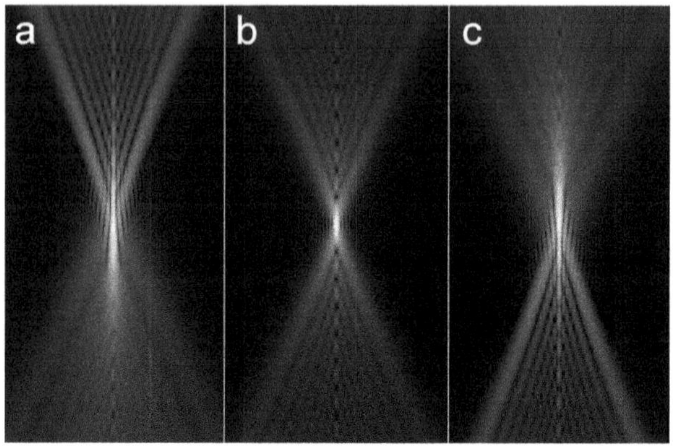

Figure 3.5 PSF distorted by spherical aberration: (a) overcorrected, (b) corrected, and (c) undercorrected. Image courtesy Wikimedia Commons.

on the lens barrel). Their SA correction is designed to be right for a specimen straight beneath the coverslip. If the specimen is in a permanent mount (refractive index approximately equal to that of glass) you can use a thinner coverslip and image deeper, but there is still no space for extensive Z-sectioning. Expensive high-NA dry objectives will have a correction collar to adjust for differences in coverslip thickness, and learning to use this (see below) will have a huge effect on image quality but the extent of useful Z-sectioning will still be limited. To get anything useful at a range of depths will require opening the pinhole to pass the aberrated PSF (Fig. 3.5), and accepting the consequent loss of resolution.

Other immersion media can be very valuable to the confocal microscopist. Glycerol immersion objectives permit aberration-free imaging of samples in common glycerol-based antifade media such as Citifluor. Water-immersion objectives offer deep imaging of living samples. These come in two forms—dipping objectives which do not use a coverslip, and coverslip objectives which do. Both will typically have correction collars—the former to adjust for refractive index, since often samples will be in saline media, the latter to adjust for coverslip thickness. Since there is water both above and below the coverslip the correction will not change with depth. Either version is very well suited to confocal optical sectioning of appropriate samples.

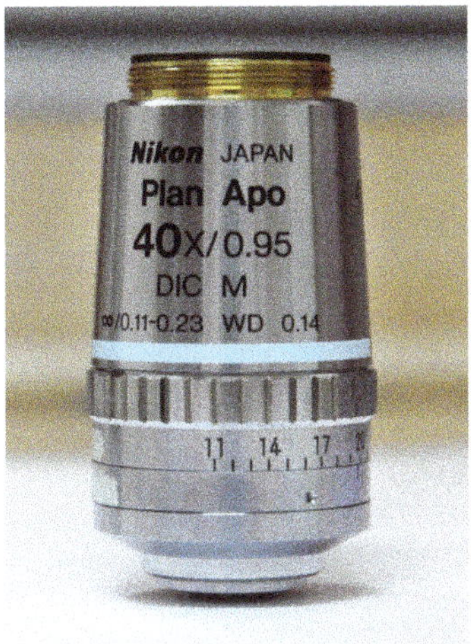

Figure 3.6 A high-power dry lens with a correction collar.

3.4.2.2.1 ***Using a correction collar*** A correction collar really needs to be adjusted by eye; the engraved markings are only a rough guide. Not only are coverslip thicknesses not very precise, adjusting by eye will enable correction for refractive index mismatches. It is a skill which needs to be learned, but learning is really not difficult. It is simplest if your sample has small fluorescent points, when the in and out of focus images will look more or less like the diagrammatic images shown in Fig. 3.7.

Suspensions and slides of fluorescent beads are widely available and make ideal training samples. Even if your specimen does not have punctate structures, once you have your eye in you will find it simple to set the collar so that the image looks the same on either side of focus. The gain in image quality will be dramatic.

3.5 High-Speed Confocal Microscopy

Point by point scanning is inevitably slow compared to widefield microscopy. This was recognized back in the early days, and Czech scientist Mojimir Petrán [12], working in conjunction with researchers at Yale in the US, devised a microscope using multiple

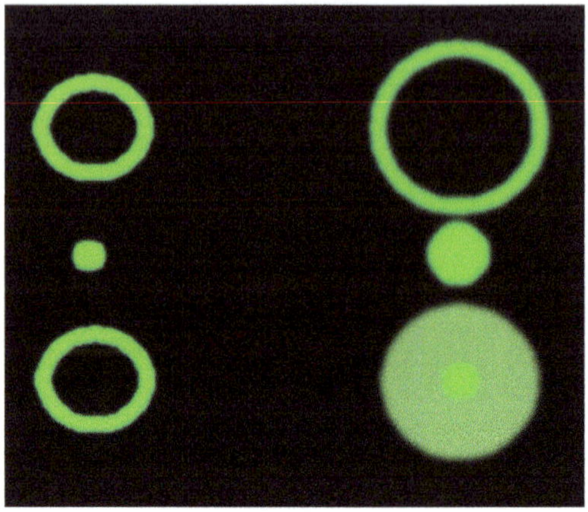

Figure 3.7 Simulated in- and out-of-focus images of a point object with and without spherical aberration.

pinholes on a spinning disk so that high-speed imaging was possible. Some confocality was inevitably sacrificed, since some of the returning light could pass through the wrong pinhole. That microscope, and its subsequent commercial derivatives, suffered from a light budget that was insufficient for fluorescent imaging, so it could only be used in reflection mode. As such, it does not really belong in this chapter.

The most successful realization of this concept came with a scan head devised by the Japanese Yokogawa company, which had a disk with micro-lenses above the pinhole disk (Fig. 3.8). This improved the illumination light budget to the extent that fluorescence imaging was possible, and since the image was collected between the two disks stray reflected excitation light was not an issue.

This was turned into a successful commercial product by the Perkin Elmer company, and marketed for many years, but it is no longer produced. The spinning disk confocal microscope is no longer a mainstream commercial product, though it survives in some niche markets. Why is this? Firstly because multipoint scanning is much better implemented in multiphoton microscopy (Chapter 4) where focal plane selectivity is achieved at the excitation

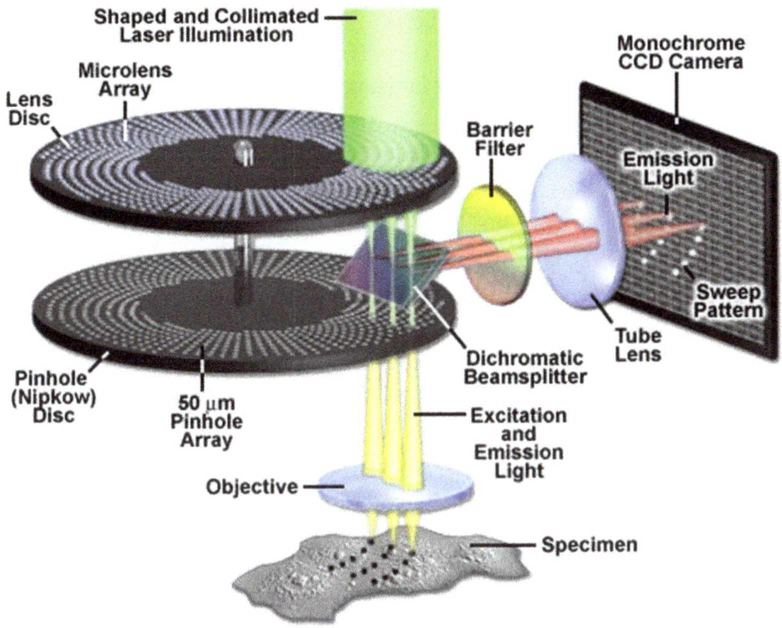

Figure 3.8 The Yokogawa scan head. Courtesy Yokogawa.

stage, so a detection pinhole is not required and crosstalk is not an issue.

The second point is that single-point scanning speeds have been improved to the point where video rate is routine, and higher speeds are possible. This brings its own problem, since such speeds are only achieved with resonant scanning [13]. Regular confocal scanning is controlled by stepper motors, so that each scan point is equivalent in spacing and dwell time. With resonant scanning the scan mirror is scanned at its resonant frequency, so the scan motion is sinusoidal but scan speeds can be much higher. The resulting image requires correction, implemented in software or software and hardware, to remove the distortion and uneven illumination that would otherwise result. Computing power is now such that this correction is simple, and all major manufacturers now offer high-speed resonant scanning.

References

1. Tsien, R. Y. (2003). Imagining imaging's future. *Nat. Rev. Mol. Cell Biol.*, **4**(Suppl), SS16–SS21.
2. Minky, M. US Patent 3,013,467, submitted 1957, granted 1961.
3. Minsky, M. (1988). Memoir on inventing the confocal scanning microscope. *Scanning*, **10**, 128–138.
4. Naora, H. (1951). Microspectrophotometry and cytochemical analysis of nucleic acids. *Science*, **114**, 279–280.
5. Naora, H. (1955). Microspectrophotometry of cell nucleus stained by Feulgen reaction. 1. Microspectrophotometric apparatus without Schwarzschild-Villiger effect. *Exp. Cell Res.*, **8**, 259–278.
6. Sheppard, C. J. R., and Wilson, T. (1978). Depth of field in the scanning microscope. *Opt. Lett.*, **3**, 115–117.
7. Brakenhoff, G. J., Blom, P., and Barends, P. (1979). Confocal scanning light microscopy with high aperture immersion lenses. *J. Microsc.*, **117**, 219–232.
8. Cox, G. C. (1993). Trends in confocal microscopy. *Micron*, **24**, 237–247.
9. Cox, G. C., and Sheppard, C. J. R. (2004). Practical limits of resolution in confocal and non-linear microscopy. *Microsc. Res. Tech.*, **63**, 18–22.
10. Zaluzec, N. J. (2003). The scanning confocal electron microscope. *Microsc. Today*, **6**, 8–12.
11. White, J. G., Amos, W. B., and Fordham, M. (1987). An evaluation of confocal versus conventional imaging of biological structures by fluorescence light microscopy. *J. Cell Biol.*, **105**, 41–48.
12. Petran, M., Hadravsky, M., Egger, M. D., and Galambos, R. (1968). Tandem scanning reflected light microscope. *J. Opt. Soc. Am.*, **58**, 661–664.
13. Tsien, R., and Backsai, B. J. (1995). Video-rate confocal microscopy. In: Pawley, J. (ed.), *Handbook of Biological Confocal Microscopy*, 2nd ed. (Plenum Press, New York), pp. 459–478.

Chapter 4

Multiphoton Microscopy

Mark B. Cannell[a] and Guy Cox[b]

[a]*Department of Physiology, Pharmacology and Neuroscience, University of Bristol, Bristol BS8 1TD, UK*
[b]*Australian Center for Microscopy and Microanalysis, Madsen Building F09, University of Sydney, NSW 2006, Australia*
mark.cannell@bristol.ac.uk, guy.cox@sydney.edu.au

Conventional (visible light) fluorescence microscopy was revolutionized by the sectioning ability provided by the confocal microscope. There are, however, several limitations associated with visible light confocal microscopy: (i) Signal loss arising either directly or indirectly from chromatic aberrations. (ii) Signal loss due to scattering. (iii) Photodamage (especially at shorter wavelengths). (iv) Possible "inner filtering" effects due to the absorption of the excitation light as it passed through the sample. These problems are especially acute at near- UV wavelengths, where several popular calcium indicators (Fura-2, Indo-1), DNA labels (DAPI, Hoechst), and NADH are excited. Furthermore, UV lasers are tricky to align and, due to longitudinal chromatic aberration, it may be difficult to ensure confocality (note that most apochromats are not corrected for near UV, and some do not even transmit it). The development of multiphoton microscopy in the 1990s offered a potential solution to these problems as well as providing the

Fundamentals of Fluorescence Imaging
Edited by Guy Cox

ISBN 978-981-4774-85-7 (Hardcover), 978-1-351-12940-4 (eBook)
www.jennystanford.com

ability to probe biological responses with a three-dimensionally (3D) resolved source of excitation. In this chapter, we describe two-photon excitation microscopy (2PEM) as an alternative approach to the imaging problems faced by scientists.

4.1 Introduction

Two-photon excitation provides a radically different approach to many of the problems of confocal microscopy. While conventional confocal microscopy uses a pinhole to reject light emitted outside the focal plane the excited volume in 2PEM is determined entirely by the excitation light field. This self-sectioning ability is a nonlinear optical process that is dependent on the numerical aperture of the objective, the intensity of the light, and the probability of exciting a molecule with multiple photons of lower energy than normally associated with the transition from the ground state to the singlet state. This can be more easily understood in terms of the Jablonski diagrams for excitation shown in Fig. 4.1. (see also Chapter 1). In single-photon excitation, the energy of the photon (hv) raises the molecule to an excited singlet state (S1) from the ground state because the interaction of the photon electric field on that of the molecule can provide sufficient energy. Subsequently the molecule decays from the S1 exited state via the vibrational ladder in ~10 ps to emit fluorescence in ~10 ns or via a further transition to a triplet state and luminescence (in ~10 ms). The excited state may also be quenched by transfer of heat via molecular collisions, but this "loss" process will not be considered further. The fact that the emitted photon must be of lower energy (longer wavelength) than a single exciting photon was recognized by G. G. Stokes in 1852 [1] and is defined by the term Stokes shift, which defines the average difference in excitation and emission wavelengths. In 1931, Maria Goppert Meyer [2] proposed that the excited state might also be reached by a combination of lower energy photons and this was demonstrated 30 years later after the invention of the laser, which was needed to produce a sufficiently bright light source [3]. The idea that two-photon excitation could be used in microscopy was first proposed in 1978 [4]. This excitation process

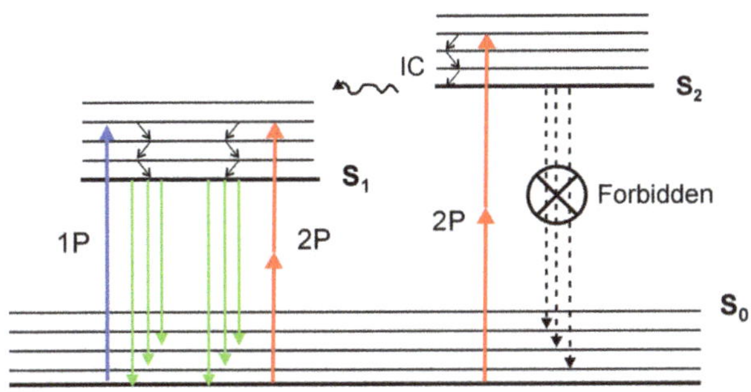

Figure 4.1 Jablonski diagram for two-photon excitation. In the simplest case (left) 2 photons each with ~half the single photon energy will excite to the same S1 state, from which fluorescence will occur. In some fluorochromes, particularly symmetrical molecules, selection rules prohibit this and excitation must occur to a higher state, typically S2. An intersystem crossing (IC) must take place to the S1 level before a fluorescent photon can be emitted.

is generated by the (near) simultaneous absorption of two (or even more) lower energy photons whose electric fields overlap in time and space. Therefore very intense light sources are needed and in 2PEM this usually achieved by focusing the light from a "mode-locked" (pulsed; see below) laser as first demonstrated by Denk et al. [5]. Although the quantum state on absorption of the two photons is different than for single photon excitation, relaxation through the vibrational ladder to the singlet excited state (from which subsequent fluorescence or other processes occur) is the same as for single-photon excitation. Therefore, although the *excitation* spectrum for TPMEM is quite different, there are no differences in the emission spectra of fluorochromes excited by a two-photon light source. In fluorescence microscopy, the most obvious change in microscope design will be in the excitation dichroic mirror which will need to be changed to a short pass mirror, rather than the conventional long pass mirror used in epifluorescence microscopes. Apart from that, few other changes in the microscope system are required, provided all ancillary optics are suitable for use with high peak power near infrared light (i.e., 700–1000 nm wavelength at generally <1 W average power).

4.2 Multiphoton Excitation

As a rule of thumb for most fluorochromes, the optimal two-photon excitation wavelength will be slightly less than twice the wavelength required for single-photon excitation. In addition, the two-photon excitation spectrum is significantly broadened and blue-shifted compared to a doubled one-photon spectrum. This may be related to quantum mechanics requiring that a *symmetrical* molecule must be two-photon excited to a different energy state than that reached by single-photon excitation (due to the combined spin of two photons), so the excitation spectrum might be expected to be different from the single photon excitation spectrum scaled by a factor of 2 (Fig. 4.1). This effect is reduced by the nonsymmetrical nature of many acceptor molecules. An additional factor is that the probability of a two-photon transition is increased if there is an allowed quantum state near the state produced by the superposition of the first photon quantum state with that of the acceptor molecule. Such virtual states are very short-lived but provide a simple explanation for the change in spectral shape between one- and two-photon excitation. In addition, the existence of a short-lived virtual state overcomes the impossibility of two photons arriving *simultaneously* at the same point in space.

To have two photons arrive almost simultaneously at a single molecule requires a very large photon flux. In 2PEM this is facilitated by two factors. The first is that the laser source can be mode-locked so that the laser energy is delivered as a rapid train of ultrashort (<1 ps) pulses. For a typical 2PEM system using a solid state titanium–sapphire (TiS) laser, the pulses are typically $\sim$100 fs (10^{-13} s) long with a repetition rate of $\sim$80 MHz. This pulse train has a duty cycle (ratio of on- to off-time) of $\sim 10^{-5}$ so the peak power (or photon flux) is approximately 10^5 times the average power. Thus pulse peak powers in the kW range can be achieved with only a few hundred mW of average laser power. Were this not the case, specimens would suffer rapid heating and destruction. This low duty cycle repetition rate is also compatible with fluorescence lifetimes of $\sim$10 ns, which gives fluorescent molecules time to decay back to the ground state between each pulse. This helps avoid further excitation to even higher levels that might result in severe phototoxic effects.

While short pulses increase the probability of two-photon excitation, having too short a pulse produces undesirable effects. First there is the possibility of three (or more) photon excitation. Since this would correspond to very short UV light, this would be highly undesirable for biological and other delicate specimens. The ratio of two- to three-photon excitation varies in inverse proportion to pulse duration so pulses no shorter than $\sim$100 fs in duration are generally used. Second, to produce short pulses, the laser spectral bandwidth must be increased. The shorter the pulse the greater the bandwidth needed (since for a given pulse shape the time/bandwidth product should be constant) and this leads to increased temporal dispersion as the pulse travels through media, since the refractive index of the medium varies with wavelength. Thus paradoxically, a very short pulse with a broad spectrum may become longer than an initially longer (narrower spectrum) pulse. To obtain a short laser pulse in the specimen, there should be minimal phase distortion of the frequency components making up the pulse at the end of the entire optical train, including the specimen itself. However, the variation in refractive index with wavelength causes the quadratic terms in the phase/wavelength relationship to change—i.e., bluer components of the pulse now arrive later (for normally dispersive media) than redder components so the pulse becomes spread in time. The entire collection of wavelengths making up the pulse travels at the speed of light within the specimen and has a "group velocity" and dispersion of these wavelengths is called *group velocity dispersion* (or GVD). It can be offset by "pre-chirping" the pulse so that the specimen recompresses the pulse to near its minimum duration [6]. Put simply, pre-chirping simply imposes the opposite wavelength-dependent delays in the pulse by ensuring that blue components of the pulse travel less distance than red components, and this can be achieved by a pair of prisms in the optical train [6]. These imposed delays are removed by the normal dispersion associated with the microscope optics (and specimen) so the pulse becomes recompressed at the focal point. The GVD is closely related to the thickness and Abbe number of the material through which the pulse passes and may be estimated by the Sellmeier equation, although for complicated microscope optics calculation of the actual GVD would be very involved as well as being specimen dependent.

However, it turns out that near optimal pre-chirping can be achieved by simply observing the change in emitted signal intensity as the pre-chirp is adjusted (i.e., by altering the distance and/or depth of the beam inside the prisms of a prism-based system). Several modern laser systems for 2PEM now include variable pre-chirping options.

The second factor that increases local light intensity to cause multiphoton excitation is the ability to focus laser light to a diffraction-limited spot. This not only increases the local photon flux but also causes the photon flux density to vary with the square of distance above and below the focal point. Since the probability of two photons arriving at a point at almost the same time depends on the probability of one photon being there multiplied by the probability of a second photon also being there, the probability of two-photon excitation depends on square of the flux density. As a result, the probability of two photon excitation depends on the 4th power of distance from the focal point in a focused beam and two-photon excitation will be limited along the optical axis to a diffraction limited region (Fig. 4.2). It is this spatial restriction of two-photon that is key to the 3D imaging capability of TPMEM. No pinhole is required to block out of focus light (as is the case in confocal microscopes), because the excitation itself is spatially restricted and little excitation occurs outside the focal region. This also limits photo-damage to the region being imaged as shown in Fig. 4.3. It has often been said that two-photon excitation is less damaging to specimens but this is only true for a spatially extended specimens; within the region being imaged damage rates can be higher than single-photon excitation—possibly related to the fact that the electric field strength may approach that of chemical bonds themselves as well as higher order multiphoton excitation effects. Thus, while volume imaging a 3D specimen with single-photon excitation leads to photodamage throughout the specimen as each plane is imaged, with 2PEM photodamage outside the focal plane being imaged does not occur. This can be clearly seen in the photo-bleached region shown in Fig. 4.3.

Finally, the spatial restriction of two-photon excitation also limits inner filtering effects due to absorption of the excitation light as it

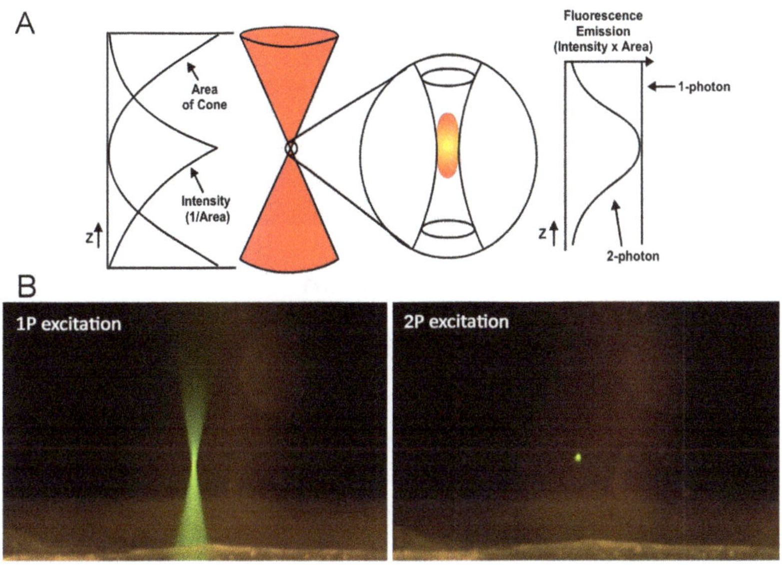

Figure 4.2 Single-photon excitation compared with two-photon excitation. (A) Diagrammatic explanation. The cone of illumination comes to a diffraction-limited spot at the focus, where the intensity is at a maximum. In the single-photon case the *total* excitation is the same at each plane, but in the two-photon case it is strongest at the focal plane. (B) Actual demonstration in a solution of fluorescein. With single-photon excitation fluorescence is excited throughout the entire illumination cone, but in the two-photon case it is confined to the focal point.

passes through the specimen. This can be seen in Fig. 4.2 as the dimming of the cone of excitation on the far (upper) side of the focal region.

4.2.1 Scattering Samples

Light scattering in biological samples is a significant problem—it was for this reason that "clearing agents" were developed early in the history of microscopy. This light scattering reduces the intensity of both the excitation light and the emitted fluorescence and appears as an overall loss of image intensity and relative contrast in fluorescence microscopy. For biological samples the scattering sources are usually lipid and protein aggregates as well as structures

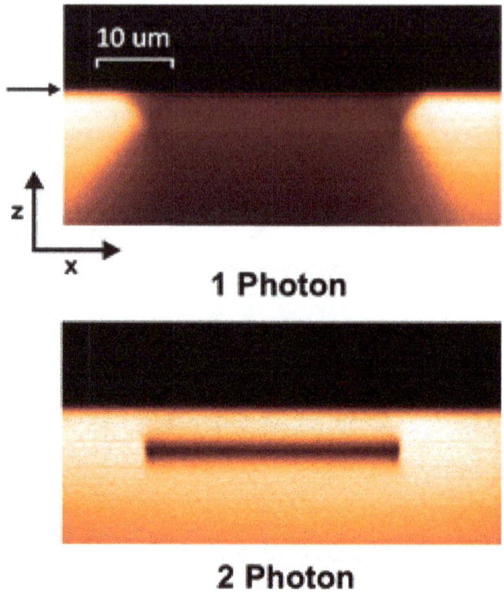

Figure 4.3 XZ scans through a fluorescently labeled gel which has previously been imaged at one plane in single- or two-photon imaging. Single-photon excitation bleaches the sample through the entire cone of illumination, while in two-photon mode only the imaged plane is bleached. Arrow indicates the cover slip.

such as nuclei and mitochondria, implying mixed Rayleigh/Mie scattering. Therefore, the scattering of light will be wavelength dependent although possibly not as strong as the 6th power wavelength dependence for pure Rayleigh scattering.

It follows that the near-IR wavelengths used for 2PEM will penetrate biological specimens more easily. In addition, scattered light will not be blocked by a confocal pinhole. This is illustrated in Fig. 4.4, which shows how the scattered photons may be blocked by a confocal pinhole. If the scanning system is bypassed in 2PEM (non-descanned detection) by placing the detectors close to the objective rear aperture, highly scattered photons can still be detected (note that the source of these photons is produced by the three-dimensionally resolved excitation point in 2PEM, so resolution is not affected).

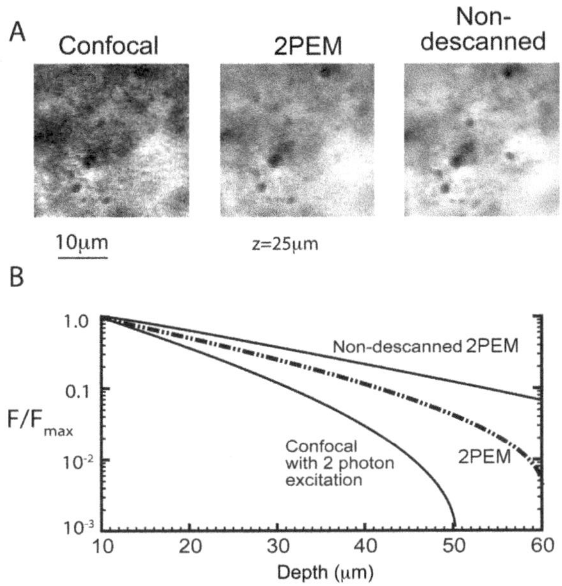

Figure 4.4 Two-photon imaging of fluorescently labeled cheese. With a confocal pinhole in place, the signal falls off very rapidly with depth. With the confocal optics but a fully open pinhole (2PEM), there is still a substantial decline since the system is not truly widefield. With a non-descanned detector placed close behind the objective, much more scattered light is detected and a usable signal is detected down to a depth of 60 μm. The images are taken at 25 μm depth.

The relative benefit of different detector modalities in a strongly scattering sample is shown in Fig. 4.4A, which shows sample images from fluorescently labeled cheese (a highly scattering sample) recorded from a plane 25 μm from the cover slip using the same two-photon excitation power. When a pinhole is present it blocks many of the photons from the sample resulting in a grainy image which is slightly improved by opening the pinhole fully. However, the long path length of the optical train (coupled with limited detector size) means that some confocality still exists (i.e., the microscope is still not fully wide field). By using non-descanned detection, the signal is further improved and details obscured by noise in the other images are clearly visible. For this sample, Fig. 4.3B shows that the signal deteriorated most quickly in the confocal mode, decreasing to less than 10% in 30 μm and was essentially undetectable by 50 μm. By comparison, when the pinhole was fully opened the 10% point occurred at a depth of 40 μm. Using non-descanned detection, the signal falls approximately exponentially within the plotted range and is still stronger than 10% of the maximum signal at a depth of 50 μm. Therefore, for this sample, non-descanned detection results in an improvement in signal strength by an order of magnitude over confocal detection at depths >40 μm. When this improvement is added to that arising from the doubling of the excitation wavelength, it is apparent that 2PEM with non-descanned detection is a superior method for deep fluorescence imaging in scattering samples.

The point spread function (PSF) is the same as for a confocal microscope except that the wavelength is longer (typically 700 nm to 1000 nm instead of 450 nm to 650 nm, assuming that we want to excite dyes that require blue or near-UV light in single-photon excitation). The longer wavelength might be expected to reduce the resolution substantially. However, in a confocal microscope a lateral resolution improvement of $\sqrt{2}$ relative to wide-field is potentially present (Chapter 3). In fluorescence imaging, this potential essentially cannot be achieved, because we cannot close the pinhole down to the tiny size required without losing most of our signal. In multiphoton mode, we do not need a pinhole, so the full $\sqrt{2}$ resolution improvement is always present. So what resolution can we expect?

$$\begin{aligned} r &= \frac{0.61\lambda}{\sqrt{2}\mathrm{NA}} \\ &= \frac{(0.61 \times 800)}{(1.414 \times 1.4)} \\ &= 247\ \mathrm{nm} \end{aligned}$$

This means that using 800 nm excitation we are getting a resolution, in X and Y, which is virtually as good as we will get with confocal excitation in the visible range. This can be seen in Fig. 4.5, which compares the point spread functions of conventional confocal microscopy with the pinhole set to one Airy unit with 2PEM without a pinhole.

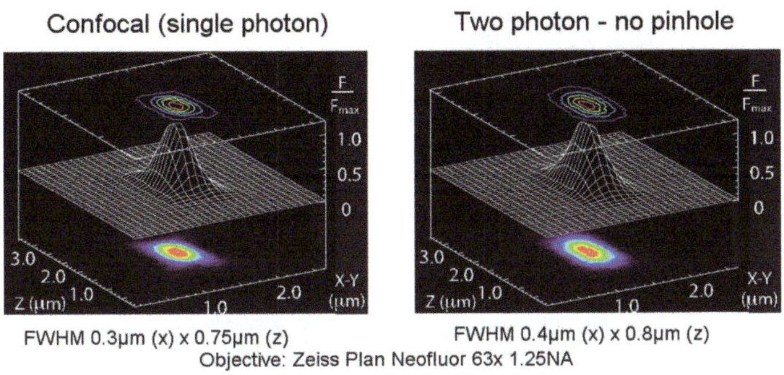

Figure 4.5 Point spread functions (PSFs) of confocal and two-photon microscopes with the same objective lens. In spite of the longer wavelength, the two-photon microscope has almost equivalent resolution.

4.3 Two-Photon Flash Photolysis

The spatially restricted nature of the excited region in 2PEM also introduces the possibility of photo-releasing a chemical from a caged precursor in a three-dimensionally restricted volume. This "flash photolysis" is facilitated by the ability of two-photon excitation to simulate the near-UV region, where photon energies are sufficient to break bonds in suitable acceptor molecules.

Microscope optics generally perform poorly in the near-UV region since the rapid change in refractive index in this region for most glasses leads to chromatic (and other) aberrations. In addition the UV light normally used for uncaging is damaging to biological specimens; it passes throughout the specimen causing damage everywhere. Since the aim of such experiments is to examine dye movement in living cells, this is a serious limitation.

2PEM reduces these problems; the source is near IR, which not only penetrates the specimen but also limits any damage to the focal plane. It is also easier to align the focal planes of the photolysis light and visible dyes. This allows one to examine biological responses with a visible wavelength probe (e.g., a Ca dye) while simultaneously and independently controlling the photolysis of a caged molecule. This is illustrated in Fig. 4.6A, which shows the photolysis of caged Ca with the resulting calcium diffusion being measured with Ca dye fluo-3.

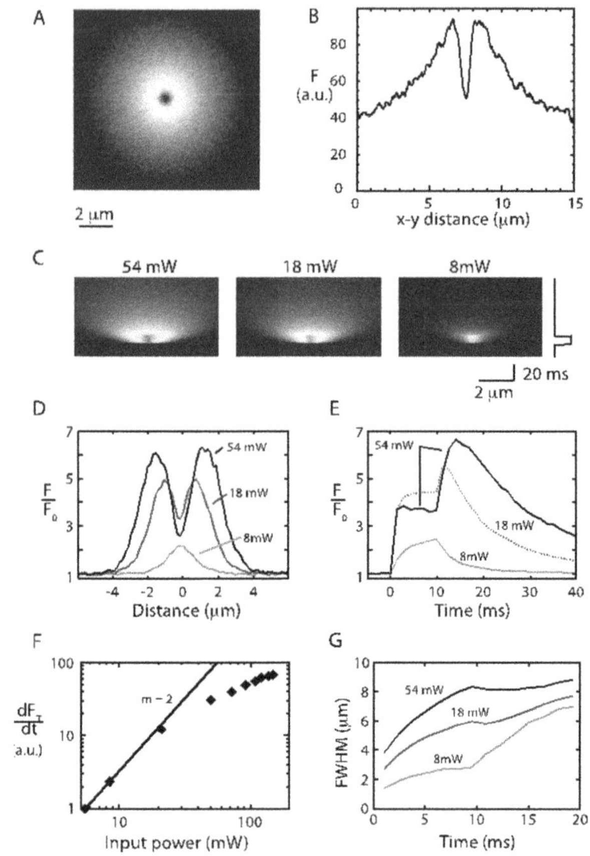

Figure 4.6 Uncaging of Ca. (A, B) After uncaging, the Ca diffuses freely but a dark spot in the center indicates, where the 2P beam has bleached the indicator dye. (C, D) Reducing the intensity of the 2P beam eliminates the bleaching. (E) The jump in fluorescence after the 2P beam is turned off is just diffusion, not a biological effect. (F) The increase on fluorescence with time is quadratic, proving that it is a two-photon effect. (G) Diffusion of calcium (measured as fluorescence full width half maximum) with time.

In this experiment the 2PEM beam was stationary (bypassing the scanning optics), while the fluo-3 response was monitored by the scanning system of the same confocal microscope. The resulting pattern of Ca release is as one would expect for diffusion from a point source, except for an unexpected dark region in the center.

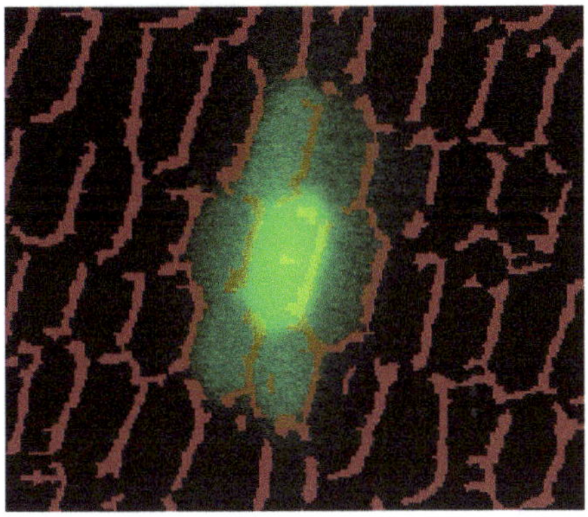

Figure 4.7 Diffusion of uncaged fluorescein in eye lens fiber cells.

This is due to the 2PEM beam also photobleaching the reporting fluo-3, and while this can be reduced at lower powers (Fig. 4.6C,D), it confirms the perfect alignment of confocal and TPMEM beams. That the photolysis is due to 2PE can be shown by examining the power dependence of the response (Fig. 6.6F), which for powers below ~20 mW has the expected quadratic behavior with saturation effects appearing at higher powers reducing the slope in a log-log plot below 2. It should also be noted that damage associated with the 2PE beam, although limited to the focal region, is of some concern. It would be easy to confuse a change after beam turn off with a real biological effect (here seen in E as a jump in fluorescence at the focal spot due to the cessation of photobleaching).

As noted above, creation of a point source of a compound within a specimen (cell) is a way of probing local and disuse wide responses. A practical example is illustrated in Fig. 4.7, where the diffusion of uncaged fluorescein between eye lens fiber cells is measured. The movement of the fluorescence between cells then gives a measure of the local gap junction permeability

4.4 Practicalities and Advantages of MPM

A two-photon microscope is essentially just a scanning confocal microscope (Chapter 3), with the addition of a pulsed infrared laser.

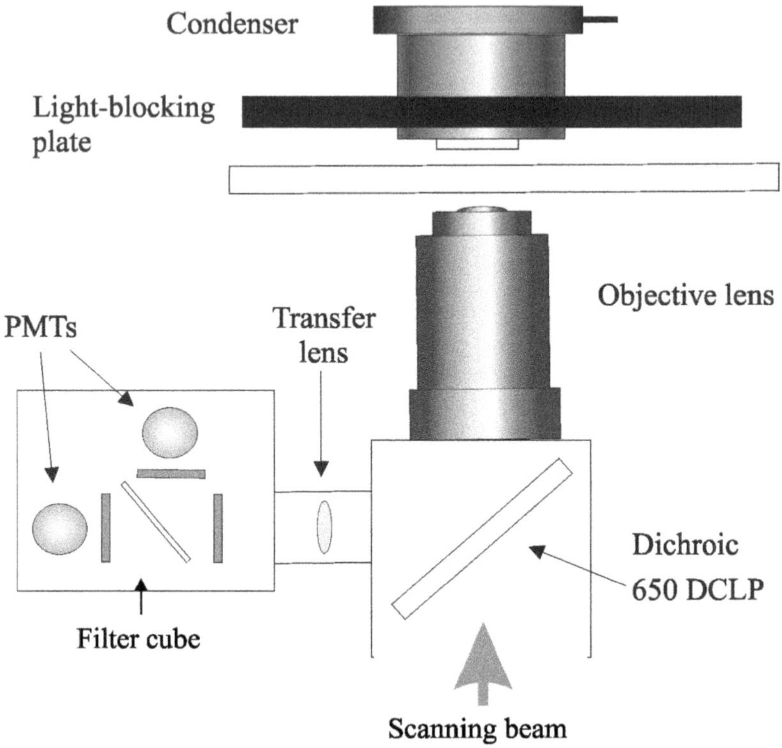

Figure 4.8 Layout of a non-descanned detector.

This is usually a titanium–sapphire laser since its output can be tuned over a wide range, but other lasers have been used. For practical reasons an inverted microscope is commonly used. The lasers and associated optics are bulky, and unlike the visible lasers used in confocal microscopy, the beam cannot be transmitted to the microscope though an optical fiber. (The fiber would spread the pulses and could also be damaged at high intensities). But upright setups are possible. The other important addition is a detector which bypasses the scanning system (Fig. 4.8). As Fig. 4.4 shows, this captures scattered light and thus greatly improves imaging at depth in biological samples.

A dichroic mirror (Chapter 2) passes the long-wavelength scanning beam, but directs visible fluorescence (<650 nm) to the detectors, here two photomultiplier tubes (Chapter 13). The transfer lens images the back focal plane of the objective on to the detectors, and a conventional filter cube (Chapter 2) sends the required wavelength to each detector.

Since no pinhole is used the detectors are sensitive to ambient room light. The blocking plate shown around the condenser in Fig. 4.8 helps to block this and so eliminates the need to work in darkness.

Two-photon imaging offers several clear advantages over single-photon imaging:

- Because two-photon events can take place only at the focus of the illuminating light, bleaching above and below the plane of focus is eliminated. In a conventional confocal system (although we are only forming an image of the focused spot formed by the laser beam), the cone of light above and below this spot is also capable of exciting—and bleaching—our fluorescent label.
- Penetration of the exciting light into tissue is increased by at least a factor of two due to its longer wavelength. Scattering of light in a turbid medium is directly related to wavelength (which is why the sky is blue), so longer wavelengths penetrate much better into cells. There is a counterbalancing effect—as we move into the infrared, water starts to absorb light—but in the range between 700 nm and 1200 nm, the gain from reduced scattering exceeds the loss due to absorption. Of course, the ability of the fluorescence to get back out of the tissue is unchanged, but we can now collect even scattered light from the fluorescence, because it no longer needs to come to a focus. Overall, the depth from which we can get useful images is substantially increased (Figs. 4.4 and 4.9).
- The exciting light is within the range in which microscope lenses perform efficiently, eliminating or simplifying optical corrections. Microscope lenses are generally designed for the range of ~700 nm to 400 nm, so that UV excitation at 350 nm lies outside the chromatic correction of the objective, but two-photon excitation at 700 nm is just about inside it. Even when we drop down into the infrared region, many lenses transmit

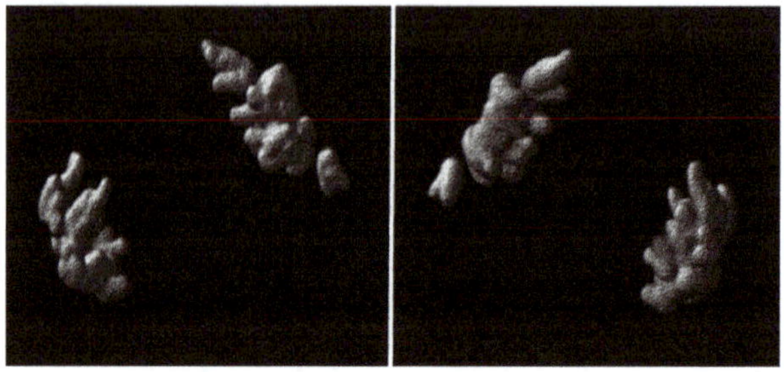

Figure 4.9 Second division of meiosis imaged at 200 μm depth in a living *Agapanthus* anther (see [8]). Two views of a full 3D reconstruction.

better than they do in the UV, and the fact that the focused spot might no longer be imaged on the confocal pinhole does not matter so much since...

- There is no need for a confocal pinhole. Because two-photon events occur only at the focus of the illuminating beam, two-photon microscopy is depth-selective even with a wide-field detector. In fact, chromatic correction becomes almost irrelevant in multiphoton because we need to bring only one wavelength to a focus.
- Long-wavelength light is much less damaging to living cells than ultraviolet or even blue light. Many experiments have shown extended cell viability during multiphoton imaging compared to equivalent single-photon imaging conditions.

Figure 4.9 shows an example of these benefits: imaging meiosis at high resolution in three dimensions 200 μm deep inside a living *Agapanthus* anther [8].

Acknowledgment

We thank Warren Zipfel for providing an initial outline for this chapter.

References

1. Stokes, G. G. (1852). Changes in the refrangibility of light. *Philos. Trans. R. Soc. London*, **142**, 463–562.
2. Göppert-Mayer, M. (1931). Über Elementarakte mit zwei Quantensprüngen. *Ann. Phys.*, **9**(3), 273–295.
3. Kaiser, W., and Garret, C. G. (1961). Two-photon excitation in CaF2: Eu2+. *Phys. Rev. Lett.*, **7**, 229–231.
4. Gannaway, J. N., and Sheppard, C. J. (1978). Second harmonic imaging in the scanning optical microscope. *Opt. Quantum Electron.*, **10**, 435–459.
5. Denk, W., Strickler, J. H., and Webb, W. W. (1990). Two-photon laser scanning fluorescence microscopy. *Science*, **246**, 73–76.
6. Soeller, C., and Cannell, M. B. (1996). Construction of a two-photon microscope and optimisation of illumination pulse duration. *Pflügers Arch.*, **432**, 555–561.
7. Cox, G., and Sheppard, C. J. R (2004). Practical limits of resolution in confocal and non-linear microscopy. *Microsc. Res. Tech.*, **63**, 18–22.
8. Feijo, J. T., and Cox, G. (2001). Visualisation of meiotic events in living anthers by means of two-photon microscopy. *Micron*, **32**, 679–684.

Chapter 5

Immunofluorescence

Anthony F. Henwood
Histopathology Department, The Children's Hospital at Westmead, Locked Bag 4001, Westmead 2145, Australia
tony.henwood@health.nsw.gov.au

5.1 Antibodies as Histochemical Markers

When a molecule is present in a vertebrate animal that is not part of the "self," it is processed by antigen-presenting T cells. These present the antigen to B cells, which bind it, and the B cells are activated to produce and secrete their own immune globulins. Immune globulins can be in several classes (IgG, IgM, IgA, IgE), which are produced at specific times and locations. However, each B cell can secrete only a single type of immune globulin, with only a single sequence at its variable site [1]. The antibody molecules produced by a single B cell are therefore identical. When a B cell is activated, it begins dividing, and all of the daughter cells in that clone also produce the same antibody [2].

Antibody preparations used in immunofluorescence can be polyclonal or monoclonal.

Fundamentals of Fluorescence Imaging
Edited by Guy Cox

ISBN 978-981-4774-85-7 (Hardcover), 978-1-351-12940-4 (eBook)
www.jennystanford.com

Because most antigens are highly complex, they present numerous epitopes that are recognized by a large number of T lymphocytes. Each subsequent activated B lymphocyte is incited to proliferate and differentiate into plasma cells, and the resulting antibody response is polyclonal. In contrast, monoclonal antibodies (MAbs) are antibodies produced by a single B lymphocyte clone. MAbs were first recognized in sera of patients with multiple myeloma in which clonal expansion of malignant plasma cells produce high levels of an identical antibody resulting in a monoclonal gammopathy. In the mid-1970s, Köhler and Milstein devised the technique for generating MAbs of a desired specificity, for which they were awarded the Nobel Prize [3]. They fused splenic B cells with myeloma cells with the resulting immortal hybridomas, each producing a unique MAb [4].

Polyclonal antibodies were the first immunoreagents used resulting in unprecedented strides in biology. Animals commonly used for antibody generation are rabbits, goats and to a lesser extent hamsters, rats, and mice. Larger animals are preferred. A single bleed from a larger animal can be used for a greater number of immunohistochemical reactions than a bleed from a smaller animal. The advantage of having a larger amount of serum per bleed is that each bleed is essentially a unique combination of antibody clones. Even when boosting the same animal with repeated immunizations with the same antigen, the antibody content in sequential bleeds may differ markedly. Hence, the lot for a polyclonal antiserum is critical, and even another batch from the same animal may have entirely different staining properties. For this reason, experienced immunohistochemists write down the lot numbers for each vial of antiserum and, when they have a good lot, buy up as much of that lot as they are likely to need in the foreseeable future to avoid inability to finish a project [2, 5].

Polyclonals frequently have excellent sensitivities because they are produced by a large number of B cell clones each generating antibodies to a specific epitope, and polyclonal sera are a composite of antibodies with unique specificities [4]. The binding affinity of an antibody for its target is variable, depending on how well the target fits the variable region. Binding affinities in the range of 10^{-10}

to 10^{-11} M are common compared with the binding affinities of enzymatic sites, which are often in the range of 10^{-6} to 10^{-8} M. This tight binding permits immunofluorescent staining, because the antibody can bind to its target so tightly that it can survive persistent washes, whereas the nonbound (or loosely bound) antibodies are washed away [2].

A polyclonal antiserum contains a mixture of different antibody isotypes as well as nonsense immunoglobulins that are normally present in the rabbit serum. Natural antibodies binding to the tissue section may occur in the antiserum as a result of unrecognized, prior antigenic stimulation. The strong staining of the epidermis, and moderate staining of vascular endothelial cells, fibroblasts, sweat glands, and arrector pili muscles of human skin described by Kobayashi et al. [6] after using some rabbit polyclonal antisera to lysozyme, myoglobin, S100 protein, and normal rabbit serum is probably due to the presence of anti-intermediate filament antibody, including an anti-keratin antibody. Such contaminating antibodies may co-purify with the specific antibody and cause false positive staining in immunofluorescence testing. Natural antibodies are usually found at low concentration and dilution may remove interference [7, 8]. It is generally accepted to use absorption controls where the antibody has been preabsorbed with the immunogen; however, this procedure does not eliminate the possibility that the antibody reacts with another protein in the tissue. Because antibodies recognize a relatively small component of an antigen, they can cross-react with similar epitopes on other antigens, but usually with less affinity [4, 9]. These low-affinity antibodies have weaker binding than specific antibodies. By incubating at low temperature, selection for the higher affinity specific antibody binding is preferred.

The problems of cross-reactivity of antisera that are associated with degeneracy and mimicry in immune recognition [10] are well established. Furthermore, the inherent drawbacks of the use of animals to produce antisera, viz., the potential presence of antibodies against contaminants of the immunogen and/or against antigens to which the animal has been exposed earlier, are well known [11].

Often it is desirable to purify polyclonal antibodies by removing non-immunoglobulin makes proteins and non-specific antibodies. Protein A/G purification takes use of the high affinity of *Staphylococcus aureus* protein A or Streptococcus protein G to the immunoglobulin Fc domain. Protein A/G purification, passing the serum through a column containing these proteins immobilized to a solid support (e.g., agarose beads), eliminates the bulk of the serum proteins from the raw antiserum. However, this does not eliminate the non-specific antibody fraction. As a result the protein A/G purified antiserum may still possess a small amount of undesirable cross-reactivity. Antigen affinity purification takes advantage of the affinity of the specific immunoglobulin fraction for the immunizing antigen against which it was generated. Affinity purification in which the antigen is coupled to a support and the polyclonal antibody is then put through the column results in the elimination of the bulk of the nonspecific immunoglobulin fraction, while enriching the fraction of immunoglobulin that specifically reacts with the target antigen. The resulting affinity-purified immunoglobulin will contain primarily the immunoglobulin of desired specificity [12].

Although in some cases high-quality and reliable results can be obtained with crude polyclonal antisera, in other cases even the most highly purified polyclonal antibody preparations (i.e., affinity-purified antibody preparations) may contain antibodies with reactivity against closely related antigenic sequences, and this cross-reactivity contributes to or even dominates the observed signal [13].

Recently chickens have been used to produce antibodies. Like mammals, chickens protect their offspring by transferring maternal antibodies from serum to egg yolk [14]. As a non-mammalian species, chickens offer a number of advantages, among them phylogenetic divergence as well as the ability to easily harvest antibodies, equivalent to mammalian IgG, from the yolk (IgY) of the egg without blood collection. The quantity of IgY harvested from a week's worth of eggs is significantly greater (up to 10-fold) than that obtained from rabbit blood collected during an equivalent period [4, 15]. The evolutionary difference between mammals and chickens should mean that more numerous sites in a mammalian antigen will evoke an antibody response. Thus, a greater number of antibodies will bind to the antigen [16].

Monoclonal antibodies (MAbs) are antibodies produced by a single clone of B cells. Köhler and Milstein [3] discovered that these cells can be immortalized by fusion with myeloma cells, resulting in hybridoma cells that are able to produce virtually unlimited quantities of MAbs. Since the Nobel Prize–winning work of these researchers, MAbs have become essential tools in basic research as well as in diagnostic testing and medical treatments [17]. A clone of hybridoma cells all produce the same, identical immune globulin, with the same variable region. These MAbs have the property that they will only bind to molecules that possess the same antigenic site [2, 4].

The principal advantages of MAbs are their homogeneity and consistency. The monospecificity provided by MAbs is useful in evaluating changes in molecular conformation, protein–protein interactions, and phosphorylation states, and in identifying single members of protein families. However, the monospecificity of MAbs may also limit their usefulness. Small changes in the structure of an epitope (e.g., as a consequence of genetic polymorphism, glycosylation, and denaturation) can markedly affect the function of a MAb. For that reason, MAbs should be generated to the state of the antigen to which it will eventually need to bind. Another key advantage of MAbs is that once the desired hybridoma has been generated, MAbs can be generated as a constant and renewable resource [4].

In 1995, Spieker-Polet et al. [18] first reported the development of a stable rabbit–rabbit hybridoma that produced rabbit monoclonal antibodies (RMAbs). According to Huang et al. [19] the availability of RMAbs has many advantages: First, rabbit antisera generally have higher affinity and recognize a greater variety of epitopes than antisera generated in mice, and conceivably RMAbs might also possess the same superior features. Second, RMAbs may be expected to recognize many antigens that are not immunogenic in mice, including mouse proteins. Last, in view of the size of the rabbit spleen, a much larger pool of hybridomas can be generated from a single rabbit with, therefore, an enhanced potential for selecting the desired RMAb.

5.1.1 Endogenous Fc Receptors

Endogenous Fc receptors (FcRs) are structures on the surface of certain cells that bind the Fc region of antibodies. Cross-linking of antibody bound by FcRs provides an important link between the cellular and humoral branches of the immune system by inducing several responses, including phagocytosis, endocytosis, antibody-dependent cytotoxicity, release of inflammatory mediators, and enhancement of antigen presentation. The nature of the response depends primarily on the cell type on which these FcRs are expressed [20].

It is thought that preincubation of a histological sample with 5–10% normal serum from the species that the secondary antibody is derived from will prevent nonspecific binding of secondary antibodies to endogenous FcRs. According to Buchwalow et al. [20] this makes little sense for the immunohistochemical staining of human cell and tissue samples, as the vast majority of secondary antibodies used in human immunofluorescence are derived from goats, and goat serum has long been reported not to bind to FcRs on human cells. Preincubation with solutions containing normal goat serum have also been assumed to prevent background staining that might result from ionic and hydrophobic interactions. Blocking the nonspecific background due to FcRs or ionic and hydrophobic interactions is considered an obligatory step prior to incubation with primary antibody [20].

Buchwalow et al. [20] concluded that endogenous FcRs do not retain the ability to bind the Fc portion of IgG Ab after fixation in formaldehyde, acetone, or alcohol. Likewise, nonspecific Ab binding to tissue proteins due to hydrophobic interaction between proteins or ionic and electrostatic interactions did not take place in routinely fixed cell and tissue samples. In contrast to the commonly accepted view, they found that the protein blocking step traditionally used in immunohistochemistry was unnecessary in the immunostaining of routinely fixed cell and tissue samples taken both from human patients and from nonhuman species [20].

5.1.2 Antibody Validation

The specificity of an antibody refers to its ability to recognize a specific epitope in the presence of other epitopes. An antibody with high specificity would result in less cross-reactivity. With respect to native protein antigens, the binding affinity of most antibodies is influenced by conformational determinants, and antibodies may not bind the same protein in a denatured state [21]. This characteristic is particularly true of MAbs, which target a single epitope. Conformation may be altered by any number of factors, including association with other proteins, posttranslational modification, temperature, pH, salt concentration, and fixation. The impact of conformational change is of less concern when using polyclonal antibodies. Polyclonal antibodies recognize multiple epitopes, some of which are likely to be linear, and conformational changes may not influence all epitopes to the same degree [4].

Unfortunately, in many cases commercial antibodies fail even the most fundamental tests of activity and/or specificity (e.g., selective staining of cells overexpressing the target protein versus those that do not express the target protein at all), and with alarming regularity vendors pass the burden for antibody validation/quality control to the end user. In some fields of neuroscience, the poor quality of antibody reagents has caused considerable frustration among investigators and led to publication and perpetuation of erroneous research results [13, 22].

Many criteria have been used to evaluate specificity [11]. Of these, a selective staining pattern, a single band of the expected size on Western blots, and the disappearance of staining after preabsorption of the antiserum with purified epitope are most often available for commercially available antisera, whereas the remaining, more solid criteria, such as absence of staining in tissues genetically deficient for the protein, identical staining patterns of antibodies raised against different epitopes on the same protein, and/or correspondence between the staining pattern after ISH and IHC are rarely available.

Evidence that the antibody binds specifically to the expected target molecule is often provided by a Western blot, which should show that the antibody stains a single band (or a set of bands) of appropriate molecular mass for that target. Presence of multiple bands or bands not at the proper molecular weight could represent the same target at different post-translational modification status, breakdown products, or splice variants. It is recommended that this antibody be used with caution [2, 22]. Since proteins are denatured when applied to the membrane, it is important to consider whether the antibody will recognize the denatured antigen or conversely if it does recognize the denatured antigen in a blot, will the antibody recognize the antigen in its native state in tissue [22].

Another control for specific staining in tissue is the preadsorption test. Mixing the diluted antibody with an excess of the immunogen should completely block staining. This shows that the staining in the tissue is against something that is at least cross-reactive with the original protein (although it does not prove that this is what the target in the tissue actually is). In general, when the original immunogen is readily available, such as for a synthetic peptide, the preadsorption test should be run as a matter of course [2].

The preabsorption control determines only the specificity of the antibody for the incubating protein or peptide and does not prove the specificity of the antibody for the protein in the tissue. An antibody can bind to any epitope that has the correct conformation, and this potentially includes the protein used for immunization, as well as any protein with a similar epitope. Therefore, the absorption control does not indicate that the protein to which the antibody binds in tissues is exactly the same protein that was used to generate the antibody [23]. The preadsorption control is meaningless for a monoclonal antibody since the immunogen will always bind it and always pass a preadsorption test, and for antibodies that have already been affinity purified (for the same reason) [2].

Recently, there has been an increase in the use of Knock-out (KO) animal tissues in which the target antigen has been genetically deleted, to validate immunohistochemical reactions [13]. Lorincz and Nusser [24] have discussed several issues with proving the specificity of an immunohistochemical reaction using KO animals, for example depending on the deletion strategy, a truncated part of the protein could be expressed in KO animals; and the expression of many other proteins (often highly homologous isoforms, which are the most likely candidates for antibody cross-reactivity) could change in conventional KO mice.

Another approach to obtaining concordant data is to compare the immunofluorescence staining patterns revealed by two different antibody preparations raised against distinct, non-overlapping epitopes on the same target protein. Because it is highly unlikely that antibodies raised against non-overlapping antigenic sequences from the same target antigen will show the same specific and nonspecific staining patterns, areas of close overlap in immunofluorescent staining for the two antibodies are very likely to represent a specific signal [2, 13, 25].

5.2 Fixation

Immunofluorescence can be done on fresh frozen cell preparations (e.g., renal and skin autoimmune diagnosis) or on tissues that have been "fixed." Fixation preserves cells and tissues in a lifelike condition, inhibits autolytic (self-digestion) and putrefaction changes, and insolubilizes tissue elements for later demonstration. It is important that fixation has minimal effect on antigenic determinates, allowing recognition by appropriate antibodies. Fixatives can be physical or chemical in action. Physical fixatives, for example heating sections or cells on to slides, alcohol, or acetone treatment, preserve tissues by dehydrating cells. Alcohol and acetone possibly also dissolve lipids from cell membranes facilitating antibody access.

Of the chemical fixatives, 4% formaldehyde in phosphate buffer is the most common. Formaldehyde crosslink proteins and thus stabilizes cell membranes and organelles, forming a lattice that permits diffusion of various substance including antibodies into cells. Aldehyde fixation is a prerequisite for immunohistochemical detection of soluble proteins, including neurotransmitters, neuropeptides, calcium-binding proteins and reporter proteins such as eGFP, to avoid washout of damaged cells during the staining procedure. However, loss of antigenicity and/or epitope masking that are due to tissue fixation are major limitations in immunohistochemistry [26]. It is appropriate to apply this to both enzyme and fluorescent immunohistochemistry since both techniques are affected equally. Clevenger et al. [27] have shown that the sequential use of formaldehyde and 0.1% Triton X-100 results in optimal permeabilization of cells for the demonstration of nuclear antigens.

A number of solutions have been proposed to resolve this issue, including various antigen-retrieval procedures, digestion of the fixed tissue with proteolytic enzymes, inhibition of proteolysis during fixation and short-fixation techniques to achieve a compromise between labeling efficiency and morphological preservation [26]. This issue is also discussed later in this chapter (1.5 formalin fixed paraffin embedded (FFPE) sections).

Lorincz and Nusser [24] have presented examples of the effect that fixation can have on the immunolocalization of antigens. Formaldehyde fixed sections of control and corresponding knockout mouse tissues often gave positive diffuse staining in both. When section were pre-treated with proteases, specific staining in the expected compartment appeared in control but was absent in corresponding knockout tissues.

Other antigens require heating, sometimes over 100°C, in acidic (pH6) or alkaline (pH9) buffers. This can be quite harsh on cells. Pavlakis et al. [28] found that several antigens were detected solely upon washings of the formalin-fixed cryosections for 15 minutes with an acidic pH buffer. This is also indicative of an antigen retrieval process on formaldehyde-fixed, frozen tissue sections, which involved no heating or boiling, and was irrelevant of the fixation time and also very much dependent on the pH of the unmasking buffer solution. They used acidic (Na_2HPO_4-citric acid buffer pH 4.0 containing 0.05% Tween) for 15 min at room temperature.

5.3 Frozen Sections (Cryotomy)

Thin sections (e.g., 4 to 15 μm) of frozen tissue can be made using a freezing microtome (cryotome) for immunofluorescence. This technique requires extensive technical expertise and the reader is recommended to refer to existing histotechnology text books and to employ the services of a good histotechnologist.

5.4 Cell Cultures

There are several methods for preparing cell cultures for immunofluorescences. Cell cultures tend to be divided into adherent and suspension cultures and this will affect how cultures are processed for immunofluorescence. Cell cultures can be stained directly or dissociated cells can be spun directly onto slides using a cytocentrifuge before immunostaining. Cell cultures can also be concentrated using centrifugation and cell blocks prepared either for cryotomy or after formalin fixation, processed to paraffin for routine microtomy and immunostaining [29].

Cell cultures that grow in an adherent manner (e.g., BT-474 and SW480 cells) will require harvesting using a protease (e.g., 0.25% trypsin, 0.53 mM EDTA) if a cell block is required. It might be easier to immunostain these adherent cell cultures on chamber slides [29, 30].

5.5 Formalin-Fixed Paraffin-Embedded (FFPE) Sections

Historically the immunohistochemical demonstration of tissue antigens has been performed on frozen tissue sections, but the preservation of antigenic epitopes comes at the expense of morphology, and frozen sections may suffer from artifactual diffusion of antigens [31]. Fixation of tissues, in particular using formaldehyde, stabilizes tissues by crosslinking proteins resulting in antigen stabilization and preserved morphology.

In the past, there were many antibodies, mainly polyclonal in type, which recognized their antigens in formalin-fixed paraffin-embedded (FFPE) tissues but, especially with the development of MAbs, researchers realized that there were many antigens that were not detectable in FFPE tissues but could be localized in nonfixed frozen sections. This impasse was breached when Shi et al. [32]

developed their antigen retrieval technique to restore an antigen's optimal immune reactivity after formaldehyde. Shi et al. employed microwave heating of paraffin-embedded sections in the presence of heavy metal solutions, such as lead thiocyanate but later it was found that a citrate buffer (pH6) or EDTA in a pH8 buffer retrieved antigens more efficiently [33, 34]. How heat-induced antigen retrieval works is subject to conjecture and several theories have been proposed. It has been suggested that heat induces reversal of various chemical modifications of the protein structure that result from formalin fixation. Other actions that produce antigen retrieval include the loosening or breakage of the cross-linkages caused by formalin fixation and hydrolysis of Schiff bases. Heat may act on multiple pathways including the extraction of diffusible blocking proteins, precipitation of proteins, and dehydration of tissue sections to allow better penetration of antibody and increased accessibility to epitopes [33, 34]. Whatever is the correct reason, heat-induced antigen retrieval works and needs to be considered in the design of immunofluorescence reactions on FFPE tissues.

5.6 Immunofluorescence

Fluorescent histotechnology relies on the use of an ultraviolet light source microscope to examine fluorescing substances (called fluorophores) in cells and tissue sections. The fluorescent substances are either natural fluorescing substances in tissue sections or attached to histochemical probes that have been allowed to react with tissue substances. The fluorescence microscope is based on the phenomenon that certain materials emit energy detectable as visible light when irradiated with the light of a specific wavelength. For example, fluorescein, a commonly used fluorophore absorbs blue light and emits green, and rhodamine, absorbs green and emits red. When a fluorescent compound absorbs a photon, its outer shell of electrons move from a "ground" state to an "excited" state. The fluorophore, when attempting to return to its "ground" state, then releases part of the energy to its molecular environment, and then emits a photon with less energy and consequently a longer wavelength. To visualize this, the emitted light is separated from the much brighter excitation light in a second filter.

Several tissue and cellular compounds are autofluorescent. Calcium gives a yellow whereas porphyrins and vitamin A give a red autofluorescence. Cartilage, cardiac muscle, and keratin give a yellow-green whereas collagen and elastic fibers give a blue-green autofluorescence. Lipofuscins give a brown-red and occasionally orange autofluorescence [35].

5.6.1 The Direct Immunofluorescent Technique (DIF)

In 1950 Coons and Kaplan developed the immunofluorescent technique whereby fluorescein was conjugated to antibodies and applied to cells and tissue sections. From this development, today's modern immunohistochemistry evolved. With today's modern immunohistochemical methods being done quite successfully on formalin-fixed paraffin-embedded sections, it is mystifying as to why immunofluorescence is still preferred for immunopathological renal and skin diagnosis. Immunofluorescence remains the most popular method of detecting immunoglobulins and complement deposits, with 83% of British laboratories using this technique [36].

Technically, frozen sections are cut, dried, rinsed in buffer, optimally diluted FITC labeled antibodies are added to sections, incubated for 30–60 minutes, nonbound antibody is rinsed from the sections with buffer, the sections are coverslipped with a nonfluorescing aqueous mountant and finally viewed using a microscope equipped with an ultraviolet light source; overall quite a simple technique (see Appendix A) [25]. Disadvantages include requirement for frozen sections and a UV microscope, impermanence of stained sections, inadequate quality control, and inferior morphology.

The detection of tissue-bound autoantibodies is rendered difficult by the presence of abundant plasma proteins, which must be removed to avoid nonspecific staining. This has resulted in a continuing popularity of frozen sections and immunofluorescence techniques in immunopathology, where they have been superseded in most other branches of immunohistochemistry by methods that are applicable to paraffin wax sections. In paraffin wax sections, the plasma proteins have been fixed in the tissues and must be removed by a digestion process rather than simple washing [37]. Unfortunately, too much digestion may destroy tissue integrity and possibly remove the immune complexes, resulting in a false negative result. Furthermore, digestion conditions are difficult to standardize. The times required vary with the degree of fixation and section thickness [38]. Frozen sections of nonfixed tissue when used for the detection of bound-immune complexes, give clean localization with minimal background without the requirement for enzymatic pre-digestion. The plasma membranes, being loosely bound, if bound at all, to the surrounding tissue elements are easily dissolved in the warm buffer pre-incubation step. Bound immune complexes seem to be unaffected.

A basic panel of antibodies should be used for the detection of tissue deposits of IgG, IgA, IgM, and complement (usually C3). This represents a minimum panel. Additional antibodies can be used at the pathologist's discretion. Widely used antibodies include those directed against C4 and/or C1q, fibrinogen, albumen, and kappa and lambda light chains [25, 37]. It is interesting to consider the appropriate optimum dilution of the FITC-labeled antibodies. A common across-the-board dilution is 1 in 10 and this has often been quoted in the literature. Dilutions up to 1 in 50 have also been used, though often dilutions are not recorded in many publications. One wonders whether many of the dilutions have been determined empirically and are based on historical data. Appendix A presents a common immunofluorescent technique.

5.6.1.1 Renal immunofluorescence

Glomerular crescents in a renal biopsy evoke certain differential diagnostic considerations and often, the need for aggressive therapeutic interventions. Extracapillary proliferation within a glomerulus, either partially or completely filling up the Bowman's space defines a crescent. It is composed of proliferating parietal epithelial cells, podocytes, macrophages, and fibroblasts [39].

Crescent formation is stimulated by the entry of fibrin and other plasma proteins from the capillary lumens and indicates glomerular basement membrane (GBM) rupture. Recruitment of T cells, macrophages, and other inflammatory cells to the site of injury follows, with cytokine release and tissue injury. These "cellular" crescents may evolve into a chronic phase of fibrocellular

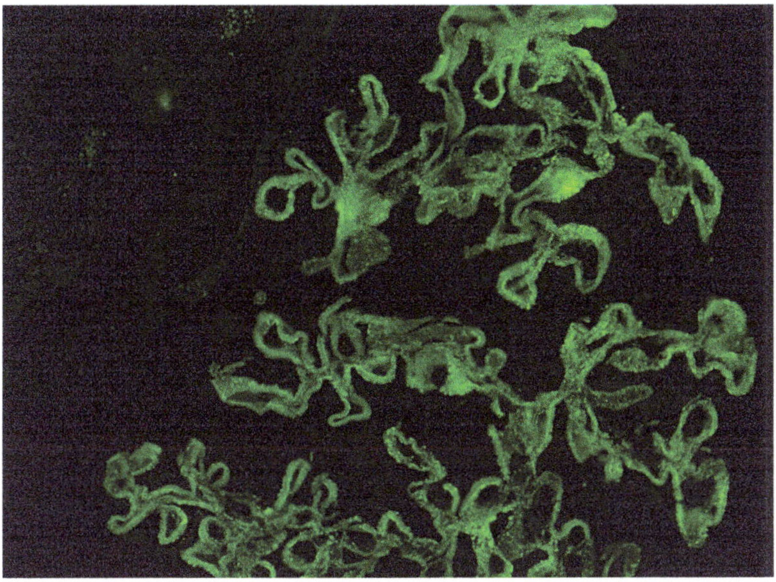

Figure 5.1 Direct immunofluorescence using FITC-conjugated goat anti-human IgA. This renal biopsy from a patient with lupus nephritis shows IgA granular deposits aligned regularly along glomerular basement membranes.

or fibrous crescents with increased extracellular matrix and collagen deposition [39] (Fig. 5.1).

Immunofluorescence, together with light microscopy and electron microscopy, allow distinction of pauci-immune glomerulonephritis, antiglomerular basement membrane nephritis and immune complex-mediated glomerulonephritis, including overlap syndromes and mimics [39].

Nonspecific or low-intensity linear IgG staining may be observed in elderly individuals, patients with diabetes mellitus and obesity, allograft biopsies, and autopsy specimens. Hence, clinicopathological correlation is of great relevance in interpreting these IF results. Also, when in doubt, comparing the intensity of linear staining of IgG with that of albumin is helpful as anti-GBM nephritis is associated with IgG intensity greater than that of albumin [39].

Not everything that fluoresces in a biopsy represents an immune event. There are autofluorescent components that require recognition. The two most common are arterial elastic fibers and lipid deposits, which may be observed in tubular cells, interstitial foam cells, and arterioles. These autofluorescent components appear yellow rather than the characteristic green of fluorescein, the usual probe used in immunofluorescence. Recognition of autofluorescent phenomena is necessary to avoid errors in interpretation but also provides useful information. They reveal underlying arterial disease and lipidemia and lipiduria characteristic of the nephrotic state. A second pitfall that must be recognized is the modest background glomerular basement membrane and tubular basement membrane staining associated with the use of a buffered transport medium. This may elicit initial concern about a linear immune reaction until the negative fibrin stain and absence of crescent formation are noted [40].

A frequent concern with renal immunofluorescence is what constitutes an adequate number of glomeruli for testing. One should be enough, but realizing that at least 7 frozen sections will need to be prepared, it seems appropriate that more than 3 glomeruli should be sought. The biopsy should be placed in a drop of Hanks or similar cell culture fluid immediately after removal from the patient and examined under a good dissecting microscope with an adequate light source. With appropriate magnification it is easy to recognize normal renal cortex and to identify glomeruli on the surface. This might be more difficult with chronic disease and fibrosis, which might cause the tissue to be uniformly pale, or with acute disease where hypercellular, bloodless glomeruli may take on the appearance of surrounding tubulo-interstitial tissues [37].

When trimming into a frozen renal sample, it is important to regularly assess the sections with a rapid H&E or toluidine blue stain for glomeruli. Sometimes sclerotic glomeruli (where the glomerulus is replaced by hyaline connective tissue) are present and these cannot be included in immunofluorescent assessment.

5.6.1.2 Skin immunofluorescence

Autoimmune bullous skin diseases represent a heterogenous group of disorders of skin and mucosa which are commonly associated with IgG or IgA autoantibodies against distinct adhesion molecules of the skin. The antibody induced loss of adhesion between

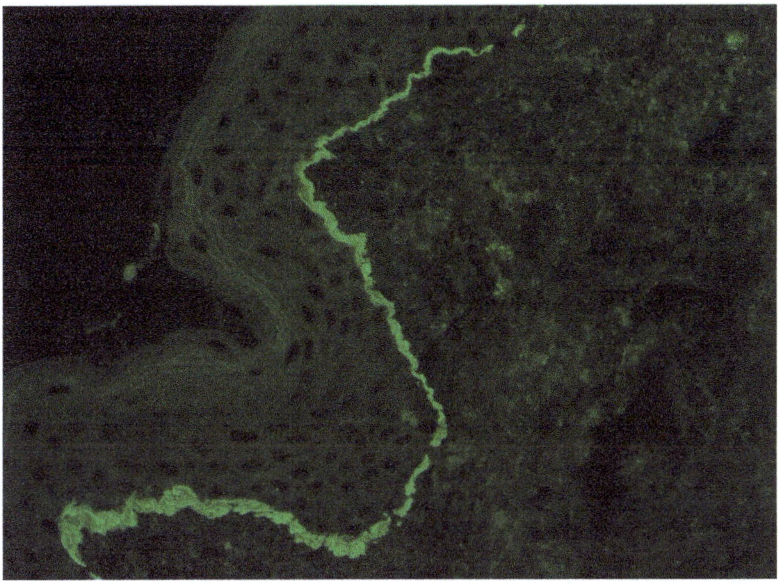

Figure 5.2 Direct immunofluorescence using FITC-conjugated goat anti-human IgA. This skin biopsy from a young patient shows linear IgA deposition along the dermoepidermal junction, consistent with vesiculobullous dermatosis, favoring linear IgA disease.

epidermis and dermis results in blister formation and extensive erosions [41] (Fig. 5.2).

The pattern of IF staining is important in diagnosis, e.g., in Pemphigus vulgaris, the pattern is "chicken wire" demarcating intercellular spaces in the epidermis, whereas in Bullous pemphigoid, there are linear deposits at the dermoepidermal junction (Chhabra et al., 2012).

The location of the biopsy site is of key importance to achieve best possible sensitivity. As immunoglobulins and complement factors are degraded by the inflammatory reaction and are thus no longer detectable directly in the blister, the biopsy should be taken from normally appearing skin in the immediate vicinity of a fresh blister [41].

Lehman and Camilleri [38] reported that lipofuscin in human sweat glands can lead to the appearance of autofluorescent granules in the secretory tubular cells of eccrine glands in IF specimens and that this phenomenon resembled the intramural fluorescent particles that can be seen in vasculitis on IF. They proposed that for skin specimens submitted for IF evaluation, non-specific eccrine gland particle autofluorescence can be differentiated from intramural granular deposits of vasculitis in the following four ways:

(1) Correlation with findings on accompanying frozen haematoxylin and eosin stained tissue may aid in differentiating blood vessels from eccrine glands when equivocal periluminal granular deposition is present on immunofluorescence.
(2) Previous studies have shown that cutaneous leukocytoclastic vasculitis involves the superficial and deep vessels in 26% and deep vessels only in 3%. Eccrine glands tended to localize to the mid to deep reticular dermis. Therefore, if periluminal granules are present in the superficial dermis, then vasculitis may be favored.
(3) Secretory granules are likely to be coarse, in contrast to the fine intramural granular deposits of cutaneous vasculitis.
(4) The presence of a periductal thick, smudged line is helpful in identifying a lumen as an eccrine gland rather than a blood vessel; its absence, however, is not helpful in excluding eccrine gland, as this study showed that it was absent in the majority of sections examined

Arbesman et al. [42] examined whether direct immunofluo rescence testing would still be accurate if performed on biopsy specimens after brief inadvertent immersion in formalin. They found that that limited exposure of skin samples to formalin may significantly impair immunogenicity for select diseases, whereas others only demonstrate reduced immunogenicity with increasing exposure. More specifically, biopsy specimens that are positive for pemphigus will not remain positive with minimal immersion in formalin. They demonstrated a spurious non-disease-specific nuclear staining or fluorochroming pattern in all diseases studied. This pattern resembles an in vivo antinuclear antibody staining pattern that is found in samples positive for lupus erythematosus (LE). This can be used as an indication that a supposedly fresh skin sample had been inadvertently placed in formalin before receipt in the laboratory.

It is important to regularly assess each section for its suitability for immunofluorescence. Since immune deposits are found in the epidermis or dermoepidermal junction, it is desirable to have a complete skin surface with no folds.

5.6.2 Indirect Immunofluorescence (IIF)

There are many issues with DIF that hinders general use [20]:

(1) Loss of antibody affinity and fluorescent activity caused by the chemical linking reaction.
(2) Covalent labeling of primary antibodies requires relatively large quantities of purified antibody
(3) Higher threshold of detection compared to indirect techniques (more antigen required for detection).
(4) Localization antisera to different antigens need to be individually labeled.

The indirect immunofluorescent (IIF) technique has been developed to circumvent many of these shortcomings. A researcher can have a panel of unlabeled localization antisera to various antigens and yet only need one labeling, fluorescence conjugated antibody to the immunoglobulin of the species from which the localization antisera were obtained. For example, if the localization antisera are all mouse monoclonals, then a rabbit or goat anti mouse immunoglobulin that has been conjugated with FITC is used to label bound mouse antibody. Labeling antibodies coupled to a large variety of fluorochromes are available, which provide satisfactory quantum yield and limited bleaching. Carbocyanine dyes (Cy2, Cy3, and Cy5) and Alexa fluorochromes (e.g., Alexa-Fluor 488, 543, etc.) are largely comparable and suitable for these applications [26].

An advantage of IIF is the increased sensitivity compared to the DIF. This results principally from the ability of the polyclonal bridging antibody to recognize multiple sites on the Fc and Fab portions of the localization antibody. The actual sensitivity, taking into account the effect of steric hindrance and other constraints is 4 or 5 times that of DIF [43].

The investigator needs to be aware of unexpected cross-reactions between the labeling antiserum and other determinates in the tissue. It is known that donkey-derived labeling antibodies can react nonspecifically with human IgG in tissue and it is recommended that preabsorbed rabbit labeling antibodies be used [44].

A concentration of 1–5 ug/ml FITC conjugated labeling antiserum has been recommended by investigators, allowing an incubation time of 30–60 minutes at room temperature.

5.6.2.1 Autoimmune sera testing

The critical function of the immune system is to discriminate self from non-self. Tolerance against self-antigens is a highly regulated process and, in order to maintain it, the immune system must be able to distinguish self-reactive lymphocytes as they develop. The presence of autoantibodies is the consequence of breakdown of tolerance and, although they are an important serological feature of autoimmune diseases, their presence is not exclusive of these conditions [45].

There are more than 80 different autoimmune diseases that affect hundreds of millions of people around the world. These systemic or organ-specific conditions are the third leading cause of morbidity and mortality in the industrial world after cancer and heart disease [46].

The IIF assay is a powerful, sensitive, and comprehensive test for screening patient's sera for autoantibodies. It is probably the single immunological test most ordered by rheumatologists after the rheumatoid factor test [47]. In 1954, Weller and Coons used a monolayer of fixed cells which were allowed to interact with immunoglobulins (autoantibodies) in the test serum. To detect a positive reaction, a fluorescent labeled antibody is used next, and the slide is read using a fluorescence microscope usually at 400× magnification. Specific staining of nuclear or cytoplasmic structures is visualized as apple-green fluorescence when using fluorescein isothiocyanate (FITC) [47] (Fig. 5.3).

IIF testing led to the detection of autoantibodies in various autoimmune diseases, including systemic lupus erythematosus, scleroderma, dermatomyositis, and mixed connective tissue disease.

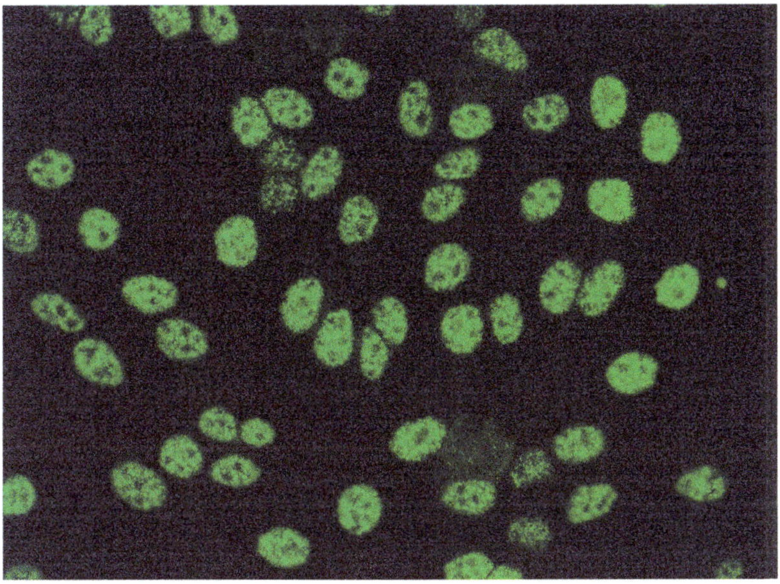

Figure 5.3 Indirect immunofluorescence (using patient's sera as the localization antibody). Speckles in the nucleus of interphase cells of Hep2000 cells often seen in cases of systemic lupus erythematosis and Sjögren's syndrome.

Certain autoantibodies produced distinct patterns of staining since they reacted with specific organelles in the nucleoplasm, nuclear membrane, nucleolus, or cytoplasm. An outstanding example is an autoantibody in the CREST (calcinosis, Raynaud's phenomenon, esophageal dysmotility, sclerodactyly, and telangiectasia) subset of scleroderma, in which immunostaining revealed a limited number of dots in the nucleoplasm of interphase cells, but a total redistribution of these dots to the centromeric regions of condensed chromosomes of cells in mitosis. It became clear that there were multiple autoantibodies of different specificities in any individual autoimmune disease, a few autoantibodies were disease specific, and different autoantibody profiles were associated with different diseases [48].

The detection of antinuclear antibodies (ANA) using IIF has been widely used to provide supporting evidence of a diagnosis of systemic lupus erythematosus (SLE). IIF for ANA detection has been in use for more than 4 decades and, over this period, the methodology has evolved to include some important changes. The use of mouse liver tissue sections as antigen substrate has been largely replaced by tissue culture cells (e.g., HEp-2). Patient sera is diluted in 0.01M phosphate buffered saline, pH 7.2 at 1:40, 1:80, 1:160, and 1:320 [49, 50]. In general, patients with SLE had higher titers of ANA than normal persons or persons with other diseases. A positive test result typically was reported only when immunofluorescent staining persisted at dilutions of 1:40 or higher. Acetone-fixed substrate slides are recommended. Ethanol and methanol fixation is discouraged as these fixations may remove Ro (SS-A) antigen. HEp-2 cells are preferable to mouse or rat tissues since the latter do not contain detectable amounts of Ro (SS-A) antigen and do not reveal antibodies to a number of organelles, most importantly, antibodies to the centromere [50, 51].

Paraneoplastic neurological syndromes (PNS) are remote effects of cancer that are not caused by invasion of the tumor or its metastases. Immunological factors appear important in the pathogenesis of PNS because antineuronal autoantibodies and T-cell responses against nervous system antigens have been defined for many of these disorders. The immunologic response is elicited by the ectopic expression of neuronal antigens by the tumor. Expression of these so-called "onconeural" antigens is limited to the tumor and the nervous system and sometimes also the testis. At the time of presentation of the neurological symptoms, most patients have not yet been diagnosed with cancer. Detection of paraneoplastic antibodies is extremely helpful in diagnosing an otherwise unexplained and often rapidly progressive neurological syndrome as paraneoplastic [52]. A gynecological carcinoma, usually ovarian or breast, is found in over 90% of patients who are seropositive for Purkinje cell cytoplasmic antibody type 1 (PCA-1 or anti-Yo). Small-cell lung carcinoma (SCLC) is found in more than 80% of patients who are seropositive for antineuronal nuclear antibody type 1 (ANNA-1 or anti-Hu). A composite substrate of mouse neural and non-neural tissue is used and in order to remove non–organ-specific autoantibodies patients' sera is preabsorbed with liver tissue [53].

5.6.3 Multi-antigen Immunofluorescence

There are many more reports in the literature using immunofluorescent double labeling (sometimes called multiplexing) rather than light microscopic double labeling:

- As Chris van de Loos and Peter Teeling [54] have remarked: "a truly generally applicable double immunolabeling method that allows for the light microscopic visualization of co-localization (two markers present at one cell, yielding a mixed color) does not exist." Unless the targets exist in different compartments of the cell (i.e., nuclear, cytoplasm, cell membrane, extracellular), this is definitely the case [54, 55].
- Diaminobenzine tetrachloride (DAB) when used as a chromogen for horseradish peroxidase activity after the first immunolocalization has been found to be quite effective in "shielding" the first staining sequence. The brown DAB reaction product covers the first localization and labeling antibodies so well that reagents from the second detection system can no longer react with components from the first staining sequence [56]. This shielding has been found effective only if the first localizing antibody is optimally titrated. To date, DAB is the only known enzymatic reaction product that shows this unique sheltering characteristic [54, 57]. DAB can also mask antigens close to the first antigen, causing failure of the second localization antibody to bind to its antigen.
- It is beyond dispute that double immunofluorescence, because of its high staining accuracy, is the most ideal method for studying co-localization [54]. In fact co-localization seems easier to resolve using fluorescence [20]. You can use one localization antibody labeled with FITC and a second labeled with Texas red resulting in yellow-green and red and where both antigens are in the same component an orange mix. Usually image processing is used where a photo is taken at each filter combination and the resulting images superimposed (see Fig. 5.4) [58].
- There are many different fluorophores now available affording a wide range of colors and with the development of quantum

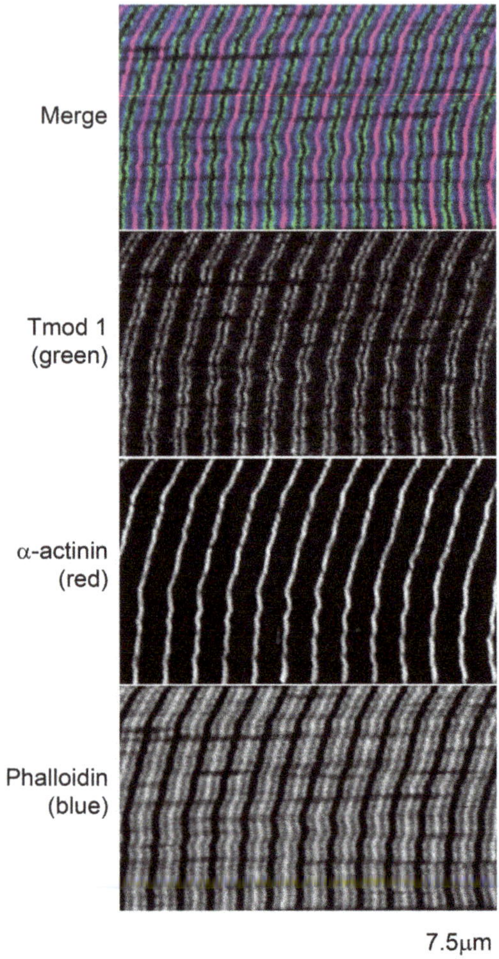

Figure 5.4 Double immunofluorescence: skeletal muscle sarcomere labeled with an antibody to tropomodulin 1 (green) and alpha-actinin (red) and the fungal toxin phalloidin (blue) showing the different parallel protein bands. Image courtesy of Michaela Yuan, the Institute of Neuroscience and Muscle Research, Kid's Research Institute, Westmead, Australia.

dots, the choices are even more enticing. Quantum dots are fluorescent nanocrystals. These are inorganic fluorophores, whose semiconductor properties endow them with relatively broad excitation but narrow emission profiles [59].

Multiple immunostaining using directly labeled antibodies are the easiest to do. Limitations, as mentioned earlier, include loss of antibody affinity and fluorescent activity caused by the chemical linking reaction, and decreased sensitivity compared to indirect techniques (more antigen needed to produce a visible reaction).

Indirect methods are commonly used for multiple immunofluorescent labeling. The localization antibodies in order to allow them to be selectively detected by the two secondary fluorescent conjugates, must differ in subclass (e.g., IgG1 or IgG2a), isotype (e.g., IgG or IgM), or species (e.g., mouse or rabbit) [20, 60]. Unfortunately most of the available primary antibodies originate mainly from two species, i.e., rabbit and mouse. When primary antibodies are raised in the same host species, the secondary species-specific antibodies can cross-react with each of the primary antibodies [20]. It is important to be aware of "inappropriate" interactions among the various primary and secondary antisera. It has been reported that sheep anti-guinea pig lgG has affinity for rabbit IgG, and that goat antirabbit IgG has affinity both for primary antisera raised in guinea pig and for primary antisera raised in monkey [61].

5.6.4 Photobleaching

One of the disadvantages of immunofluorescence is photobleaching. The fluorescence of FITC is rapidly lost after exposure to excitation light, and FITC-stained preparations mounted in conventional buffered glycerol exhibit prompt bleaching under illumination. The prevention of photobleaching is very important for obtaining stable and accurate images that allow repeated observations. It has also been suggested that it is important for quality assurance purposes. Various chemicals, including polyvinyl alcohol (PVA), p-phenylenediamine (PPD), n-propyl gallate (NPG), and 1,4-diazobicyclo[2,2,2]-octane (DABCO), have been added to the mountants to retard the fading of immunofluorescence [62]. Grimwood and Proffer [63] showed that FITC-positive slides could be preserved at room temperature for 30 months using these mountants. Dikicioglu et al. [62] found in their study that the preservation of fluorescence in FITC-stained DIF-positive slides using mounting media with an antifading reagent is possible for 2 years at room temperature in darkness. However, in daily practice, they found that storage for longer than 11 months prevented a reliable diagnosis. They concluded that a storage time of 11 months for slides is also acceptable without a colder environment.

Espada et al. [64] have shown that fluorochromes are very well preserved after dehydration (or simply air drying) followed by permanent mounting in non-aqueous synthetic media, particularly DePeX. They also found that fluorescence had not faded after one year of storage at room temperature.

5.6.5 Autofluorescence

Plant and animal tissue contain naturally fluorescent compounds. This autofluorescence is mainly due to the presence of endogenous flavins, reduced NADH, lipofuscins, reticulin fibers, collagen, and elastin. The degree of natural fluorescence relates to the tissue type. Tissues such as kidneys and liver can be very challenging, as these organs have extremely high metabolic rates with constant NADP-NADPH turnover, high accumulation of flavins and lipofuscins, and are highly vascularized, thus containing high amounts of collagen and elastin, and their parenchyma, especially that of liver, are rich in reticular fibers. These characteristics are not shared with tissues, such as the pancreas, that have low accumulation of both flavins and lipofucins, and reduced vascularization. The molecules that contribute to intrinsic and induced autofluorescence are responsible for an emission of fluorescence between 450 and 650 nm that overlaps the emission wavelength of fluorophores widely used in immunofluorescence studies, such as fluorescein isothiocianate (FITC) and phycoerithrin (PE) (from 488 nm to 568 nm) [65].

Depending on the cause of autofluorescence, different methodologies have been used [65]:

- Chemical procedures, such as treatment with ammonia/ethanol
- Quenching using dyes, such as Sudan Black B, Pontamine Sky Blue, and Trypan Blue

- Use of photochemical methods, such as photobleaching, through which, the molecular structure of a fluorophore is changed so that it loses its ability to fluoresce
- Use of filter sets which is particularly suited for the reduction of NADH-associated autofluorescence

Van de Lest et al. [66] demonstrated that autofluorescence in immunofluorescence microscopy can be efficiently eliminated by subtraction of the autofluorescence image captured using an excitation wavelength outside the excitation spectrum of the fluoroprobe. Autofluorescence in tissue sections treated for immunofluorescence microscopy can be effectively eliminated by digitalization of the fluorescence signal at two different wavelengths. Correction for section thickness and minimal fading of the fluoroprobe make this method suitable for (semi)quantitative analysis of stained sections. The method is based on the fact that most autofluorescent compounds have broad-banded excitation and emission spectra, whereas specific fluorescent probes have narrow spectra.

5.6.5.1 Autofluorescence in formalin-fixed sections

It is not surprising that fixation and processing (using formaldehyde, ethanol, xylene [or its replacements] and heated wax) would have an effect on autofluorescence. Autofluorescence, either intrinsic or induced by fixation and tissue processing, may either mask specific fluorescent signals or be mistaken for fluorescent labels [67]. Autofluorescence interferes with routine specific fluorescent labeling by red, green, and blue fluorescent probes, making it very difficult to visualize or co-localize multiple proteins of interest [68]. Baschong et al. [67] have listed three different strategies can be used to control autofluorescence: extraction of the autofluorescent constituent by dissolution (ammonia-ethanol), chemical modification of the fluorochrome (borohydride), and masking of the autofluorescent structure by appropriate staining (Sudan Black). Their experience highlights the fact that there is no general recipe available for the control of autofluorescence. Success was found in a tactical approach that can be summarized as choice of the appropriate reagent(s) by trial and error, either single or combined, mutually designed for the specific tissue type, fixation solution, processing technique, and wavelength of excitation light.

5.6.5.2 Formalin-induced fluorescence

Biogenic amines and some of their precursors may be found in a variety of human neoplasms including pheochromocytomas, neuroblastomas, nonchromaffin paragangliomas, carcinoid tumors, melanomas, and a variety of polypeptide producing endocrine tumors [69]. The formaldehyde-induced fluorescence technique has been used to identify low concentrations of biogenic amines in chromaffin cells of the adrenal gland, argentaffin cells, and in neural tissue. The fluorescence phenomenon itself is based on the reaction of formaldehyde vapor with intracellular biogenic amines such as dopamine, epinephrine, and norepinephrine. This reaction forms 3,4-dihydroisoquinolines, a group of compounds which fluoresce with an excitation peak of 390 nm and an emission maximum of 480 nm [70].

The active principles in pigmented nevi and malignant melanoma responsible for the formaldehyde condensation are thought to be DOPA (3, 4-dihydrosyphenylalanine) and dopamine. A positive green fluorescence is seen [69]. Sztramska et al. [71], using microscopic digital imaging to quantitatively determine the fluorescence intensity of primary melanomas and benign pigmented lesions, found that the FIF emitting molecules (most probably 5-S-cysteinyldopa) were present in melanomas at the concentration significantly higher than in benign lesions. They suggest that FIF analysis might be useful in supporting the diagnosis of melanoma in cases of problematic lesions which are difficult to diagnose using histologic criteria [71].

The staining method is quite simple. After dewaxing sections, dehydrate, clear, and mount using a nonfluorescing mountant.

Acknowledgments

I would like to thank Amanda Charlton and Andrew Williams from the Children's Hospital at Westmead for their advice and support.

References

1. Neuberger, M. S. (2008). Antibody diversification by somatic mutation: from Burnet onwards. *Immunol. Cell Biol.*, **86**(2), 124–132.
2. Saper, C. B. (2009). A guide to the perplexed on the specificity of antibodies. *J. Histochem. Cytochem.*, **57**(1), 1–5.
3. Köhler, G., and Milstein, C. (1975). Continuous cultures of fused cells secreting antibody of predefined specificity. *Nature*, **256**(5517), 495–497.
4. Lipman, N. S., Jackson, L. R., Trudel, L. J., and Weis-Garcia, F. (2005). Monoclonal versus polyclonal antibodies: distinguishing characteristics, applications, and information resources. *ILAR J.*, **46**(3), 258–268.
5. Hjelm, B., Forsström, B., Igel, U., Johannesson, H., Stadler, C., Lundberg, E., et al. (2011). Generation of monospecific antibodies based on affinity capture of polyclonal antibodies. *Protein Sci.*, **20**(11), 1824–1835.
6. Kobayashi, M., Yanagihara, E., and Hayashi, T. (1988). Specificity problem of polyclonal rabbit antibody. *J. Clin. Pathol.*, **41**(6), 705–706.
7. Henwood, A. F., (1982). The importance of optimum titre in immunoperoxidase techniques. *Aust. J. Med. Lab. Sci.*, **3**(1), 21–27.
8. Andersen, D. C., Kliem, A., Schrøder, H. D., and Jensen, C. H. (2011). Newly formed skeletal muscle fibers are prone to false positive immunostaining by rabbit antibodies. *Acta Histochem.*, **113**(1), 68–71.
9. Scott, G. S. and Henwood, A. F. (1984). Crossreactivity in immunoperoxidase: prostate specific antigen. *Aust. J. Med. Lab. Sci.*, **5**(2), 73–75.
10. Cohn, M. (2005). Degeneracy, mimicry and crossreactivity in immune recognition. *Mol. Immunol.*, **42**(5), 651–655.
11. Pradidarcheep, W., Labruyère, W. T., Dabhoiwala, N. F., and Lamers, W. H. (2008). Lack of specificity of commercially available antisera: better specifications needed. *J. Histochem. Cytochem.*, **56**(12), 1099–1111.
12. KPL (2013). Guide to purification of polyclonal antibodies. http://www.kpl.com/docs/techdocs/Purification%20of%20Polyclonal%20Antibodies%20Guide.pdf
13. Rhodes, K. J., and Trimmer, J. S. (2006). Antibodies as valuable neuroscience research tools versus reagents of mass distraction. *J. Neurosci.*, **26**(31), 8017–8020.
14. Gassmann, M., Thömmes, P., Weiser, T. H. O. M. A. S., and Hübscher, U. (1990). Efficient production of chicken egg yolk antibodies against a conserved mammalian protein. *FASEB J.*, **4**(8), 2528–2532.

15. Hanly, W. C., Artwohl, J. E., and Bennett, B. T. (1995). Review of polyclonal antibody production procedures in mammals and poultry. *ILAR J.*, **37**(3), 93–118.
16. Larsson, A., Karlsson-Parra, A., and Sjöquist, J. (1991). Use of chicken antibodies in enzyme immunoassays to avoid interference by rheumatoid factors. *Clin. Chem.*, **37**(3), 411–414.
17. Leenaars, M., and Hendriksen, C. F. (2005). Critical steps in the production of polyclonal and monoclonal antibodies: evaluation and recommendations. *ILAR J.*, **46**(3), 269–279.
18. Spieker-Polet, H., Sethupathi, P., Yam, P. C., and Knight, K. L. (1995). Rabbit monoclonal antibodies: generating a fusion partner to produce rabbit-rabbit hybridomas. *Proc. Natl. Acad. Sci. U.S.A.*, 92(20), 9348–9352.
19. Huang, Z., Zhu, W., Szekeres, G., and Xia, H. (2005). Development of new rabbit monoclonal antibody to estrogen receptor: immunohistochemical assessment on formalin-fixed, paraffin-embedded tissue sections. *Appl. Immunohistochem. Mol. Morphol.*, **13**(1), 91–95.
20. Buchwalow, I., Samoilova, V., Boecker, W., and Tiemann, M. (2011). Non-specific binding of antibodies in immunohistochemistry: fallacies and facts. *Sci. Rep.*, **1**, 28.
21. Nelson, P., Westwood, O., Jefferies, R., Goodall, M., and Hay, F. 1997. Characterization of anti-IgG monoclonal antibody A57H by epitope mapping. *Biochem. Soc. Trans.*, **25**, 373S.
22. Bordeaux, J., Welsh, A. W., Agarwal, S., Killiam, E., Baquero, M. T., Hanna, J. A., et al. (2010). Antibody validation. *Biotechniques*, **48**(3), 197.
23. Burry, R. W. (2000). Specificity controls for immunocytochemical methods. *J. Histochem. Cytochem.*, **48**(2), 163–165.
24. Lorincz, A., and Nusser, Z. (2008). Specificity of immunoreactions: the importance of testing specificity in each method. *J. Neurosci.*, **28**(37), 9083–9086.
25. Henwood, A. F., Harmata, P., and Scott, G. S. (1983). A control for routine immunofluorescent histochemistry. *Aust. J. Med. Lab. Sci.*, **4**, 187–188.
26. Gasser, E. M. S., Straub, C. J., Panzanelli, P., Weinmann, O., Sassoè-Pognetto, M., and Fritschy, J. M. (2006). Immunofluorescence in brain sections: simultaneous detection of presynaptic and post-synaptic proteins in identified neurons. *Nat. Protoc.*, **1**(4), 1887–1897.
27. Clevenger, C., Bauer, K., and Epstein, A. (1985). A method for simultaneous nuclear immunofluorescence and DNA content quantitation using monoclonal antibodies and flow cytometry. *Cytometry*, **6**, 206–214.

28. Pavlakis, E., and Chalepakis, G. (2008). pH-dependent antigen unmasking in paraformaldehyde-fixed tissue cryosections. *Appl. Immunohistochem. Mol. Morphol.*, **16**(5), 503–506.

29. Gruber, H. E., Ingram, J., Zinchenko, N., and Hanley Jr., E. N. (2010). Practical histological methods for use with cultured cells. *Biotech. Histochem.*, **84**(6), 283–286.

30. Cutler, A. H., Cockayne, S., Liddell, J. D., Herman, O. M., Gill, S. D., Margaritov, E., et al. (2014). Cultured cell lines prepared by novel dual-freeze cryoembedded method mimic frozen tissue blocks in maintaining complement components. *J. Histotechnol.*, **37**(1), 26–32.

31. Randall, K. J., and Pearse, G. (2008). A dual-label technique for the immunohistochemical demonstration of T-lymphocyte subsets in formalin-fixed, paraffin-embedded rat lymphoid tissue. *Toxicol. Pathol.*, **36**(6), 795–804.

32. Shi, S. R., Key, M. E., and Kalra, K. L. (1991). Antigen retrieval in formalin-fixed, paraffin-embedded tissues: an enhancement method for immunohistochemical staining based on microwave oven heating of tissue sections. *J. Histochem. Cytochem.*, **39**(6), 741–748.

33. Leong, T. Y., and Leong, A. S. (2007). How does antigen retrieval work?. *Adv. Anat. Pathol.*, **14**(2), 129–131.

34. D'Amico, F., Skarmoutsou, E., and Stivala, F. (2009). State of the art in antigen retrieval for immunohistochemistry. *J. Immunol. Meth.*, **341**(1), 1–18.

35. Elias, J. M. (2000). Mechanics of fluorescence microscopy. *J. Histotechnol.*, **23**(3), 193–202.

36. Furness, P. N. and Boyd, S. (1996). Electron microscopy and immunocytochemistry in the assessment of renal biopsy specimens: actual and optimal practice. *J. Clin. Pathol.*, **49**(3), 233–237

37. Furness, P. N. (2000). ACP best practice no. 160. Renal biopsy specimens. *J. Clin. Pathol.*, **53**(6), 433–438.

38. Lehman, J. S. and Camilleri, M. J. (2013). Potential diagnostic pitfall in direct immunofluorescence evaluation for vasculitis in skin biopsies: fluorescent secretory granules within eccrine glands. *J. Cutaneous Pathol.*, **40**, 603–605.

39. Kambham, N. (2012). Crescentic glomerulonephritis: an update on pauci-immune and anti-GBM diseases. *Adv. Anat. Pathol.*, **19**(2), 111–124.

40. Bonsib, S. M. (2002). Differential diagnosis in nephropathology: an immunofluorescence-driven approach. *Adv. Anat. Pathol.*, **9**(2), 101–114.

41. Kneisel, A., and Hertl, M. (2011). Autoimmune bullous skin diseases. Part 2: diagnosis and therapy. *J. Dtsch. Dermatol. Ges.*, **9**(11), 927–947.

42. Arbesman, J., Grover, R., Helm, T. N., and Beutner, E. H. (2011). Can direct immunofluorescence testing still be accurate if performed on biopsy specimens after brief inadvertent immersion in formalin?. *J. Am. Acad. Dermatol.*, **65**(1), 106–111.

43. Swanson, P. E. (1988). Foundations of immunohistochemistry. A practical review. *Am. J. Clin. Pathol.*, **90**(3), 333.

44. Greka, A., Weins, A., and Mundel, P. (2014). Abatacept in B7-1-positive proteinuric kidney disease reply. *N. Engl. J. Med.*, **370**(13), 1263–1266.

45. Lleo, A., Invernizzi, P., Gao, B., Podda, M., and Gershwin, M. E. (2010). Definition of human autoimmunity—autoantibodies versus autoimmune disease. *Autoimmun. Rev.*, **9**(5), A259–A266.

46. Sfriso, P., Ghirardello, A., Botsios, C., Tonon, M., Zen, M., Bassi, N., et al. (2010). Infections and autoimmunity: the multifaceted relationship. *J. Leukocyte Biol.*, **87**(3), 385–395.

47. von Mühlen, C. A., and Tan, E. M. (1995). Autoantibodies in the diagnosis of systemic rheumatic diseases. *Semin. Arthritis Rheumat.*, **24**(5), 323–358.

48. Tan, E. M. (2012). Autoantibodies, autoimmune disease, and the birth of immune diagnostics. *J. Clin. Invest.*, **122**(11), 3835–3836.

49. Tan, E. M., Feltkamp, T. E. W., Smolen, J. S., Butcher, B., Dawkins, R., Fritzler, M. J., et al. (1997). Range of antinuclear antibodies in "healthy" individuals. *Arthritis Rheumat.*, **40**(9), 1601–1611.

50. Kavanaugh, A., Tomar, R., Reveille, J., Solomon, D. H., and Homburger, H. A. (2000). Guidelines for clinical use of the antinuclear antibody test and tests for specific autoantibodies to nuclear antigens. *Arch. Pathol. Lab. Med.*, **124**(1), 71–81.

51. Tozzoli, R., Bizzaro, N., Tonutti, E., Villalta, D., Bassetti, D., Manoni, F., et al. (2002). Guidelines for the laboratory use of autoantibody tests in the diagnosis and monitoring of autoimmune rheumatic diseases. *Am.J. Clin. Pathol.*, **117**(2), 316–324.

52. De Beukelaar, J. W., and Smitt, P. A. S. (2006). Managing paraneoplastic neurological disorders. *Oncologist*, **11**(3), 292–305.

53. Vernino, S., and Lennon, V. A. (2000). New Purkinje cell antibody (PCA-2): Marker of lung cancer–related neurological autoimmunity. *Ann. Neurol.*, **47**(3), 297–305.

54. van der Loos, C. M., and Teeling, P. (2008). A generally applicable sequential alkaline phosphatase immunohistochemical double staining. *J. Histotechnol.*, **31**(3), 119–127.
55. Robertson, D., Savage, K., Reis-Filho, J. S., and Isacke, C. M. (2008). Multiple immunofluorescence labelling of formalin-fixed paraffin-embedded (FFPE) tissue. *BMC Cell Biol.*, **9**(1), 13.
56. Sternberger, L. A., and Joseph, S. A. (1979). The unlabeled antibody method. Contrasting color staining of paired pituitary hormones without antibody removal. *J. Histochem. Cytochem.*, **27**, 1424–1429.
57. Pirici, D., Mogoanta, L., Kumar-Singh, S., Pirici, I., et al. (2009). Antibody elution method for multiple immunohistochemistry on primary antibodies raised in the same species and of the same subtype. *J. Histochem. Cytochem.*, **57**, 567–575.
58. Buscone, S., Argentieri, M. C., Pilla, D., and Cattoretti, G. (2014). Whole-slide, quadruple immunofluorescence labeling of routinely processed paraffin sections. *Appl. Immunohistochem. Mol. Morphol.*, **22**(4), e1–e7.
59. Yuste, R. (2005). Fluorescence microscopy today. *Nat. Methods*, **2**(12), 902–904.
60. Gellrich, S., Ventura, R., Jones, M., Tan, S. Y., and Mason, D. Y. (2004). Immunofluorescent and FISH analysis of skin biopsies. *Am. J. Dermatopathol.*, **26**(3), 242–247.
61. Wessendorf, M. W., and Elde, R. P. (1985). Characterization of an immunofluorescence technique for the demonstration of coexisting neurotransmitters within nerve fibers and terminals. *J. Histochem. Cytochem.*, **33**(10), 984–994.
62. Dikicioglu, E., Meteoglu, I., Okyay, P., Culhaci, N., and Kacar, F. (2003). The reliability of long-term storage of direct immunofluorescent staining slides at room temperature. *J. Cutaneous Pathol.*, **30**(7), 430–436.
63. Grimwood, R. E., and Proffer, L. H. (2000). Long-term preservation of direct immunofluorescence staining in slides stored at room temperature. *J. Cutaneous Pathol.*, **27**(5), 224–227.
64. Espada, J., Juarranz, Á., Galaz, S., Cañete, M., Villanueva, Á., Pacheco, M., and Stockert, J. C. (2005). Non-aqueous permanent mounting for immunofluorescence microscopy. *Histochem. Cell Biol.*, **123**(3), 329–334.
65. Viegas, M. S., Martins, T. C., Seco, F., and Do Carmo, A. (2007). An improved and cost-effective methodology for the reduction of autofluorescence in direct immunofluorescence studies on formalin-fixed paraffin-embedded tissues. *Eur. J. Histochem.*, **51**(1), 59–66.

66. Van de Lest, C. H., Versteeg, E. M., Veerkamp, J. H., and Van Kuppevelt, T. H. (1995). Elimination of autofluorescence in immunofluorescence microscopy with digital image processing. *J. Histochem. Cytochem.*, **43**(7), 727–730.
67. Baschong, W., Suetterlin, R., and Laeng, R. H. (2001). Control of autofluorescence of archival formaldehyde-fixed, paraffin-embedded tissue in confocal laser scanning microscopy (CLSM). *J. Histochem. Cytochem.*, **49**(12), 1565–1571.
68. Sun, Y., Yu, H., Zheng, D., Cao, Q., Wang, Y., Harris, D., and Wang, Y. (2011). Sudan black B reduces autofluorescence in murine renal tissue. *Arch. Pathol. Lab. Med.*, **135**(10), 1335–1342.
69. Delellis, R. A. (1971). Formaldehyde-induced fluorescence technique for the demonstration of biogenic amines in diagnostic histopathology. *Cancer*, **28**(6), 1704–1710.
70. Dunn, D. R. and Barth, R. F. (1974). Identification of melanoma cells by formaldehyde-induced fluorescence. *Cancer*, **33**(3), 701–706.
71. Sztramska, A., Piwinski, M., and Chwirot, B. W. (2010). Quantitative measurements of formalin-induced fluorescence for differential diagnostics of melanomas and lesions of human skin. *Melanoma Res.*, **20**(5), 408–416.

Appendix A. Direct Immunofluorescence

Solutions

Tris or PBS buffer can be used depending on laboratory preference

10× TRIS Buffer:

- TRIS(hydroxymethyl)aminomethane (Sigma T-1378) 30.29 g
- Sodium Chloride 43.05 g
- Distilled Water 450 ml
- pH to 7.6 with concentrated Hydrochloric Acid
- Make up to 500 ml with Distilled water
- For use, add 10 ml concentrate buffer to 90 ml distilled water

10× PBS Buffer:

- Distilled Water 800 ml
- Sodium Chloride 80 g

- Potassium Chloride 2 g
- DiSodium Hydrogen Phosphate (Na_2HPO_4) 14.4 g
- Potassium diHydrogen Phosphate (KH_2PO_4) 2.4 g
- Adjust pH to 7.2
- Make up to 1 liter with distilled water
- For use, add 10 ml concentrate buffer to 90 ml distilled water

Antibody diluent:

- Tris or PBS Buffer (dilute from 10× above) 100 ml
- Bovine Serum Albumen 1 g
- Sodium Azide 0.1 g
- Add a few drops of green or blue food coloring
- Keeps for several months in the fridge.
- Caution Harmful: contains Sodium Azide

Mountant:

(http://www.jacksonimmuno.com/technical/anti-fade.asp) or use commercial supply.

(1) Prepare a stock solution of 20%(w/v) *n*-propyl gallate (Sigma P3130) in dimethyl formamide or dimethyl sulfoxide. (Note: *n*-propyl gallate does not dissolve well in water-based solutions.)
(2) Thoroughly mix 1 ml 10× Tris or 10× PBS with 9 ml glycerol (ACS grade 99–100% purity) and slowly add 0.1 ml 20% *n*-propyl gallate dropwise with rapid stirring.

Procedure:

(1) Cut frozen sections, assessing for the presence of glomeruli or complete skin surface
(2) Mark back of slide, circling the sections, with a xylene-resistant marker
(3) Air-dry for 15 minutes or more
(4) Place in Buffer at 37°C for 5–7 minutes
(5) Drain excess buffer, add optimally diluted FITC labelled antibody and incubate in the dark for 30–60 minutes
(6) Rinse excess antibody from sections with buffer
(7) Place in Buffer at 37°C for 5–7 minutes

(8) Drain and coverslip with fade-resistant mountant (alternatively, dehydrate with alcohol, clear in xylene and coverslip with DePeX
(9) Assess staining with a fluorescent microscope

Appendix B. Procedures to Reduce Autofluorescence

Sudan Black B Method to Reduce Autofluorescence

(Viegas et al., 2007; Oliveira et al., 2010, 68)

(1) This technique can be applied either before or after immunofluorescence staining.
(2) Tissue sections were immersed in 0.1% Sudan Black B (Sigma) in 70% ethanol for 20 minutes, at room temperature
(3) Rinse slides in water 25 minutes

Ammoniacal Ethanol Method to Reduce Autofluorescence

(Viegas et al., 2007; Oliveira et al., 2010)

(1) Place the slide in 0.25% ammonia in 70% ethanol for 1 hour at room temperature
(2) Rinse well in water

Borohydride Reduction Method to Reduce Autofluorescence

This is especially useful for formaldehyde fixed tissues (Baschong et al., 2001)

(1) Place slides in freshly prepared 1% sodium borohydride in PBS at 4°C for 40 minutes (the solution should be fizzy like carbonated water)
(2) Rinse slides well in water

Chapter 6

Fluorescent Proteins

Anya Salih
Confocal Bio-Imaging Facility S8, Office of Deputy Vice-Chancellor, Research and Development, Western Sydney University, NSW 1797, Australia
a.salih@westernsydney.edu.au

6.1 Introduction

The increasing need to understand the function of genes and proteins in cells and organisms and the advances in imaging platforms to image these events in radically new ways have contributed to the explosive development of fluorescent protein technologies in the last quarter of the century. Numerous proteins exhibit fluorescence, but the term *fluorescent proteins* (FPs) generally defines a diverse group of proteins found in marine organisms that are homologues of the prototypical jellyfish (*Aequorea victoria*) green fluorescent protein (GFP, avGFP, or wild-type wtGFP). This 28-kDa protein has bright green fluorescence at UV or blue light excitation. The main advantage of GFP-like proteins over other fluorescent proteins, stains or tags is their ability to form an internal chromophore without requiring any accessory cofactors, enzymes or substrates other than molecular oxygen [1–4]. This unique property makes

Fundamentals of Fluorescence Imaging
Edited by Guy Cox

ISBN 978-981-4774-85-7 (Hardcover), 978-1-351-12940-4 (eBook)
www.jennystanford.com

possible the formation of the chromophore in live organisms, tissues and cells. Consequently, one of the most common uses for GFPs is as gene tags fused to proteins or polypeptides of interest so that the expression of the target protein becomes accompanied by the fluorescent glow of GFP. What was previously invisible becomes visible, and standard tools of fluorescence and confocal microscopy can be used to observe the target protein in a living cell as it performs its function. GFP-like FPs have fueled a new era in cell biology and biomedicine and are widely used as genetically encoded fluorescent labels to visualize structures and dynamic processes in living cells and organisms [2, 5–8].

Originally, in the 1990s, only green GFP-type proteins were available to scientists, but a range of mutations of the wild-type GFP expanded the color palette to cyan and yellow FPs. Soon, many other FPs, such as the red fluorescent DsRed, were discovered in reef corals and other marine organisms [9, 10]. Now there are hundreds of GFP homologues available, with diverse spectral and molecular properties, covering almost the entire visible spectrum. Reef organisms were also the source of the photoactivatable FPs (PAFPs) that spontaneously alter their spectral properties after irradiation by certain wavelengths and thus enable tracking of labeled molecules and cells in space and time [11].

One of the most common uses for GFP variants is as tags fused to proteins or polypeptides. This large family of multi-colored GFP homologues is now used in thousands of laboratories around the world to track where and when certain proteins are expressed in cells or in whole organisms in real time. The protein fusion constructs are designed so as not to disrupt the structure and function of the studied protein and to produce efficient fluorescent labeling of proteins in living cells. Importantly, differently colored GFP-like proteins fused to different target proteins provide opportunities to simultaneously observe multiple cellular events and to unravel the fine details of how cells and organisms function. When fused to appropriate targeting sequences, GFP-type proteins can be expressed in subcellular organelles. GFP-type proteins have been expressed in bacteria, yeast, algae, plants, mammalian cells, embryos, worms, insects, and living vertebrates, including fish, mice, and even monkeys.

6.2 Historical Perspectives

6.2.1 Early Discoveries

A brief description of the history of GFP discoveries is presented here since so many scientists contributed to its development and success that it is impossible to fully review their contributions in this short chapter. Excellent reviews can be found in books by Chalfie and Kain [8] and Shimomura [12].

The first step in what has often been termed as the "GFP revolution" in cell biology was made by Osamu Shimomura in 1961, when he discovered a green fluorescent protein accidentally in a bioluminescent jellyfish *Aequorea victoria* (avGFP) while studying aequorin, a chemoluminescent calcium-activated photoprotein [13]. Aequorin and GFP were found localized in the jellyfish light organs and blue fluorescent emissions by aequorin were shown to excite green GFP fluorescence by Förster resonance energy transfer between the two molecules [14]. Another GFP-type protein was found in a bioluminescent relative of corals, the sea pansy *Renilla reniformis* [15], and it likewise transforms blue light emitted by luciferase into green light. Since both proteins form components of a bioluminescent system, it was assumed that GFPs were restricted to bioluminescent organisms.

On its own, GFP was found to fluoresce bright green under ultraviolet light, the color that led to its name. Its fluorescence excitation spectrum has a dominant maximum at 390 nm and a smaller maximum at 470 nm, while the emission spectrum is at 505 nm, with a shoulder around 540 nm [2] (Fig. 6.1).

Wt-GFP was shown to be made of 238 amino acids, and was cloned by Prasher et al. [16]). Residues 65–67 (Ser-Tyr-Gly) in its sequence were found to spontaneously form the fluorescent chromophore *p*-hydroxybenzylideneimidazolinone [17] (Fig. 6.2). In 1996, the crystal structure of the wt-GFP was resolved, revealing that the cyclic tripeptide chromophore is buried in the center of a molecule that resembled a can in shape (β-barrel structure) and proportion, with an 11-stranded sheet wrapped into a cylinder measuring 42 by 24 Å (Fig. 6.3) [18, 19]. The rigid and cylindrical structure of wt-GFP is conserved in all the FPs yet discovered,

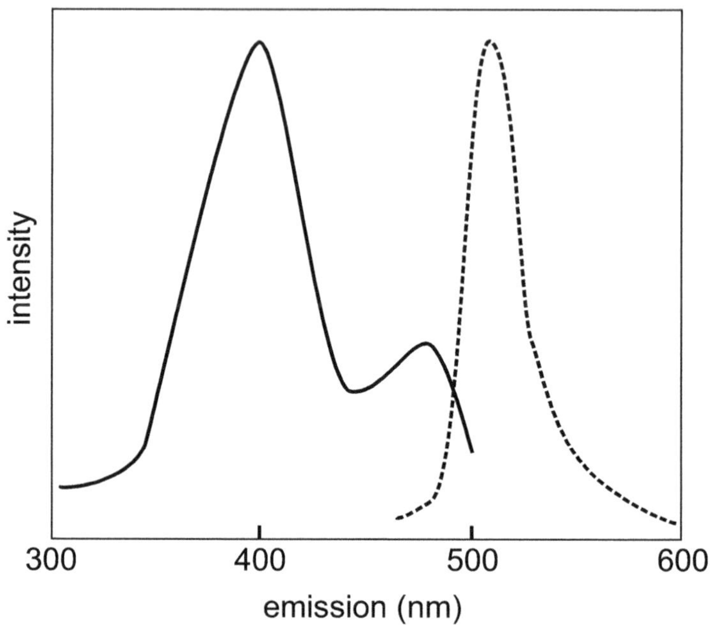

Figure 6.1 Excitation and emission spectra of wild-type GFP.

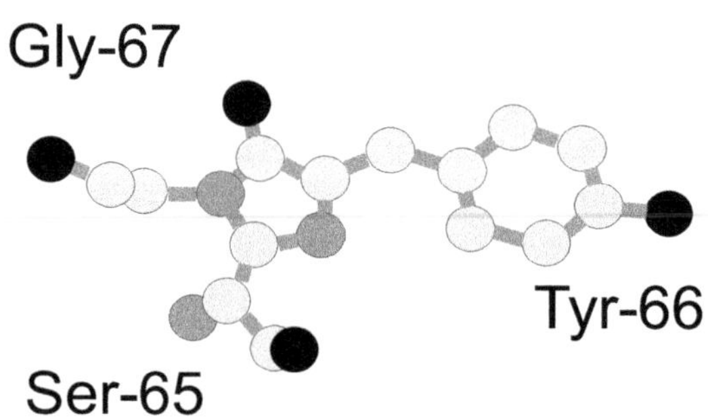

Figure 6.2 The GFP chromophore. Image courtesy of Guy Cox.

possibly having evolved to protect the chromophore. The first heterologous expression was by Chalfie et al. [1].

Major achievements were made by Tsien and his colleagues, who contributed much to our understanding of how GFP works and engineered new mutants [2, 4, 7]. They investigated how

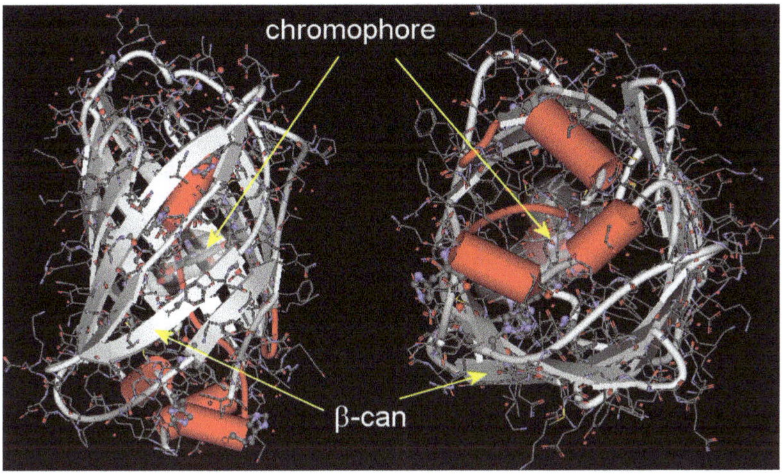

Figure 6.3 The structure of GFP shown in side view (left) and top view (right). The β-can and chromophore are indicated. Image courtesy of Alexander Savitsky.

GFP's structure produces the green fluorescence and experimented with the structure to produce GFP variants that emitted light at different wavelengths. Subsequently, once GFP-like proteins were discovered in nonbioluminescent organisms, his group added other GFP variants to the FP toolkit.

The mutants were brighter and had additional spectral variants, fluorescing in cyan, blue, and yellow wavelengths [2]. One of the most widely used GFP mutants, named enhanced GFP (eGFP), was developed by Tsien's group with an excitation/emission spectrum that closely matches the 488 nm argon-ion laser line and the green 509 nm emission matching filter sets designed for fluorescein (FITC) [7]. It is bright and photostable, and one of the most popular variants. Various other early modifications of GFP included codon optimization for expression in mammalian cells, bacteria, and yeast and improved folding mutations [1].

However, despite considerable efforts to generate red FPs (RFPs), none were discovered or created from bioluminescent hydrozoan jellyfish. The most red-shifted avGFP variant that was created was green-yellow, with emission at 529 nm [18]. Red-shifted GFP-like proteins were required for several reasons: they provide an additional color for multi-color labeling; they penetrate deeper than other visible light wavelengths due to limited absorption by tissues, and their excitation is less phototoxic to cells.

6.2.2 FPs from Nonbioluminescent Organisms

The search for red-emitting FPs continued until a group of Russian scientists headed by Lukyanov discovered that GFP-like FPs are the key color determinants in reef-building corals and related organisms (class Anthozoa, order Scleractinia), ranging in colors from cyan to red, and cloned two orange-red RFPs emitting at 583 and 593 nm [9]. The 583 nm red FP cloned from an anemone-like organism *Discosoma* was named DsRed [9] and became one of the most widely used RFPs to date.

Finally, the red color barrier was broken, and thereafter a stream of discoveries of variously colored GFP-type proteins were made in a variety of nonbioluminescent marine organisms. FPs were found to be immensely diverse and abundant, and up to 98% of all surveyed corals at sites of Great Barrier Reef were found to contain FPs [20]. Even the nonfluorescent, but strongly purple-blue pigmented chromoproteins (CPs) common to coral tissues and anemone tentacle tips were found to belong to the GFP family [21].

Mutagenesis of CPs generated some of the most red-shifted FPs, with an emission maximum at 645 nm [22]. Over the last two decades, GFP-like FPs and CPs have been found in a wide range of marine invertebrates including jellyfish, corals, anemones, zoanthids, copepods, and even in lancelets, primitive chordate organisms (e.g., [9, 23–26]). Molecular phylogeny showed that the colors evolved from a single, most likely green, fluorescent ancestral protein by variation of the p-HBI chromophore [25, 27]. A colorless, nonfluorescent protein known as nidogen found in mammalian basement membranes was found to have a 10% sequence homology to wtGFP, suggesting that the FP family is even more wide-spread than we currently know. Spectra of some of the available FPs are shown in Fig. 6.4, and their applications in Figs. 6.5 and 6.6.

The search for the far-red GFP-like proteins continues on today. However, screening of corals and other cnidarians revealed that

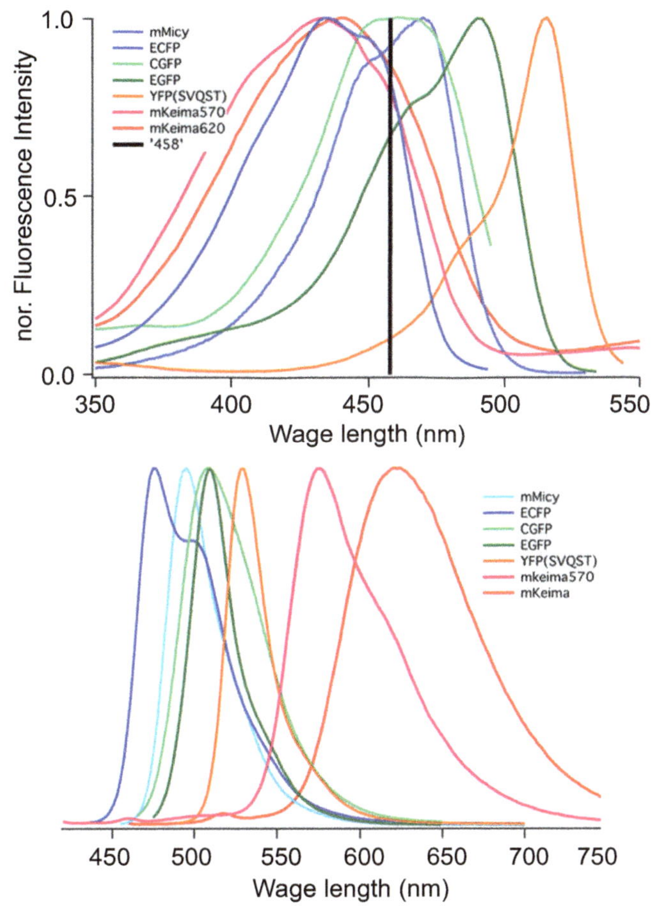

Figure 6.4 Spectra of a range of currently available fluorescent proteins. Top: excitation. Bottom: emission. Image courtesy of Atsushi Miyawaki.

emissions of FPs do not exceed the 650–700 nm waveband (e.g., [25, 28]. Since many anthozoans, such as reef corals, are symbiotic with photosynthetic microalgae whose chlorophylls absorb in these wave-lengths, the lack of the far-red-shifted FPs absorbing light beyond 650 nm may be due to the necessity to avoid interfering with photosynthesis. However, GFP-type proteins also occur in organisms that lack symbiotic microalgae, and their FP emissions also fall well under the 650 nm limit.

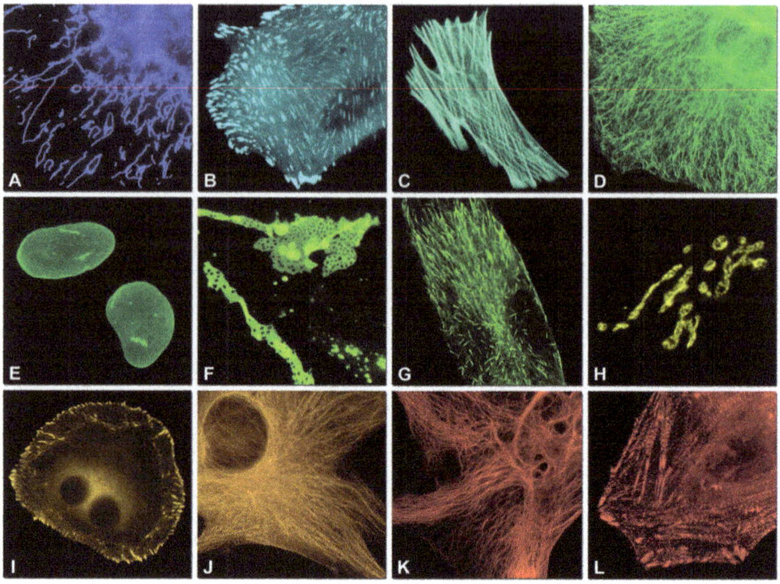

Figure 6.5 (A–L) Selected monomeric FP fusions with targeting proteins imaged in wide-field fluorescence. FP fusion terminus and number of linker amino acids is indicated after the name of the targeted organelle or fusion protein. (A) EBFP2-mito-N-7 (human cytochrome C oxidase subunit VIII; mitochondria); (B) mCerulean-paxillin-N-22 (chicken; focal adhesions); (C) mTFP1-actin-C-7 (human β-actin; filamentous actin); (D) mEmerald-keratin-N-17 (human cytokeratin 18; intermediate filaments); (E) superfolder GFP-lamin B1-C-10 (human lamin B1; nuclear envelope); (F) mVenus-Cx43-N-7 (rat α-1 connexin-43; gap junctions); (G) YPet-EB3-N-7 (human microtubule-associated protein; RP/EB family); (H) mKO-Golgi-N-7 (N-terminal 81 amino acids of human β-1,4-glactosyltransferase; Golgi complex); (I) tdTomato-zyxin-N-7 (human zyxin; focal adhesions); (J) TagRFP-tubulin-C-6 (human α-tubulin; microtubules); (K) mCherry-vimentin-N-7 (human vimentin; intermediate filaments); (L) mPlum–actinin-N-19 (human nonmuscle; cytoskeleton). Image courtesy of Nathan Shaner [29].

6.2.3 The Green Glow of the Nobel Prize

In 2008, the Nobel Prize in Chemistry was awarded jointly to Osamu Shimomura, Martin Chalfie, and Roger Y. Tsien "for the discovery and development of the green fluorescent protein, GFP." It is a shining example of how fundamental research in an obscure topic can lead

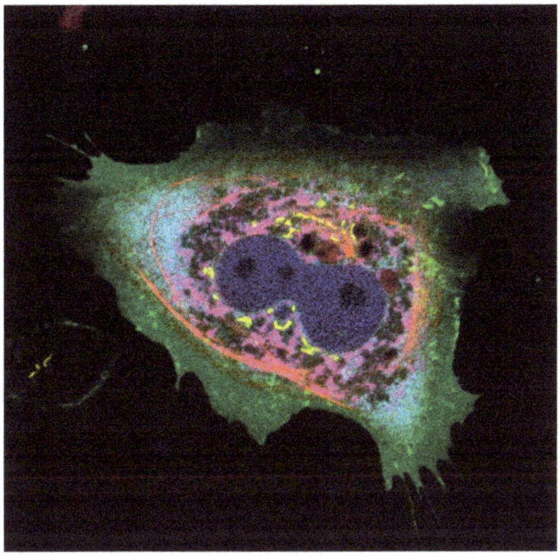

Figure 6.6 Simultaneous six-color imaging of subcellular structures in a Vero cell using a single laser line (458 nm). CFP localized on the plasma membrane (green), mMiCy in the endoplasmic reticulum (cyan), EGFP in the Golgi (yellow), YFP along the microtubules (red), dKeima570 in the nucleus (dark blue), and mKeima in the mitochondria (purple). Image courtesy of Atsushi Miyawaki.

to spectacularly beneficial applications that extend across many disciplines.

Within a few years, research on GFP-type proteins contributed to yet another outstanding development in science: the nano-microscopic realm of super-resolution microscopy (Chapters 14 and 16), enabling visualization of proteins in cells at nanometer scales. The 2014 Nobel Prize in Chemistry was awarded to three scientists who developed super-resolution techniques: Stefan Hell, William Moerner, and Eric Betzig.

6.3 General Properties of Fluorescent Proteins

With hundreds of FPs now available researchers face a difficult task in selecting the right variant for a specific experiment. They may well simply choose something that has worked in their lab previously.

Several reviews have covered the topic of what properties to look for in FPs [7, 30], and here some basic characteristics of FPs, their spectral properties, and their important parameters relevant to fluorescence imaging will be addressed briefly.

6.3.1 General Structure

All GFP-type proteins, regardless of their origin or the degree of genetic mutagenesis, are ~25 kD in size (reviewed Tsien [2]). They differ from various low molecular weight fluorophores, since not only are they genetically encodable, but their emissions are tuned by the surrounding protein matrix, enabling rational genetic manipulations to optimize their optical properties [7, 31].

The FP structure consists of an extremely rigid β-barrel fold comprising 11 β-sheets that surround central α-helices [18]. Together with the α-helices at either end of the molecule, a cylinder is produced (42 Å long and 24 Å diameter) [18] creating what is referred to as a "β-can" formation (Fig. 6.3). All GFP-like proteins consist of ~220–240 amino acid residues, which fold into a barrel with an internal helix and the chromophore at the center [2].

The chromophore is formed by a unique posttranslational modification of the three amino acid residues X-Tyr-Gly at positions 65–67 (Fig. 6.2). The amino acid side chains inside the barrel are important in chromophore formation and in influencing spectral properties [2].

Each protein monomer can assemble into a dimer or tetramer structure [25, 32] and the majority of nonbioluminescent anthozoan FPs are tetrameric. This is a disadvantage in functional imaging since the large size of tetramer can interfere with normal cell functioning. All GFP-like proteins have a tendency to dimerize and even monomeric ones can form dimers at high concentrations [4, 33]. A great deal of effort has therefore been expended in producing bright and stable monomeric variants. For avGFP and its variants, a single-point mutation, A206K, is sufficient to prevent dimerization [4]. The most effective approach to reduce oligomerization has been a combination of rational protein engineering and directed evolution and has been used to develop a number of non-oligomerizing variants: RFP from *Discosoma* sp. [34], a GFP, Azami-Green, from the coral *Galaxia* sp.; an orange FP from *Fungia concinna*; green-to-red photoconvertible FPs from *Lobophyllia hemprichii* [35] and *Dendronephthya* sp. [28]; a photoactivatable GFP from Pectiniidae; a chromoprotein from *Montipora* sp. to create a red FP with a large Stokes shift.

6.3.1.1 Blue and cyan FPs

The use of blue and cyan FPs for cellular imaging has been more limited than for greens because many require near-UV light for excitation which is phototoxic to mammalian cells. Other shortcomings included low quantum yield, increased cellular autofluorescence, light scattering and more specialized microscope systems for imaging at near-UV wavelengths [36, 37], although most microscopes nowadays are appropriately configured.

Cyan FPs developed from avGFP bearing a tryptophan substitution in the chromophore have been among the most frustrating to end users due to their low brightness. One of the earliest color variants derived from avGFP was a blue FP (BFP), with a broad absorption band at 380 nm and emission peaking at 448 nm [38]. It is imaged using standard DAPI fluorescence filter sets, but it has fallen out of use as it is not very bright [37]. An enhanced, brighter variant was subsequently developed, EBFP, which has been frequently used as a donor FP, with the acceptor EGFP, in FRET microscopy [39]. The best BFP variant, EBFP2, is the brightest and most photostable; it forms an excellent donor for FRET and is useful for long-term imaging in living cells [37].

An enhanced version of CFP (ECFP), another mutant of avGFP, had been generated with a bimodal spectra, with excitation peaks at 433 and 445 nm and emission maxima at 475 and 503 nm [33]. It has been frequently used in FRET applications, has improved brightness and photostability, but still only about 40% of the brightness of EGFP [33, 37]. The complex spectral characteristics of ECFP were improved by mutagenesis to produce Cerulean, leading to a higher extinction coefficient, increased quantum yield, and which when coupled with yellow-emitting FPs, such as Venus, has been one of the most widely used donors for FRET applications [37].

Naturally occurring cyan FPs in anthozoans have emission peaks between 475–499 nm and are some of the most common reef FPs, particularly in shallow reefs [25, 28]). They possess the same chromophore as the green FPs. Their excitation peaks are usually between 440 and 465 nm, with some rare exceptions of more blue-shifted peaks. They are readily excited by near-UV light, but can be excited by blue-violet light around 450 nm. Cyan FPs tend to be bright and photostable, have lower extinction coefficients of all colors, and have wider excitation and emission curves than greens (full width at half maximum (FWHM) of 55 nm for the former compared to ~35 nm for the latter).

One example of an anthozoan bright and stable cyan addition to the FP toolkit is the monomeric teal FP (mTFP1) from *Clavularia* sp. [29, 40] (Fig. 6.5C). A great deal of recent effort has gone into engineering highly improved cyan FP variants [41, 42], culminating in the development of mTurquoise2 whose quantum yield of 0.93 is among the highest described for any known FP [43].

6.3.1.2 Green FPs

The wild-type avGFP has a complex absorption spectrum, with maximal excitation at 397 nm and a minor peak at 476 nm (Fig. 6.1) which arise from its deprotonated and protonated states [41]. This complex absorption spectrum, with high absorption by UV wavelengths and low quantum yield, limited its utility. Protein engineering approaches have generated many different spectral variants from the avGFP, known collectively as "enhanced" (e) GFPs: improving protein maturation and expression in many different heterologous cell systems, as for example, improvement of expression in mammalian cells, improved maturation at physiological temperatures [7].

The most commonly used version, EGFP, has the major absorption peak at 488 nm, but it is also efficiently excited by UV at 400 nm [33] and this can complicate multicolor imaging or fluorescence resonance energy transfer (FRET) experiments with donors excited by UV. Another avGFP variant, mWasabi (emission maximum = 509 nm), has very narrow excitation and emission peaks, is 1.6-fold brighter than EGFP, is photostable and has almost no UV (below 400 nm) excitation. This makes it superior to EGFP for use in multicolor imaging in combinations with UV–violet light excitable FPs and for FRET experiments [37]. Emerald (Fig. 6.5D), also an avGFP variant, excitation 487 nm/emission 509 nm, is a weak dimer and has an initial fast bleaching component which subsequently stabilizes to photostability equivalent to EGFP [2]. Another very useful avGFP variant, the "superfolder" GFP (Fig. 6.5E), folds even when fused to insoluble proteins and is brighter and more acid-resistant than avGFP and EGFP [42].

A novel monomeric green FP, mNeonGreen, has been recently developed from the cephalochordate *Branchiostoma lanceolatum*, with high photostability and very high quantum yield, demonstrating that the use of fresh starting material can lead to the discovery of substantially improved variants [30].

Green fluorescent FP-like proteins are the most abundant in reef corals and other anthozoans and are very conspicuous on reefs. Their emissions are at ~500–520 nm, similar to wtGFP and eGFP, and they usually have narrower emissions and excitation spectra at FWHM than the cyans. Reef FPs hold promise for the development of novel green emitters for bioimaging and a variety of cloned green FPs are becoming available for imaging applications.

A very bright Azami-Green (excitation 492 nm/emission 505 nm) FP has been cloned from a coral *Galaxia* sp. It has a higher quantum yield than eGFP and has acid stability while being very similar spectrally to EGFP, although sharing only 5.7% amino acid homology with it. It is tetrameric but is has been converted to a monomeric form mAzami-Green (mAG), which proved to be very successful as a label.

6.3.1.3 Yellow FPs

Yellow FPs (YFPs) are useful in multicolor labeling applications since they are readily spectrally separated from cyan and red FPs, as well as from the more blue-shifted green FPs, as EGFP. They are also of importance for the design of yellow-red FRET pairs. Up to now, there are still relatively few of these compared to other color classes.

The first GFP-type YFPs were engineered from avGFP and are defined by the T203Y mutation, an anionic phenolate tyrosine-derived chromophore, and an emission peak ~20 nm red-shifted from that of EGFP [18]. The enhanced version, EYFP, has become one of the brightest and most widely utilized FPs, even though it is pH sensitive [7]. A brighter and more acid-stable variant, Citrine, with absorption and emission maxima at 516 and 529 nm, respectively, was subsequently developed, which is more efficiently expressed in mammalian cells and has increased photostability [46]. Yet another improved version of EYFP, Venus [47], was developed with even more acid tolerance but reduced photostability, causing a significant problem for long-term imaging experiments [37] (Fig. 6.5F).

Naturally occurring YFPs are the least abundant color variants in reef Anthozoa and commonly have emission maxima at ~525–565 nm, although we have found them present in several Great Barrier Reef corals and related species, including *Discosoma* spp. and zoanthids (personal observations). Two yellow FPs have been cloned and used in imaging, although infrequently: zYFP538 from a zoanthid *Zoanthus* sp., with maximal emissions at 538 nm [9], and phiYFP, with emission maximum at 537 nm, from the jellyfish *Phialidium* [24]. While phiYFP contains a GFP-like chromophore in a modified environment, zoanYFP possesses a Lys66-Tyr67-Gly68 chromophore-forming sequence, arranged into a unique three-ring chromophore structure [9].

6.3.1.4 Red FPs

Red and particularly far-red emitting FPs are useful for imaging deep into tissues. All wild-type RFPs originate from anthozoans and are tetrameric, so a key goal of research was to develop monomeric forms [7, 48].

The majority of monomeric RFPs in use in imaging are derived from only a handful of species. The soft coral *Discosoma* sp. gave wt-DsRed, the first RFP to be isolated. As well as being a tetramer, it takes some time to mature after expression, fluorescing in the green until maturation, which is a problem for dual labeling [49, 50]. Twenty years of refinement has led to a range of bright, monomeric (m-) versions: mCherry (Fig. 6.5K), mStrawberry, and other "mFruits," e.g., tdTomato (Fig. 6.5I,L), with emission maxima ranging from 540 nm to 610 nm [7, 48].

The bubble-tip anemone *Entacmaea quadricolor* has also given rise to a family of RFPs. The native protein, eqFP611, has a large Stokes shift [51]. It was successfully used in a number of applications, e.g., in dual-labeling experiments with GFP in plants [52]. Its monomeric variant named mRuby, with excitation and emission maxima at 558 nm and 605 nm, and a Stokes shift of 47 nm, was found very useful for imaging applications and was about 10-fold brighter than EGFP [53]. One of the most recent additions to the far-red FP toolkit and the most red-shifted RFP has been the monomeric mGarnet, mutated from eqFP611, with fluorescence peaking at 670 nm [54]). It can be efficiently excited with 640 nm light and has been demonstrated to be an excellent live-cell fusion marker in STED microscopy (Chapter 16), using 640 nm excitation and 780 nm depletion wavelengths.

TagFP is another member of the *E. quadricolor* family, emitting at 635 nm (Fig. 6.5J). Derivatives from it include mKate2, a very bright far-red monomeric RFP with fast maturation and high pH stability [55]. With excitation/emission maxima at 588 and 633 nm, it is one of the most far-red RFPs, is very photostable, and is ~3-fold brighter than TagFP635. Its usefulness for deep imaging has been demonstrated in transgenic *Xenopus* embryos. Another mutant of TagFP635, named Katushka2S, is another far-red RFP with high brightness and very fast maturation [22, 55].

Another far red RFP was discovered through site-directed and random mutagenesis of a nonfluorescent chromoprotein from the anemone *Heteractis crispa*. HcRed has its emission maximum at 645 nm, but did not achieve popular use in imaging since it is a tetramer. Further mutagenesis produced a brighter dimeric variant, HcRed-2 (commercially traded as HcRed1), but no monomeric variant has yet been developed.

6.3.2 Photoactivatable and Photoconvertible FPs

A chance accident opened the door to a whole new range of FP applications. Researchers in Atsushi Miyawaki's lab in Japan isolated a green fluorescent protein from the coral *Trachyphllia geoffroyi*. It

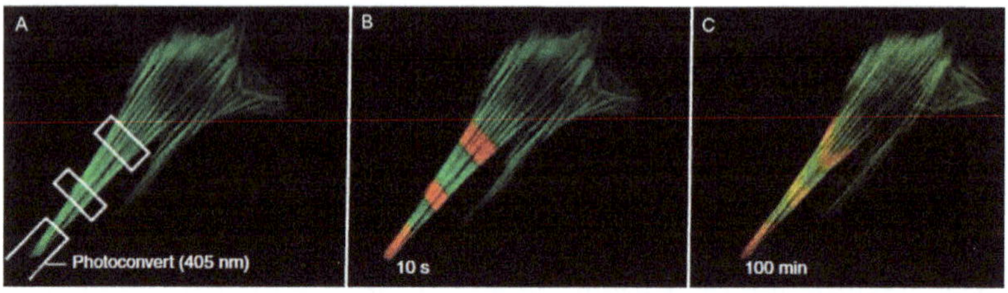

Figure 6.7 (A) A single FoLu fibroblast cell expresses t2EosFP–actin as part of the filamentous actin cytoskeletal network. The fusion protein was photoconverted using 405 nm diode laser irradiation for 10 s. The three regions irradiated in a single scan are marked by white polygons. (B) 10 s later, the field was imaged simultaneously with a 488 nm (green fluorescence) and a 543 nm laser (red fluorescence). (C) After 100 minutes, actin movement is revealed by the red fluorescence. Image courtesy of Jörg Wiedenmann. Reproduced by permission from Wiedenmann et al. [57].

was excited in the blue at 490 nm, with bright and stable green fluorescence at 520 nm. One aliquot was accidentally left on a bench exposed to daylight when the others were put away. Next day it had changed to having an excitation maximum in the green and emission in the red. Investigation showed that UVA irradiation converted the protein from a "GFP" to an "RFP," with both forms equally stable [56]. The protein was named "Kaede" (maple) since its color change mimicked maple leaves in autumn.

Other examples soon followed. Eos (Greek for dawn) has a green to red conversion like Kaede, and comes from the coral *Lobophyllia hemprichii* [57]. The conversion mechanism involves the photocleavage of the protein backbone and the formation of a double bond at His65, resulting in an extended π-electron conjugation of Tyr66 to the imidazole ring of His65 [58]. This was the first such conversion to be fully understood (Fig. 6.7).

On a different tack, a sea-anemone nonfluorescent GFP homologue was found by a Russian team to become stably fluorescent after activation by intense light [11]. This was the father of a line of "kindling," or photoactivatable FPs, some even derived from *A. victoria* GFP.

The significance of these activatable and convertible FPs is that they can be used to follow dynamic processes in the cell. If a labeled protein is expressed in the cell, we can then activate or convert a particular region and follow the subsequent fate of this protein. Previously such studies could only be done by selective photobleaching, involving damaging levels of irradiation. See Chapter 11 for more on this topic.

A further development was the discovery that is some cases conversion or kindling could be reversible. Dronpa (from the Japanese for appear and disappear), is normally excited in the blue. Green light destroys the fluorescence, but near-UV restores it, and the process can be repeated for hundreds of cycles without obvious damage [59]. This is significant for photo-localization microscopy, a super-resolution technique which locates single molecules as they switch on or off (Chapter 14). It has also been proposed as a technique for data storage, but this has not got beyond a "proof of concept."

References

1. Chalfie, M,. Tu, Y., Euskirchen, G., Ward, W., and Prasher, D. (1994). Green fluorescent protein as a marker for gene expression. *Science*, **263**, 802–805.
2. Tsien, R., (1998). The green fluorescent protein. *Ann. Rev. Biochem.*, **67**, 509–544.
3. Reid, B. G., and Flynn, G. C. (1997). Chromophore formation in green fluorescent protein. *Biochemistry*, **36**, 6786–6791.
4. Zacharias, D. A., Baird, G. S., and Tsien, R. Y. (2000). Recent advances in technology for measuring and manipulating cell signals. *Curr. Opin. Neurobiol.*, **10**, 416–421.
5. Miyawaki, A. (2003). Fluorescence imaging of physiological activity in complex systems using GFP-based probes. *Curr. Opin. Neurobiol.*, **13**, 591–596.
6. Lippincott-Schwartz, J., and Patterson, G. H. (2003). Development and use of fluorescent protein markers in living cells. *Science*, **300**, 87–91.
7. Shaner, N. C., Steinbach, P. A., and Tsien, R. Y. (2005). A guide to choosing fluorescent proteins. *Nat. Methods*, **2**, 905–909.
8. Chalfie, M., and Kain, S. R., (eds.) (2005). *Green Fluorescent Protein: Properties, Applications and Protocols*. Wiley, p. 488.

9. Matz, M. V., Fradkov, A. F., Labas, Y. A., Savitsky, A. P., Zaraisky, A. G., Markelov, M. L., and Lukyanov, S. A. (1999). Fluorescent proteins from nonbioluminescent Anthozoa species. *Nat. Biotechnol.*, **17**, 969–973.
10. Matz, M. V., Lukyanov, K. A., and Lukyanov, S. A. (2002). Family of the green fluorescent protein: journey to the end of the rainbow. *Bioessays*, **10**, 953–959.
11. Lukyanov, K. A., Chudakov, D. M., Lukyanov, S., and Verkhusha, V. V. (2005). Photoactivatable fluorescent proteins. *Nat. Rev. Mol. Cell Biol.*, **6**, 885–890.
12. Shimomura, O. (2006). *Bioluminescence: Chemical Principles and Methods*. World Scientific, p. 500.
13. Shimomura, O., Johnson, F. H., and Saiga, Y. (1962). Extraction, purification and properties of aequorin, a bioluminescent protein from the luminous hydromedusan, *Aequorea. J. Cell. Comp. Physiol.*, **59**, 223–239.
14. Morise, H., Shimomura, O., Johnson, F. H., and Winant, J. (1974), Intermolecular energy transfer in the bioluminescent system of *Aequorea. Biochemistry*, **13**, 2656–2662.
15. Ward, W. W., and Cormier, M. J. (1979). An energy transfer protein in coelenterate bioluminescesnce – characterization if the *Renilla* green fluorescent protein. *J. Biol. Chem.*, **254**, 781–788.
16. Prasher, D. C., Eckenrode, V. K., Ward, W. W., Prendergast, F. G., and Cormier, M. J. (1992). Primary structure of the *Aequorea victoria* green-fluorescent protein. *Gene*, **111**, 229–233.
17. Shimomura, O. (1979). Structure of the chromophore of Aequorea green fluorescent protein. *FEBS Lett.*, **104**, 220–222.
18. Ormö, M., Cubitt, A. B., Kallio, K., Gross, L. A., Tsien, R. Y., and Remington, S. J. (1996). Crystal structure of the *Aequorea victoria* green fluorescent protein. *Science*, **273**, 1392–1395.
19. Yang, F., Moss, L. G., and Phillips, Jr., G. N. (1996). The molecular structure of green fluorescent protein. *Nat. Biotechnol.*, **14**, 1246–1251.
20. Salih, A., Larkum, A. W. D., Cox, G., Kühl, M., and Hoegh-Guldberg, O. (2000). Fluorescent pigments in corals are photoprotective. *Nature*, **408**, 850–853.
21. Lukyanov, K. A., Fradkov, A. F., Gurskaya, N. G., Matz, M. V., Labas, Y. A., Savitsky, A. P., Markelov, M. L., Zaraisky, A. G., Zhao, X., Fang, Y., Tan, W., and Lukyanov, S. A. (2000). Natural animal coloration can be determined by a nonfluorescent green fluorescent protein homolog. *J. Biol. Chem.*, **275**, 25879–25882.

22. Gurskaya, N. G., Fradkov, A. F., Terskikh, A., Matz, M. V., Labas, Y. A., Martynov, V. I., Yanushevich, Y. G., Lukyanov, K. A., Lukyanov, S. A. (2001). GFP-like chromoproteins as a source of far-red fluorescent proteins. *FEBS Lett.*, **507**, 16–20.

23. Labas, Y. A., Gurskaya, N. G., Yanushevich, Y. G., Fradkov, A. F., Lukyanov, K. A., Lukyanov, S. A., and Matz, M. V. (2002). Diversity and evolution of the green fluorescent protein family. *Proc. Natl. Acad. Sci. U.S.A.*, **99**(7), 4256–4261.

24. Shagin, D. A., Barsova, E. V., Yanushevich, Y. G., Fradkov, A. F., Lukyanov, K. A., Labas, Y. A., Semenova, T. N., Ugalde, J. A., Meyers, A., Nunez, J. M., Widder, E. A., Lukyanov, S. A., and Matz, M. V. (2004). GFP-like proteins as ubiquitous metazoan superfamily: evolution of functional features and structural complexity. *Mol. Biol. Evol.*, **21**, 841–850.

25. Alieva, N. O., Konzen, K. A., Field, S. F., Meleshkevitch, E. A., Hunt, M. E., Beltran-Ramirez, V., Miller, D. J., Wiedenmann, J., Salih, A., and Matz, M. V. (2008). Diversity and evolution of coral fluorescent proteins. *PLoS One*, **3**, e2680.

26. Li, G., Zhang, Q.-J., Zhong, J., and Wang, Y.-Q. (2009). Evolutionary and functional diversity of green fluorescent proteins in cephalochordates. *Gene*, **446**, 41–49.

27. Ugalde, J. A., Chang, B. S., and Matz, M. V. (2004). Evolution of coral pigments recreated. *Science*, **305**, 1433.

28. Salih, A. (2012). Screening reef corals for novel GFP-type fluorescent proteins by confocal imaging - in vivo cellular imaging using fluorescent proteins. *Methods Mol. Biol.*, **872**, 217–233.

29. Shaner, N. C., Patterson, G. H., and Davidson, M. W. (2007). Advances in fluorescent protein technology. *J. Cell Sci.*, **120**, 4247–4260.

30. Shaner, N. C. (2014). Fluorescent proteins for quantitative microscopy: Important properties and practical evaluation. *Methods Cell Biol.*, **123**, 95–111.

31. Pakhomov, A. A., and Martynov, V. I. (2008). GFP family: structural insights into spectral tuning. *Cell Chem. Biol.*, **15**, 755–764.

32. Wall, M. A., Socolich, M., and Ranganathan, R. (2000). The structural basis for red fluorescence in the tetrameric GFP homolog DsRed. *Nat. Struct. Biol.*, **7**, 1133–1138.

33. Cubitt, A. B., Heim, R., Adams, S. R., Boyd, A. E., Gross, L. A., and Tsien, R. Y. (1995). Understanding, improving and using green fluorescent proteins. *Trends Biochem. Sci.*, **20**, 448–455.

34. Campbell, R. E., Tour, O., Palmer, A. E., Steinbach, P. A., Baird, G. S., Zacharias, D. A., and Tsien, R. Y. (2002). A monomeric red fluorescent protein. *Proc. Natl. Acad. Sci. U.S.A.*, **99**, 7877–7882.

35. Wiedenmann, J., Ivanchenko, S., Oswald, F., Schmitt, F., Röcker, C., Salih, A., Spindler, K.-D., and Nienhaus, G. U. (2004). EosFP, a fluorescent marker protein with UV-inducible green-to-red fluorescence conversion. *Proc. Natl. Acad. Sci. U.S.A.*, **101**, 15905–15910.

36. Rizzo, M. A., and Piston, D. W. (2005). In *Live Cell Imaging: A Laboratory Manual*. Goldman, R. D., Spector, D. L. (eds.), Cold Spring Harbor Laboratory Press; ColdSpring Harbor, pp. 3–23.

37. Day, R. N., and Davidson, M. W. (2009). The fluorescent protein palette: tools for cellular imaging. *Chem. Soc. Rev.*, **38**, 2887–2921.

38. Heim, R., Cubitt, A. B., and Tsien, R. Y. (1995). Improved green fluorescence. *Nature*, **373**, 663–664.

39. Patterson, G. H., Piston, D. W., and Barisas, B. G. (2000). Förster distances between green fluorescent protein pairs. *Anal. Biochem.*, **284**, 438–440.

40. Ai, H. W., Henderson, J. N., Remington, S. J., and Campbell, R. E. (2006). Directed evolution of a monomeric, bright and photostable version of *Clavularia* cyan fluorescent protein: structural characterization and applications in fluorescence imaging. *Biochem. J.*, **400**, 531–540.

41. Goedhart, J., van Weeren, L., Hink, M. A., Vischer, N. O., Jalink, K., and Gadella, T. W., Jr. (2009). Bright cyan fluorescent protein variants identified by fluorescence lifetime screening. *Nat. Methods*, **7**, 137–139.

42. Markwardt, M. L., Kremers, G.-J., Kraft, C. A., Ray, K., Cranfill, P. J. C., Wilson, K. A., et al. (2011). An improved cerulean fluorescent protein with enhanced brightness and reduced reversible photoswitching. *PLoS One*, **6**, e17896.

43. Goedhart, J., von Stetten, D., Noirclerc-Savoye, M., Lelimousin, M., Joosen, L., Hink, M. A., van Weeren, L., Gadella, T. W. J., Jr., and Royant, A. (2012). Structure-guided evolution of cyan fluorescent proteins towards a quantum yield of 93%. *Nat. Commun.*, **3**, 751 (1–9).

44. Chattoraj, M., King, B. A., Bublitz, G. U., and Boxer, S. G. (1996). Ultra-fast excited state dynamics in green fluorescent protein: multiple states and proton transfer. *Proc. Natl. Acad. Sci. U.S.A.*, **93**, 8362–8367.

45. Pédelacq, J. D., Cabantous, S., Tran, T., Terwilliger, T. C., and Waldo, G. S. (2006). Engineering and characterization of a superfolder green fluorescent protein. *Nat. Biotechnol.*, **24**, 79–88.

46. Griesbeck, O., Baird, G. S., Campbell, R. E., Zacharias, D. A., and Tsien, R. Y. (2001). Reducing the environmental sensitivity of yellow fluorescent

protein. Mechanism and applications. *J. Biol. Chem.*, **276**, 29188–29194.

47. Nagai, T., Ibata, K., Park, E. S., Kubota, M., Mikoshiba, K., and Miyawaki, A. (2002). A variant of yellow fluorescent protein with fast and efficient maturation for cell-biological applications. *Nat. Biotechnol.*, **20**, 87–90.
48. Shaner, N. C. (2013). The mFruit collection of monomeric fluorescent proteins. *Clin. Chem.*, **59**, 440–441.
49. Bevis, B. J., and Glick, B. S. (2002). Rapidly maturing variants of the *Discosoma* red fluorescent protein (DsRed). *Nat. Biotechnol.*, **20**, 83–87.
50. Strack, R. L., Strongin, D. E., Bhattacharyya, D., Tao, W., Berman, A., Broxmeyer, H. E., Keenan, R. J., and Glick, B. S. (2008). A noncytotoxic DsRed variant for whole-cell labelling. *Nat. Methods*, **5**, 955–957.
51. Wiedenmann, J., Schenk, A., Röcker, C., Girod, A., Spindler, K. D., and Nienhaus, G. U. (2002). A far-red fluorescent protein with fast maturation and reduced oligomerization tendency from *Entacmaea quadricolor* (Anthozoa, Actinaria). *Proc. Natl. Acad. Sci. U.S.A.*, **99**, 11646–11651.
52. Forner, J., and Binder, S. (2007). The red fluorescent protein eqFP611: application in subcellular localization studies in higher plants. *BMC Plant Biol.*, **6**, 28.
53. Kredel, S., Oswald, F., Nienhaus, K., Deuschle, K., Röcker, C., Wolff, M., Heilker, R., Nienhaus, G. U., and Wiedenmann, J. (2009). mRuby, a bright monomeric red fluorescent protein for labeling of subcellular structures. *PLoS One*, **4**, e4391.
54. Hense, A., Prunsche, B., Gao, P., Ishitsuka, Y., Nienhaus, K., and Nienhaus, G. U. (2015). Monomeric Garnet, a far-red fluorescent protein for live-cell STED imaging. *Sci. Rep.*, **5**, 18006.
55. Shcherbo, D., Merzlyak, E. M., Chepurnykh, T. V., Fradkov, A. F., Ermakova, G. V., Solovieva, E. A., Lukyanov, K. A., Bogdanova, E. A., Zaraisky, A. G., Lukyanov, S., and Chudakov, D. M. (2007). Bright far-red fluorescent protein for whole-body imaging. *Nat. Methods*, **26**, 741–746.
56. Ando, R., Hama, H., Yamamoto-Hino, M., Mizuno, H., and Miyawaki, A. (1999). An optical marker based on the UV-induced green-to-red photoconversion of a fluorescent protein. *Proc. Natl. Acad. Sci. U.S.A.*, **99**, 12651–12656.
57. Wiedenmann, J., Ivanchenko, S., Oswald, F., Schmitt, F., Röcker, C., Salih, A., Spindler, K.-D., and Nienhaus, G. U. (2004). EosFP, a fluorescent

marker protein with UV-inducible green-to-red fluorescence conversion. *Proc. Natl. Acad. Sci., U.S.A.*, **101**, 15905–15910.

58. Nienhaus, K., Nienhaus, G. U., Wiedenmann, J., and Nar, H. (2005). Structural basis for photon-induced protein cleavage and green to red conversion of fluorescent protein EosFP. *Proc. Natl. Acad. Sci., U.S.A.*, **102**, 9156–91569.
59. Ando, R., Mizuno, H., and Miyawaki, A. (2004). Regulated fast nucleocytoplasmic shuttling observed by reversible protein highlighting. *Science*, **306**, 1370–1373.

Chapter 7

Dyes for Labeling Organelles and Compartments

Iain Johnson
Lumencor, Inc., 14940 NW Greenbrier Parkway, Beaverton, OR 97006, USA
iain.johnson@lumencor.com

In this chapter, I will describe the properties and applications (primarily in mammalian cells) of fluorescent probes for labeling organelles and subcellular compartments. I will begin by outlining some general considerations that apply to all these probes and then move on to descriptions of individual probes arranged according to their labeling targets. The inclusion of this material in a volume on fundamentals of fluorescence imaging signifies the fact that probes are contrast-generating reagents and off-target localization of the probe is just as much of a problem as optical aberrations in compromising the fidelity of image data. Furthermore, the spectroscopic and biochemical properties of probes impact many aspects of the design and performance of imaging experiments (Table 7.1).

All of the subcellular targets discussed below can be labeled using fluorescent protein (FP) fusions (Chapter 6) or antibodies (Chapter 5), in addition to the organic dyes and dye conjugates that

Fundamentals of Fluorescence Imaging
Edited by Guy Cox

ISBN 978-981-4774-85-7 (Hardcover), 978-1-351-12940-4 (eBook)
www.jennystanford.com

Table 7.1 Spectroscopic and biochemical properties of fluorescent probes

Property	Impact on Experimentation
Biochemical Properties	
Translocation and binding to molecular target	Correct identification of organelle or other subcellular structure of interest
Stability of localization (retention)	Length of available image acquisition time, feasibility of time-lapse data collection
Toxicity[a]	Artifactual perturbation of cellular structure and/or function
Aqueous solubility	Requirement for water-miscible co-solvents[b] or carriers.[c] Penetration of probe into multilayer tissues
Membrane permeability	Capacity for labeling intracellular targets in live cells without permeabilization
Spectroscopic Properties	
Property	**Impact**
Photobleaching rate	Length of available image acquisition time, feasibility of time-lapse data collection
Excitation spectrum	Capacity for generating fluorescence from available microscope light source wavelengths
Emission spectrum	Emission spectral separation from other probes and autofluorescence background is essential for multiplexing applications[d]
Extinction coefficient, fluorescence quantum yield	Product is proportional to the available fluorescence photon output of a dye[e]
Photoswitching	Suitability for super-resolution microscopy via single-molecule localization (e.g., PALM, STORM)
Dual emission or excitation[f]	Ratio measurements can compensate for cell-to-cell variation of dye concentration. Increased occupation of spectral bandwidth limits multiplexing with other probes
Fluorogenic response[g]	Avoids need to remove excess unbound dye by washing before imaging
Fluorescence self-quenching[h]	Nonlinear relationship between dye concentration and fluorescence output

a. Includes both chemical toxicity and phototoxicity. *b.* e.g. dimethylsulfoxide (DMSO) *c.* e.g. 2-hydroxypropyl-β-cyclodextrin, bovine serum albumin (BSA). *d.* e.g. two-color colocalization. *e.* In absence of photobleaching. *f.* A shift in excitation or emission wavelength of the probe upon interaction with its biomolecular target. Examples include JC-1, BODIPY FL ceramide and FRET probes. *g.* A probe with a fluorogenic response is nonfluorescent except when bound to its biomolecular target. *h.* Due to dye–dye association promoted by accumulation in small compartments.

are the primary topic for consideration in this chapter. It is useful at the outset to contrast the utilities of these alternative labeling approaches. Labeling with FP fusions generally requires about 24 hours after transient transfection before sufficient expression is obtained for microscopic visualization. Once attained, expression in living cells can generally be maintained for several days. Dye-based labeling protocols produce specimens that are ready for imaging in less than an hour (Section 7.1), but the specificity of labeling is typically retained for 24 hours or less, unless post-labeling fixation is applied. In many cases, dye-based probes depend on particular physiological activities of live cells such as mitochondrial membrane potential or lysosomal acidity to attain specific localization. This is sometimes referred to as vital staining. They can therefore sometimes provide useful read-outs of these activities in addition to marking the location of the target subcellular structure. This is somewhat less so in the case of FP fusions, which use the cellular machinery of protein synthesis, trafficking, and degradation to achieve localization. For the same reason, FP fusions are generally less susceptible to off-target localization than dye-based probes, which, like low-molecular-weight drugs, are targeted via ligand-binding or partition equilibria [1]. Labeling using antibodies (immunofluorescence; Chapter 5) is largely restricted to fixed specimens.

7.1 Labeling Protocols

This chapter does not seek to offer a compendium of labeling protocols. Such protocols are widely available in the primary research literature, laboratory manuals such as the *Methods in Enzymology* series, and in package inserts from commercial vendors. However a brief outline of generally applicable considerations is appropriate and useful. Most of the dye-based probes described in this chapter are loaded into cells via spontaneous uptake, with very simple practical requirements. Cells are immersed for 5–30 minutes in a staining solution prepared by dilution (1:100 or more) of a dye stock solution in DMSO or water into minimal culture medium. At the end of the incubation period, the staining solution is removed and the cells are washed. The latter step is necessitated by the fact that labeling protocols invariably call for applying the probe in vast excess relative to the abundance of its cellular target. Large nonspecific background signals are the inevitable result of incomplete washing, unless the probe is fluorogenic (Table 7.1), in which case the wash step is generally not required. The following notes serve to highlight the important variables in the labeling process:

(1) *Probe concentration*. It is axiomatic that biologically relevant image data must be independent of probe concentration (or FP expression level). Applied probe concentrations in cellular labeling protocols are typically in the range of 10 nM^{-1} μM. It is important to recognize that these concentrations relate to the extracellular staining solution volume and not to the (much smaller) intracellular volumes in which the probe ultimately resides. Thus, even allowing for fact that only a fraction (usually <10%) of the added probe actually penetrates into the cell, intracellular concentrations can easily reach 10 μM in the cytoplasm and 1 mM or more in compact organelles such as mitochondria. Intracellular concentrations of this magnitude are readily capable of perturbing cellular structures and functions represented in the final image data. From the perspective of avoiding nonspecific staining and toxic side-effects, the probe concentration should be as low as possible within the constraint of providing adequate signal-to-noise characteristics under the prevailing image acquisition conditions. Therefore, it is always advisable to test a range of probe concentrations when establishing protocols for new probe–specimen combinations.

(2) *Incubation medium*. Serum-free media should be used for loading of organic dye probes. Many dyes bind appreciably to serum proteins, thereby lowering the free concentration available for loading. Absence of serum-derived growth nutrients such as biotin for the short duration of the incubation period is not usually deleterious. Simple buffered salt solutions such as Hank's balanced salt solution (HBSS) are often quite adequate

temporary substitutes, provided attention is paid to pH control (see below).

(3) *Incubation time.* For monolayer cell cultures, incubation periods of more than 30 minutes are almost never necessary, nor desirable. For tissue specimens, longer incubation periods ($\sim$1 hour) may be required to allow for dye penetration into subsurface cell layers. The limited aqueous solubility of many organic dyes can result in exhaustion of the staining solution before it reaches subsurface layers. This situation can be improved by pre-complexing the dye with a solubilizing carrier such as 2-hydroxypropyl-β-cyclodextrin.

(4) *Environment: temperature and pH.* Insufficient attention to these apparently prosaic factors is the most common source of problems in live cell specimen preparation. Conventional bicarbonate buffered media will alkalinize to a significant and damaging extent within a few minutes of removal from a 5% CO_2 incubator. If it is not operationally possible to maintain the CO_2 environment during the manipulations required for cell loading, then rapid ("seconds not minutes") manipulation and return to the incubator is essential. The CO_2 sensitivity liability of bicarbonate buffered media can also be circumvented using alternative pH buffers such as HEPES. Staining solutions and media for washing should be pre-warmed to the incubation temperature of the specimen (usually 37°C), to avoid unnecessary and potentially harmful thermal shocks. Pre-warming of bicarbonate buffered media should, of course, be carried out in a 5% CO_2 incubator (rather than an ambient atmosphere water bath).

The above considerations apply only to cell loading with small (i.e., molecular weight <1000) molecules that have some degree of plasma membrane permeability. Loading of larger molecules such as proteins, peptides, and dextrans into live cells requires techniques that transiently permeabilize or physically penetrate the plasma membrane, or delivery via endogenous transport processes such as endocytosis (Section 7.6).

7.2 Nucleus

7.2.1 Hoechst 33258, Hoechst 33342, and DAPI

The nucleic acid-binding dyes Hoechst 33258, Hoechst 33342, and 4′,6-diamidino-2-phenylindole (DAPI) are the most widely used nuclear stains for fluorescence microscopy. They are fairly interchangeable, having similar structural and spectroscopic properties, although Hoechst 33342 is preferred for applications in living cells due to its greater membrane permeability. All three dyes bind to DNA via hydrogen-bonded interactions with AT base pairs in the minor groove. Although they bind to RNA, the resulting complexes are effectively nonfluorescent [2], resulting in fluorescence signals that are predominantly nuclear specific. Extranuclear staining, although rarely observed, may originate from mitochondrial DNA (Fig. 7.1), mycoplasma contamination, or lipid transfection complexes. All three dyes are fluorogenic (Table 7.1) and have fluorescence excitation/emission peaks at about 350 nm/460 nm. Excitation at 405 nm, even though it is sub-optimal in terms of the dye excitation spectrum, still produces satisfactory fluorescence signal levels in most cases. This is due to the fact that nuclear DNA is so abundant in mammalian cells that optimal excitation of the dye can be sacrificed for the operational convenience of diode laser or LED light sources.

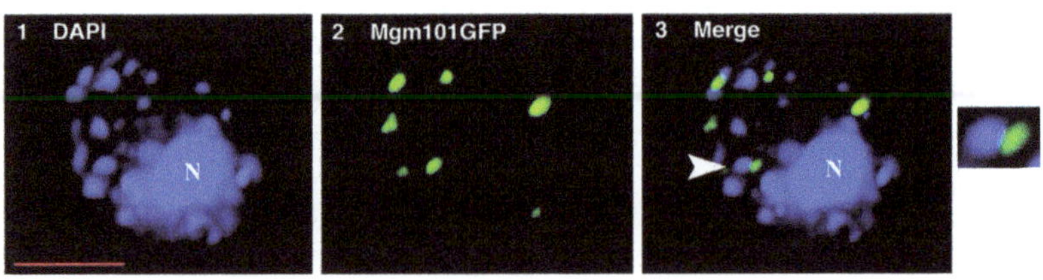

Figure 7.1 DAPI labeling of mitochondrial and nuclear DNA in *Saccharomyces cerevisiae*. Wild-type cells expressing Mgm101GFP were grown overnight in YPG to early log phase and shifted to YPDGal containing 1 μg/ml DAPI for 20 min at 25°C to visualize mtDNA-containing nucleoids. Cells were washed into SD and imaged using a Deltavision deconvolution microscope. Panel A is a whole cell projection. N indicates nuclear DAPI labeling. Arrows indicate the region magnified 250% in the corresponding inset. Scale bars = 2 μm. From Meeusen and Nunnari (2003). *J. Cell Biol.*, **163**, 503–510.

Hoechst 34580, which has an excitation peak at about 370 nm, provides an alternative should more efficient 405 nm excitation be required [3]. The most common applications of Hoechst 33258, Hoechst 33342, and DAPI are as fiducial markers in multicellular specimens rather than for imaging nuclear structure per se. In this respect, detection of near-UV-excited blue fluorescence emission is advantageous, leaving the rest of the visible range from 500–650 nm clear for detection of protein targets by FPs and labeled antibodies. However, photoconversion of the dyes to species with red-shifted spectra [4, 5] has the potential to cause increased levels of channel crosstalk when performing multiplexed imaging of two or more probes. Hoechst 33342 is often used for time-lapse imaging of mitotic cell division, although considerable care must be exercised to avoid impacts of dye-mediated phototoxicity [6]. Consequently, FP-histone fusion proteins are now generally preferred for this application [7].

7.2.2 Long-Wavelength Dyes

Because there is a practical benefit to reserving emission channels in 500–650 nm wavelength region for detection of lower abundance protein targets by FPs and labeled antibodies, the main alternatives to Hoechst 33258, Hoechst 33342, and DAPI are the anthraquinone derivative DRAQ5 [8] for live cells and the cationic cyanine dye TO-PRO-3 for fixed cells. Both dyes have excitation/emission spectral characteristics similar to Cy5. DRAQ5 has an extremely low fluorescence quantum yield (0.004), but its strong binding affinity for DNA coupled with the high cellular abundance of its target result in good signal-to-background characteristics in fluorescence microscopy applications. DRAQ5 does not bind appreciably to RNA [9]. The primary DNA binding site of DRAQ5 overlaps that of Hoechst 33342, indicated by displacement of the latter in competition experiments [8]. Both DRAQ5 and Hoechst 33342 induce cellular DNA damage responses at typical working concentrations [10]. TO-PRO-3 is a cyanine dye that has largely superseded the phenanthridine derivative propidium iodide (PI) for nuclear staining applications, although PI remains in widespread use for cell viability analysis (see Chapter 8). Both TO-PRO-3 and PI form fluorescent complexes with

RNA as well as DNA, resulting in additional utility for Nissl staining (Section 7.7).

7.3 Cytoplasm

Labeling of the cytoplasm is required for measurement of cell volume and shape [11] and for analysis of cellular migration (e.g., “wound-healing assays”) and adhesion. Cellular expression of unfused monomeric FPs provides a simple and effective way of achieving this objective. Calcein AM (detected in the FITC channel) remains the most useful dye-based probe for cytoplasmic labeling. Calcein AM labeling has little impact on cellular motility, growth, adhesion, and other functional properties, and exhibits little cell-to-cell staining variability, except in cells expressing multidrug resistance transporters (for which the dye is a substrate). Calcein AM labeling persists for at least 24 hours in most mammalian cell types, but is not compatible with paraformalehyde fixation. Dyes with protein-reactive functionality such as CellTracker Green CMFDA or CFSE are suitable alternatives for experiments requiring label retention after fixation. Calcein AM, CMFDA, or CFSE require activation by cellular esterases and so cannot be used to label cells that have already been fixed. For this type of specimen, uniform cytoplasmic labeling can be obtained using thiol-reactive dyes such as fluorescein-5-maleimide or Alexa Fluor 488 C_5 maleimide [11].

7.4 Cytoskeleton

Although the cytoskeleton is not conventionally categorized as an organelle or subcellular compartment, it is discussed here because of its role in controlling the morphology and localization of other organelles [12] and transport and secretory vesicles (Section 7.6). FP-actin and FP-tubulin fusions have become standard tools for imaging actin filaments and microtubules in living cells [13]. In some cases, FP appendages may cause perturbation of actin and tubulin structure and dynamics. Such perturbations can become particularly problematic when imaging higher-order structures such as actin patches, cables and rings. Alternatively, FP fusions of actin and tubulin binding proteins such as Abp140 [13], Utr-CH [14], and microtubule-associated protein (MAP4) can be used. As many of these proteins are structural stabilizers, their FP fusions may be unsuitable for imaging actin and tubulin assembly and disassembly dynamics [14]. High-specificity staining of f-actin in fixed cells by dye-labeled conjugates of the hepatotoxic cyclic peptide phalloidin is primarily used in combination with immunofluorescence labeling of proteins (Chapter 5). Phalloidin (molecular weight = 789) does not spontaneously penetrate live cells, except in the case of hepatocytes, the target of its cytotoxicity in vivo [15]. Tubulin Tracker Green, comprising Oregon Green 488 dye coupled to the anti-neoplastic drug paclitaxel, provides an alternative to FP fusions for labeling tubulin in live cells. As paclitaxel is a potent stabilizer of tubulin polymerization, this probe is unsuitable for imaging microtubule instability dynamics.

7.5 Mitochondria

Specific labeling of mitochondria in living cells is generated by electrophoretic uptake of cationic rhodamine, rosamine, and cyanine dyes (Fig. 7.2) driven by the respiration-coupled membrane potential. Because the mitochondrial membrane potential (typically $\sim$–180 mV) is much larger than the plasma membrane potential ($\sim$–90 mV), cationic dyes added extracellularly progressively accumulate first in the cytoplasm and then in mitochondria. Although some dyes (e.g., MitoTracker Green, nonyl acridine orange) are often described as "potential insensitive," this is true only in a relative, not an absolute sense. The mitochondrial association of all these lipophilic cation dyes is diminished following dissipation of the membrane potential caused by treatment with uncoupling reagents [16, 17]. Thus "potential insensitive" is a description of measurement performance but not the mitochondrial staining mechanism. Because mitochondrial fission–fusion dynamics are coupled to variations in membrane potential, mitochondrially targeted FP fusions are preferable to potential sensitive dyes for time-lapse imaging of mitochondrial morphology [18]. In principle, the distribution of cationic dyes across the mitochondrial membrane can be quantitatively related to the transmembrane electrical potential by application of the Nernst potential [19–21]. However, in the complex environment of a living cell, numerous factors in addition to the mitochondrial membrane potential impact intracellular dye distribution, making potentiometric quantitation a challenging undertaking requiring extensive controls [19]. The best dyes for quantitative applications are the tetramethylrhodamine esters TMRE and TMRM, used at concentrations below 10 nM to avoid complications from aggregation and accompanying self-quenching of fluorescence. Low labeling concentrations also serve to minimize dye-mediated phototoxicity, to which mitochondria (like nuclei; Section 7.2), are particularly susceptible [18]. Cationic cyanine dyes such as JC-1 and $DiOC_6(3)$ are more susceptible to aggregation and therefore less amenable to quantitative applications than TMRE and TMRM. Cyanine dyes are also potent inhibitors of mitochondrial respiration [22]. Intramitochondrial aggregation of JC-1 generates a fluorescence emission peak shift from ~525–590 nm. JC-1 is a useful qualitative indicator of mitochondrial membrane potential changes induced by pharmacological agents, but quantitative interpretation of the dual-emission potentiometric response is not straightforward [23]. MitoTracker dyes are cationic rosamines and cyanines with thiol-reactive chloromethyl substituents (Fig. 7.2). Mitochondrial localization is achieved via same electrophoretic uptake mechanism

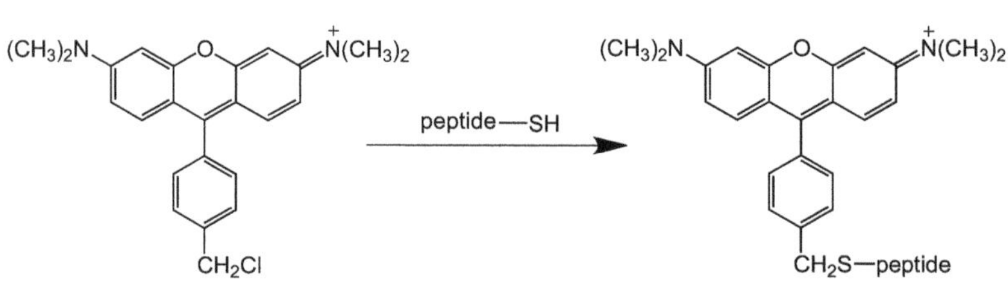

Figure 7.2 MitoTracker Orange CMTMRos is a cell-permeant cationic dye that is sequestered by actively respiring in the mitochondria. Reaction with thiols (SH) on proteins and peptides results in intramitochondrial retention after dissipation of the mitochondrial membrane potential. Consequently, MitoTracker Orange staining is also dependent on reduced thiol levels determined by the intramitochondrial redox potential.

as TMRE and TMRM, requiring that cells are actively respiring during labeling process. The capacity of MitoTracker dyes to react with protein thiols [24] preserves mitochondrial localization when the membrane potential is dissipated following fixation of cells. Thus they are primarily useful in conjunction with immunofluorescence labeling (Chapter 5). By the same token, the capacity to reversibly respond to potential changes in live cell applications is diminished [17]. Furthermore, derivatization of internal residues in proteins inevitably perturbs their physiological function [24], so there is a greater liability for toxicity than there is with freely diffusible dyes such as TMRM and TMRE. MitoTracker dyes exhibit photo-switching behavior without auxiliary thiol or phosphine additives, enabling super-resolution imaging of mitochondrial fission–fusion dynamics using STORM single-molecule localization [25]. Other organelle probes with demonstrated utility for STORM (Chapter 14) include LysoTracker Red (Section 7.6), ER-Tracker Red (Section 7.7) and DiI (Section 7.8) (Fig. 7.3). Autofluorescent constituents of mitochondria such as reduced nicotinamide adenine dinucleotide (NADH) and flavin adenine dinucleotide (FAD) are primarily useful for functional imaging of redox activity rather than as structural markers [26].

7.6 Intracellular Vesicles

7.6.1 Endosomes

Endocytosis in mammalian cells is conventionally classified as either clathrin-dependent (receptor-mediated) or clathrin-independent (caveolar) [27]. Specific labeling of endosomes in the clathrin-dependent pathway can be achieved using fluorescent conjugates of protein ligands, the most ubiquitous of which is the iron transporter transferrin. Conjugates of transferrin and other protein ligands typically incorporate standard immunofluorescent labeling dyes such as FITC, Cy dyes and the Alexa Fluor series. Markers used to visualize caveolar endocytosis include BODIPY dye-labeled glycosphingolipids and fluorescent albumin conjugates [28]. The lipid raft marker cholera toxin B subunit (Section 7.8) is sometimes advocated

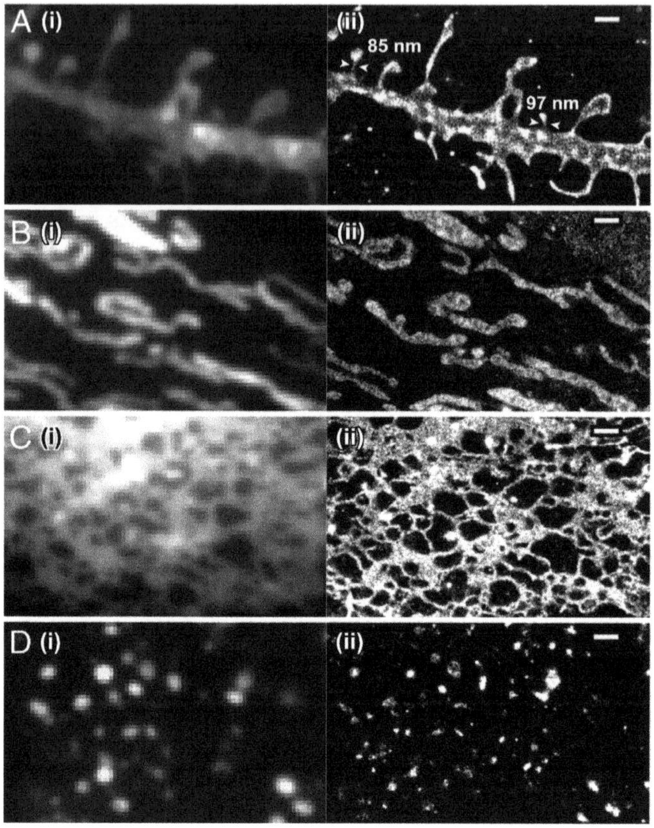

Figure 7.3 STORM images of four membrane organelles/structures in live cells. Conventional (i) and STORM (ii) images of (A) the plasma membrane labeled with DiI in a hippocampal neuron, (B) mitochondria labeled with MitoTracker Red in a BS-C-1 cell, (C) the ER labeled with ER-Tracker Red in a BS-C-1 cell, and (D) lysosomes labeled with LysoTracker Red in a BS-C-1 cell. The conventional fluorescence images were taken immediately before STORM imaging with a low excitation intensity to avoid switching off the dyes appreciably. The STORM images were acquired in 15 sec (A), 10 sec (B and C) and 1 sec (D). Arrowheads in A indicate spine necks along with their measured widths. Scale bars = 1 μm. From Shim et al. [25].

as a tracer for caveolar endocytosis but has been shown to follow the clathrin-dependent endocytic pathway in some cell types. Pathway-specific inhibitors such as chloropromazine, dynasore, genistein, filipin, and nystatin are useful accessory reagents for unraveling mechanistic complexities in the endocytic uptake of fluorescent ligands. Labeling endosomes via uptake of fluid phase tracers such as fluorescent dextrans cannot discriminate between endocytic pathways. Typical protocols use nontoxic, highly water-soluble 10 kDa or 70kDa dextrans conjugated to standard immunofluorescent labeling dyes. Dextrans conjugated to pH-sensitive dyes such as pHrodo Red or Oregon Green 488 generate fluorescence that increases or decreases with progressive endosomal acidification [29, 30]. Cells are incubated with 1–10 mg/ml dextran for 10–60 minutes, with the duration depending on whether labeling of early or late endosomes is desired. Endosomal membrane labeling is most commonly performed using the styryl dyes FM 1-43 and FM 4-64 (also referred to as SytaptoGreen and SynaptoRed). In contrast to lipophilic plasma membrane labeling dyes such as DiI and DiD (Section 7.8), FM 1-43 and FM 4-64 are highly soluble in aqueous media and their association with membrane lipids is relatively weak. Internalized endosomal membranes can therefore be visualized without plasma membrane background staining by simply washing away dye associated with the latter [31]. Wash removal of noninternalized dye background is more difficult in tissue preparations than in disseminated cell cultures. This problem can be circumvented by addition of extracellular fluorescence masking dyes [32] or cyclodextrin adsorption of non-internalized styryl dye [33]. Retention of labeling following paraformaldehyde fixation can be accomplished by substituting the primary amine-terminated FM 1-43FX and FM 4-64FX dyes for FM 1-43 and FM 4-64, respectively. Phagocytosis is a type of endocytosis performed by specialized cells such as macrophages, neutrophils, and dendritic cells. Typical strategies for visualization of phagocytosis are analogous to those for receptor-mediated endocytosis, using fluorescent conjugates of protein ligands, immune complexes, or microbial pathogens, with additional plasma membrane labeling in a second fluorescent color (Section 7.8) to provide morphological context [34] (Fig. 7.4). Preparation of these conjugates with dyes that generate increased

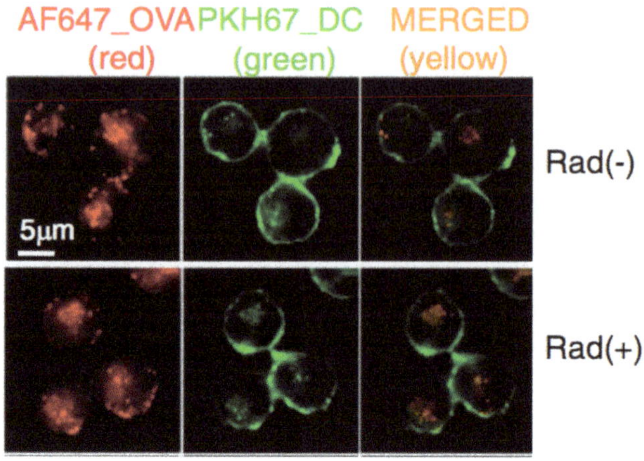

Figure 7.4 Imaging of ovalbumin phagocytosis by dendritic cells. DC2.4 cells were stained with the green-fluorescent lipophilic carbocyanine dye PKH67 on ice for visualization of plasma membranes and endosomal membranes, then pulsed with red-fluorescent Alexa Fluor 647-labeled ovalbumin (AF647_OVA) for 5 min at 37°C and then fixed. The images demonstrate increased endosomal retention of ovalbumin in cells treated with the HSP90 inhibitor radicicol (Rad(+)). From Imai et al. [34].

fluorescence output upon acidification or oxidation (e.g., pHrodo Red or DCHF respectively) results in detection biased towards internalized over extracellular probe populations.

7.6.2 Lysosomes

Four distinct methods are available for labeling lysosomes: (i) acidotropic dye staining, (ii) pulse-chase loading with fluorescent dextrans, (iii) expression of FP-LAMP fusions, and (iv) in situ coupling to cysteine residues of cathepsin [35]. Acidotropic dyes are designed to be entrapped in lysosomes via protonation of weak base substituents in the highly acidic (pH = 4.8) lysosome interior, thereby converting the initially neutral dye to a cationic species. Dyes originally adopted for this purpose such as neutral red and acridine orange have now been largely replaced by the LysoTracker and LysoSensor series marketed by Molecular Probes/Life Technologies and proprietary formulations offered by other commercial vendors. LysoTracker Red DND-99 is currently the most commonly used of these acidotropic dyes. It has some known deficiencies including photoconversion to a green-emitting species [36] and a weak base pK_a (7.4) that is much higher than internal lysosomal pH [37], resulting in labeling of vesicles that are less acidic than lysosomes. LysoSensor Green DND-189 has better specificity for lysosomal sequestration as its pK_a is much lower (5.2). LysoSensor dyes are fluorogenic (Table 7.1), producing significant fluorescence output only in their protonated (lysosomally sequestered) form. They are more often used as pH indicators rather as structural markers. Lysosomal sequestration of both LysoTracker and LysoSensor dyes in live cells is rapid (<5 minutes), but typically does not persist for more than 1–2 hours. For longer-term experiments, pulse-chase loading with inert, nontoxic fluorescent dextrans via the endocytic pathway (Section 7.6.1) produces lysosomal staining that can persist for several days. Loading is initiated by incubation for 30–60 minutes with the fluorescent dextran ("pulse") followed by a "chase" (i.e., incubation in dextran-free medium) for about 4 hours to allow complete maturation of early endosomes to lysosomes.

Peroxisomes, like lysosomes, are primarily repositories of catabolic processes and are similar in size (0.5–1.5 μm). Although they are ubiquitous in mammalian (and other eukaryotic) cells, they are particularly abundant in the liver and kidney. Electrostatically neutral dyes of the BODIPY family coupled to peroxisome-targeting PTS1 tripeptides spontaneously incorporate into live cells and accumulate in peroxisomes within minutes [38]. The same peptide sequences are also utilized for peroxisome targeting of FPs [39].

7.7 Endoplasmic Reticulum and Golgi

Labeling of the endoplasmic reticulum (ER) is typically accomplished using moderately lipophilic, membrane-permeant dyes such as $DiOC_6(3)$ or $DiIC_1(5)$ for detection in the FITC or Cy5 channels, respectively. This labeling method has little specificity for the ER versus other cellular membranes. ER labeling is predominant merely because ER is by far the most abundant eukaryotic cellular membrane, comprising about 50–60% of the total. In fact with $DiOC_6(3)$ or $DiIC_1(5)$, very low applied dye concentrations (<10 nM) will usually result in primarily mitochondrial labeling. Working concentrations of around 1–5 μM are needed to label the ER, usually inducing mitochondrial toxicity as a side effect. ER-Tracker Green and Red (Molecular Probes/Life Technologies) are derivatives of the antidiabetic drug glibenclamide, nominally targeted to plasma membrane ATP-sensitive potassium channels. Labeling of the ER is the result of residual nonspecific membrane binding of the lipophilic probe after saturation of the primary target [40]. However the collateral mitochondrial staining exhibited $DiOC_6(3)$ and $DiIC_1(5)$ is absent. These examples illustrate the limitations of ER labeling via hydrophobic partitioning of dyes. Superior labeling specificity can generally be achieved using FP fusions carrying ER targeting sequences such as KDEL (Lys-Asp-Glu-Leu). The ribosomes of the rough ER in fixed tissues can be visualized using a wide variety of RNA-binding fluorogenic dyes such as propidium iodide, TO-PRO-3 (Section 7.2.2) or ethidium homodimer-1 (EthD-1). Individual ribosomes are too small (~30 nm diameter) to be resolved by conventional wide-field or confocal microscopy but are discernible using super-resolution imaging techniques [41]. Lipid droplets are cytosolic inclusions of neutral lipids such as triglycerides and cholesteryl esters that originate from the ER. Labeling of lipid droplets can be accomplished via hydrophobic partitioning of neutral dyes such as Nile Red, BODIPY 493/503 [42] or LD540 [43].

The Golgi apparatus is the primary site for post-synthetic glycosylation of proteins and lipids. This characteristic is exploited for targeting of both dye-based and FP-fusion labeling of this organelle. NBD and BODIPY ceramides are biosynthetic precursors of glycosphingolipids. Incubation of live cells with NBD ceramide or BODIPY ceramide results in uptake and metabolic conversion to glycosphingolipids in the Golgi. The concentration of accumulated BODIPY ceramide-derived glycosphingolipids results in a shift of the dye emission from green from (515 nm) to red (620 nm) [44]. This effect is useful in color-discriminating Golgi staining from background labeling of other cellular membranes. On the other hand, it presents a potential complication in the interpretation of multiple-label colocalization experiments. Similar emission spectral shifts are generated by BODIPY dye-based probes used in other contexts [42, 44]. The most generic method for labeling the Golgi apparatus in fixed and permeabilized cells and tissues is based on glycoprotein binding by fluorescent conjugates of lectins such as wheat germ agglutinin (WGA).

7.8 Plasma Membrane

In addition to defining cellular boundaries, labeling of the plasma membrane has particular utility for providing morphological context for total internal reflection fluorescence (TIRF) imaging of cell surface processes [45]. The plasma membrane can be labeled by modification of either its lipid or protein structural components. In living cells, labeling does not usually remain specifically associated with the plasma membrane for more than 1–2 hours, due to the rapid internalization of both proteins and lipids (Fig. 7.5). Labeling of membrane lipids is accomplished via hydrophobic partitioning of fluorescent analogs of phospholipids such as DiI and DiD [45], and cholesterol [46]. These fluorescent lipid analogs are highly insoluble in water; direct addition of stock solutions in ethanol or DMSO to aqueous media often results in precipitation and uneven labeling across cell populations. Delivery in the form of cyclodextrin complexes [46, 47] or fusogenic liposomes [48] can ameliorate these problems. Because fluorescent lipid analogs are susceptible to extraction by detergents used for permeabilization, use in con-

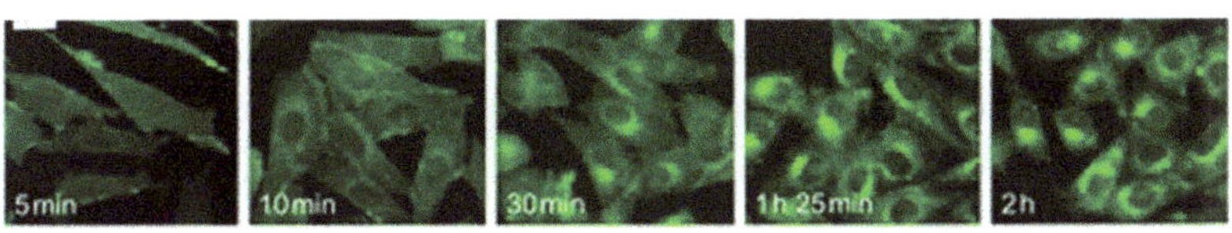

Figure 7.5 Internalization of fluorescent lipid analogs exemplified by wild-type chinese hamster ovary (CHO) cells labeled with cyclodextrin-complexed BODIPY-cholesterol for 2 min and imaged live at the indicated time points over the following 2 h. Scale bar = 20 μm. *Source*: Figure 3A from Hölttä-Vuori et al. [46].

junction with immunofluorescent labeling is not straightforward. In this case, the most generally applicable alternatives for plasma membrane labeling are fluorescent conjugates of WGA that bind to cell surface *N*-acetyl-D-glucosamine and sialic acid residues. WGA labeling must be applied before permeabilization to avoid collateral labeling of the Golgi apparatus (Section 7.7) and other intracellular structures. Lipid rafts are plasma membrane substructures that are postulated to function as hubs for signal transduction and protein trafficking. They can be labeled using fluorescent analogs of their glycosphingolipid and cholesterol components [44] or fluorescent conjugates of cholera toxin B subunit, which binds to the pentasaccharide substituent of the glycosphingolipid ganglioside GM1. Because lipid rafts are small (10–200 nm) and transient, their structure and function remains somewhat controversial. Some of these controversies may eventually be resolved through the application of super-resolution microscopy methods [49].

Abbreviations

UV	ultraviolet
nm	nanometers
BODIPY	difluoro-4-bora-3a,4a-diaza-*s*-indacene
CFSE	5-(and-6)-carboxyfluoresceindiacetate, succinimidyl ester
DCHF	2′,7′-dichlorodihydrofluorescein
DiI	1,1′-dioctadecyl-3,3,3′,3′-tetramethylindocarbocyanine
DiD	1,1′-dioctadecyl-3,3,3′,3′-tetramethylindodicarbocyanine
DNA	deoxyribonucleic acid
FP	fluorescent protein
FRAP	fluorescence recovery after photobleaching
FRET	Förster resonance energy transfer
kDa	kilodaltons
LAMP	lysosome-associated membrane protein
LED	light-emitting diode
NBD	nitrobenzoxadiazole
PALM	photoactivated localization microscopy
STORM	stochastic optical reconstruction microscopy

References

1. Tarcsay, Á., and Keserű, G. M. (2013). Contributions of molecular properties to drug promiscuity. *J. Med. Chem.*, **56**, 1789–1795.
2. Kapuscinski, J. (1995). DAPI: a DNA-specific fluorescent probe. *Biotech. Histochem.*, **70**, 220–233.
3. Shapiro, H. M., and Perlmutter, N. G. (2001). Violet laser diodes as light sources for cytometry. *Cytometry*, **44**, 133–136.
4. Zurek-Biesiada, D., Kędracka-Krok, S., and Dobrucki, J. W. (2013). UV-activated conversion of Hoechst 33258, DAPI, and Vybrant DyeCycle fluorescent dyes into blue-excited, green-emitting protonated forms. *Cytometry A*, **83**, 441–445.
5. Jež, M., Bas, T., Veber, M., Košir, A., Dominko, T., Page, R., and Rožman, P. (2013). The hazards of DAPI photoconversion: effects of dye, mounting media and fixative, and how to minimize the problem. *Histochem. Cell Biol.*, **139**, 195–204.
6. Purschke, M., Rubio, N., Held, K. D., and Redmond, R. W. (2010). Phototoxicity of Hoechst 33342 in time-lapse fluorescence microscopy *Photochem. Photobiol. Sci.*, **9**, 1634–1639.
7. Zink, D., Sadoni, N., and Stelzer, E. (2003). Visualizing chromatin and chromosomes in living cells. *Methods*, **29**, 42–50.
8. Njoh, K. L., Patterson, L. H., Zloh, M., Wiltshire, M., Fisher, J., Chappell, S., Ameer-Beg, S., Bai, Y., Matthews, D., Errington, R. J., and Smith, P. J. (2006). Spectral analysis of the DNA targeting bisalkylaminoanthraquinone DRAQ5 in intact living cells. *Cytometry A*, **69**, 805–814.
9. Martin, R. M., Leonhardt, H., and Cardoso, M. C. (2005). DNA labeling in living cells. *Cytometry A*, **67**, 45–52.
10. Zhao, H., Traganos, F., Dobrucki, J., Wlodkowic, D., and Darzynkiewicz, Z. (2009). Induction of DNA damage response by the supravital probes of nucleic acids. *Cytometry A*, **75**, 510–519.
11. Elliott, J. T., Tona, A., and Plant, A. L. (2003). Comparison of reagents for shape analysis of fixed cells by automated fluorescence microscopy. *Cytometry A*, **52**, 90–100.
12. Terasaki, M., Chen, L.-B., and Fujiwara, K. (1986). Microtubules and the endoplasmic reticulum are highly interdependent structures. *J. Cell Biol.*, **103**, 1557–1568.
13. Stringham, E. G., Marcus-Gueret, N., Ramsay, L., and Schmidt, K. L. (2012). Live cell imaging of the cytoskeleton. *Methods Enzymol.*, **505**, 203–217.

14. Burkel, B. M., von Dassow, G., and Bement, W. M. (2007). Versatile fluorescent probes for actin filaments based on the actin-binding domain of utrophin. *Cell Motil. Cytoskel.*, **64**, 822–832.

15. Anderl, J., Echner, H., and Faulstich, H. (2012). Chemical modification allows phallotoxins and amatoxins to be used as tools in cell biology. *Beilstein J. Org. Chem.*, **8**, 2072–2084.

16. Jan, F., Keij, J. F., Bell-Prince, C., and Steinkamp, J. A. (2000). Staining of mitochondrial membranes with 10-nonyl acridine orange, MitoFluor Green, and MitoTracker Green is affected by mitochondrial membrane potential altering drugs. *Cytometry*, **39**, 203–210.

17. Buckman, J. F., Hernández, H., Kress, G. J., Votyakova, T. V., Pal, S., and Reynolds, I. J. (2001). MitoTracker labeling in primary neuronal and astrocytic cultures: influence of mitochondrial membrane potential and oxidants. *J. Neurosci. Methods*, **104**, 165–176.

18. Mitra, K., and Lippincott-Schwartz, J. (2010). Analysis of mitochondrial dynamics and functions using imaging approaches. *Curr. Protoc. Cell Biol.*, 4.25.1–4.25.21.

19. Gerencser, A. A., Chinopoulos, C., Birket, M. J., Jastroch, M., Vitelli, C., Nicholls, D. G., and Brand, M. D. (2012). Quantitative measurement of mitochondrial membrane potential in cultured cells: calcium-induced de- and hyperpolarization of neuronal mitochondria. *J. Physiol.*, **590**, 2845–2871.

20. Lemasters, J. J., and Ramshesh, V. K. (2007). Imaging of mitochondrial polarization and depolarization with cationic fluorophores. *Methods. Cell Biol.*, **80**, 283–295.

21. Perry, S. W., Norman, J. P., Barbieri, J., Brown, E. B., and Gelbard, H. A. (2011). Mitochondrial membrane potential probes and the proton gradient: a practical usage guide. *BioTechniques*, **50**, 98–115.

22. Conover, T. E., and Schneider, R. F. (1981). Interaction of certain cationic dyes with the respiratory chain of rat liver mitochondria. *J. Biol. Chem.*, **256**, 402–408.

23. Di Lisa, F., Blank, P. S., Colonna, R., Gambassi, G., Silverman, H. S., Stern, M. D., and Hansford, R. G. (1995). Mitochondrial membrane potential in single living adult rat cardiac myocytes exposed to anoxia or metabolic inhibition. *J. Physiol.*, **486**, 1–13.

24. Dong, H., Cheung, S. H., Liang, Y., Wang, B., Ramalingam, R., Wang, P., Sun, H., Cheng, S. H., and Lam, Y. W. (2013). "Stainomics": identification of mitotracker labeled proteins in mammalian cells. *Electrophoresis*, **34**, 1957–1964.

25. Shim, S. H., Xia, C., Zhong, G., Babcock, H. P., Vaughan, J. C., Huang, B., Wang, X., Xu, C., Bi, G. Q., and Zhuang, X. (2012). Super-resolution fluorescence imaging of organelles in live cells with photoswitchable membrane probes. *Proc. Natl. Acad. Sci. U.S.A.*, **109**, 13978–13983.
26. Heikal, A. A. (2010). Intracellular coenzymes as natural biomarkers for metabolic activities and mitochondrial anomalies. *Biomarkers Med.*, **4**, 241–263.
27. Doherty, G. J., and McMahon, H. T. (2009). Mechanisms of endocytosis. *Annu. Rev. Biochem.*, **78**, 857–902.
28. Sharma, D. K., Choudhury, A., Singh, R. D., Wheatley, C. L., Marks, D. L., and Pagano, R. E. (2003). Glycosphingolipids internalized via caveolar-related endocytosis rapidly merge with the clathrin pathway in early endosomes and form microdomains for recycling. *J. Biol. Chem.*, **278**, 7564–7572.
29. Haggie, P. M., and Verkman, A. S. (2009). Unimpaired lysosomal acidification in respiratory epithelial cells in cystic fibrosis. *J. Biol. Chem.*, **284**, 7681–7686.
30. Deriy, L. V., Gomez, E. A., Zhang, G., Beacham, D. W., Hopson, J. A., Gallan, A. J., Shevchenko, P. D., Bindokas, V. P., and Nelson D. J. (2009). Disease-causing mutations in the cystic fibrosis transmembrane conductance regulator determine the functional responses of alveolar macrophages. *J. Biol. Chem.*, **284**, 35926–35938.
31. Gaffield, M. A., and Betz, W. J. (2006). Imaging synaptic vesicle exocytosis and endocytosis with FM dyes. *Nat. Protoc.*, **1**, 2916–2921.
32. Pyle, J. L., Kavalali, E. T., Choi, S., and Tsien, R. W. (1999). Visualization of synaptic activity in hippocampal slices with FM1-43 enabled by fluorescence quenching. *Neuron*, **24**, 803–808.
33. Kay, A. R., Alfonso, A., Alford, S., Cline, H. T., Holgado, A. M., Sakmann, B., Snitsarev, V. A., Stricker, T. P., Takahashi, M., and Wu, L. G. (1999). Imaging synaptic activity in intact brain and slices with FM1-43 in C. elegans, lamprey, and rat. *Neuron*, **24**, 809–817.
34. Imai,T., Kato, Y., Kajiwara, C., Mizukami, S., Ishige, I., Ichiyanagi, T., Hikida, M., Wang, J. Y., and Udono, H. (2011). Heat shock protein 90 (HSP90) contributes to cytosolic translocation of extracellular antigen for cross-presentation by dendritic cells. *Proc. Natl. Acad. Sci. U.S.A.*, **108**, 16363–16368.
35. Fan, F., Nie, S., Yang, D., Luo, M., Shi, H., and Zhang, Y. H. (2012). Labeling lysosomes and tracking lysosome-dependent apoptosis with a cell-permeable activity-based probe. *Bioconjugate Chem.*, **23**, 1309–1317.

36. Freundt, E. C., Czapiga, M., and Lenardo, M. J. (2007). Photoconversion of Lysotracker Red to a green fluorescent molecule. *Cell Res.*, **17**, 956–958.
37. Duvvuri, M., Gong, Y., Chatterji, D., and Krise, J. P. (2004). Weak base permeability characteristics influence the intracellular sequestration site in the multidrug-resistant human leukemic cell line HL-60. *J. Biol. Chem.*, **279**, 32367–32372.
38. Pap, E. H., Dansen, T. B., and Wirtz, K. W. (2001). Peptide-based targeting of fluorophores to peroxisomes in living cells. *Trends Cell Biol.*, **11**, 10–12.
39. Wiemer, E. A., Wenzel, T., Deerinck, T. J., Ellisman, M. H., and Subramani, S. (1997). Visualization of the peroxisomal compartment in living mammalian cells: dynamic behavior and association with microtubules. *J. Cell Biol.*, **136**, 71–80.
40. Zünkler, B. J., Wos-Maganga, M., and Panten, U. (2004). Fluorescence microscopy studies with a fluorescent glibenclamide derivative, a high-affinity blocker of pancreatic beta-cell ATP-sensitive K+ currents. *Biochem. Pharmacol.*, **67**, 1437–1444.
41. Gahlmann, A., and Moerner W. E. (2014). Exploring bacterial cell biology with single-molecule tracking and super-resolution imaging. *Nat. Rev. Microbiol.*, **12**, 9–22.
42. Ohsaki, Y., Shinohara, Y., Suzuki, M., and Fujimoto, T. (2010). A pitfall in using BODIPY dyes to label lipid droplets for fluorescence microscopy. *Histochem. Cell Biol.*, **133**, 477–480.
43. Spandl, J., White, D. J., Peychl, J., and Thiele, C. (2009). Live cell multicolor imaging of lipid droplets with a new dye, LD540. *Traffic*, **10**, 1579–1584.
44. Marks, D. L., Bittman, R., and Pagano, R. E. (2008). Use of Bodipy-labeled sphingolipid and cholesterol analogs to examine membrane microdomains in cells. *Histochem. Cell Biol.*, **130**, 819–832.
45. Anantharam, A., Onoa, B., Edwards, R. H., Holz, R. W., and Axelrod, D. (2010). Localized topological changes of the plasma membrane upon exocytosis visualized by polarized TIRFM. *J. Cell Biol.*, **188**, 415–428.
46. Hölttä-Vuori, M., Uronen, R. L., Repakova, J., Salonen, E., Vattulainen, I., Panula, P., Li, Z., Bittman, R., and Ikonen, E. (2008). BODIPY-cholesterol: a new tool to visualize sterol trafficking in living cells and organisms. *Traffic*, **9**, 1839–1849.
47. Tanhuanpää, K., and Somerharju, P. (1999). Gamma-cyclodextrins greatly enhance translocation of hydrophobic fluorescent phospholipids from vesicles to cells in culture. *J. Biol. Chem.*, **274**, 35359–35366.

48. Csiszár, A., Hersch, N., Dieluweit, S., Biehl, R., Merkel, R., and Hoffmann, B. (2010). Novel fusogenic liposomes for fluorescent cell labeling and membrane modification. *Bioconjugate. Chem.*, **21**, 537–543.
49. Owen, D. M., Magenau, A., Williamson, D., and Gaus, K. (2012). The lipid raft hypothesis revisited–new insights on raft composition and function from super-resolution fluorescence microscopy. *Bioessays*, **34**, 739–747.

Chapter 8

Indicator and Reporter Dyes

Iain Johnson
Lumencor, Inc., 14940 NW Greenbrier Parkway, Beaverton, OR 97006, USA
iain.johnson@lumencor.com

In contrast to the probes discussed in Chapter 7, which are primarily structural markers, indicator and reporter dyes are sensors of functional activities of cells and tissues. Consequently, their applications are primarily in live specimens. An essential prerequisite is that probes must be delivered to the region of interest in the specimen and retained there for the duration of the imaging experiment. The technical challenges involved in satisfying this prerequisite increase as specimen complexity increases from cells to tissues to intact organisms. Requirements for subcellular localization of these probes depends on the type of activity being measured and the physiological context of the experiment. For example, the experimental objective of measuring calcium fluxes in the vicinity of the plasma membrane can be achieved by localization of fluorescent indicators incorporating peptide sequences targeted for post-translational myristoylation. Alternatively, an unlocalized fluorescent indicator can be used in combination with plane-selective imaging techniques such as SPIM or total internal reflection fluorescence (TIRF) to confine data collection to the membrane and

Fundamentals of Fluorescence Imaging
Edited by Guy Cox

ISBN 978-981-4774-85-7 (Hardcover), 978-1-351-12940-4 (eBook)
www.jennystanford.com

its immediate environs. The contents of this chapter are organized primarily according to molecular targets such as calcium ions and reactive oxygen species (ROS). In most cases alternative approaches using small molecule dye-based probes and fluorescent protein (FP) fusions are available.

8.1 Calcium and Other Cations

Ion indicators consist of fluorophores coupled to ion binding functional groups known as chelators in such a way that ion binding to the chelator is transduced into a proportional change in the fluorescence of the fluorophore. There is a fundamental assumption here that fluorescence is proportional to ion concentration that is not necessarily valid when the indicator is used in the complex molecular environment of a cell or tissue. In particular, the concentration of the indicator can vary from cell to cell due to non-uniform loading (or expression in the case of FP-based indicators) or photobleaching, resulting in fluorescence signal variations that have no relationship to intracellular ion concentrations. Thus, the validity of this assumption must always be critically examined when inferring biological significance from fluorescent indicator image data. The same structural paradigm (Fig. 8.1) applies for both organic dye-based small molecule indicators and FP-based genetically encoded calcium indicators (GECI). Indicators for intracellular calcium based on coupling the chelator BAPTA (1,2-bis(*o*-aminophenoxy)ethane-*N*,*N*,*N*′,*N*′-tetraacetic acid) to various fluorescent dyes were largely developed in the period from 1985–2000 [1–3]. Development of GECI, where the chelator function is fulfilled by a calcium-binding protein such as calmodulin or troponin C, dates from about 1997 [4] and continues to be very active [5, 6]. The key functional attributes of these indicators (Fig. 8.1) are the ability of the chelator to selectively bind the target ion at physiologically relevant concentrations and transduce that binding to the dye, thereby generating a proportional fluorescence response. The dissociation constant for the ion-binding equilibrium (K_d) is the key factor in determining the suitability of a particular indicator in relation to the ion concentration range under investigation. For the

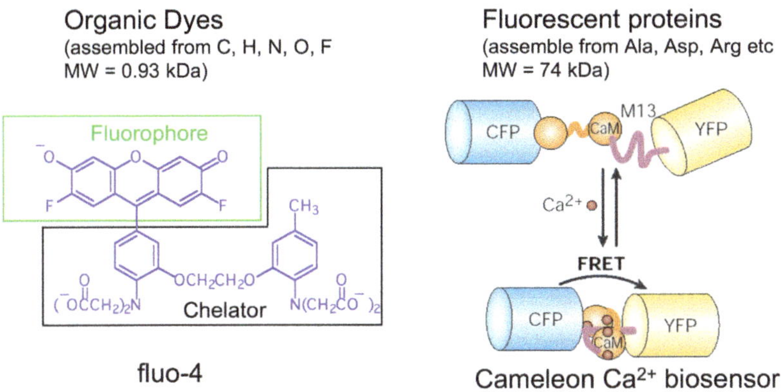

Figure 8.1 Comparison of structures of small molecule (fluo-4) and genetically encoded (Cameleon) calcium indicators. MW, molecular weight.

mammalian cell cytoplasm, with a typical basal Ca^{2+} level of 100 nM, K_d values of 150–500 nM are appropriate. In other subcellular domains, the optimal indicator K_d may be higher or lower [7]. When faced with unexpected results, it should be kept in mind that no indicator has perfect selectivity for its target ion. For example, fura-2 has about 100-fold greater binding affinity for Zn^{2+} than for Ca^{2+}. The ability of fura-2 to detect Ca^{2+} is predicated on the fact that in most (but not all) cellular contexts, concentrations of free Zn^{2+} are negligible. Following the fluorophore+chelator paradigm, indicators for cations other than calcium (Mg^{2+}, Zn^{2+} K^+, Na^+, etc.) have been synthesized by replacing the BAPTA component with a different chelator with appropriate selectivity. However, calcium indicators are the focus of the present discussion because they are the most technically optimized and widely used. This status derives from the ubiquitous and central role of free calcium in signal transduction pathways and the large fluctuations of calcium concentration that encode the signals. Thus intracellular calcium concentration fluctuations that have ranges of as much as 100-fold are easier to measure than those of other cations whose intracellular concentration ranges rarely exceed twofold. Note that the relevant concentrations are those of free ions. As implied by the detection mechanism outlined above, fluorescent ion indicators are generally

Table 8.1 Characteristics of calcium ion indicators

Indicator type	Excitation 350–500 nm	Excitation 500–650 nm	Key characteristics
dye, single ex/em	fluo-4, fluo-8, Cal520, Oregon Green 488 BAPTA	rhod-4, X-rhod-1	Visible range excitation Single cell measurements Fast response
dye dual ex/em ratio	fura-2, fura-8, indo-1	none	Ultraviolet excitation Cell population measurements Fast response
FP, single ex/em	GCaMP6	R-CaMP2	Cell type-specific expression in tissues Organelle-specific expression in cells Slower response than dyes
FP, dual ex/em ratio	Twitch [12], YC3.6*	none	Cell type-specific expression in tissues Slower response than dyes

*YC, Yellow Cameleon.

only sensitive to free ions and not those bound to proteins,[a] nucleic acids or other biomolecules.

In comparing dye-based small molecule indicators and FP-based genetically encoded calcium indicators (Table 8.1), by far the most compelling advantage of the latter is the capacity to localize them to specific cell types in tissues or specific molecular targets in cells [7, 8]. This advantage is offset by generally lower fractional fluorescence change per calcium ion detected (represented by SNR or $\Delta F/F_0$, both terms referring to the enhancement of fluorescence on calcium binding [ΔF] relative to the calcium free [F_0] indicator species) and slower response kinetics [9–12]. However, the SNR of small molecule indicators in cells and tissues is typically far lower than in cell-free solutions, primarily due to detection of signals from unresponsive and irrelevant cells. The localization specificity of FP-based sensors rectifies this deficiency. A similar balance of

[a]Unless the protein concerned is deliberately introduced as a component of a GECI.

characteristics is also found for other FP-based and small molecule indicators discussed in subsequent sections of this chapter.

Both BAPTA-based small molecule indicators and GECI can be subdivided into two groups: those that provide ratiometric calcium-binding responses and those that do not (Table 8.1). Ratiometric indicators generate fluorescence signals measured at two different excitation (or emission) wavelengths whose ratio is calcium-dependent but is independent of intracellular dye concentration. Ratio measurements therefore provide a means for avoiding false-positive indications of calcium variations that are actually due to unstable or non-uniform intracellular loading, photobleaching and other indicator concentration artifacts.[b] Generally speaking, the motivation for using ratiometric measurements is more compelling when attempting to compare calcium levels in two or more cells as opposed to measurements where observation is restricted to a single cell. Ratiometric small molecule indicators such as fura-2 require ultraviolet (or two-photon) excitation; longstanding efforts to develop ratiometric calcium indicators with the capacity for excitation at >400 nm have yielded few successes [13]. Ratiometric GECIs are most usually CFP/YFP FRET constructs (Fig. 8.1; [4]). Production of stable cell lines is sometimes impeded by CFP/YFP recombination, resulting in expression of inactive indicators [14].

Indicators with long-wavelength excitation (500–650 nm) are necessary for spectral separation in cells expressing GFP or optogenetic photostimulators such as channelrhodopsin-2 (ChR2) [15, 16]. Long-wavelength excitation also provides benefits of decreased phototoxicity and increased tissue transparency. Dye-based small molecule indicators and GECIs both provide long-wavelength excitation capabilities (Table 8.1). Rhod-2 and other long-wavelength small molecule indicators tend to exhibit mitochondrial localization when loaded as into cells, as their AM ester derivatives are structurally similar to TMRE and other mitochondrial markers described in Chapter 7. This does not preclude successful applications for cytoplasmic calcium detection

[b]Subtract background signals from component fluorescence intensity values *before* calculating ratios.

[15, 16], but may require a time investment for optimization of the loading conditions. In a recent comparative evaluation of dye-based small molecule indicators for detection of inositol trisphosphate-evoked Ca^{2+} signals in cultured human neuroblastoma cells, Cal520 and rhod-4 were found to provide the best $\Delta F/F_0$ characteristics among green- and red-fluorescent indicators respectively [11].

Cell loading and response calibration are the two most prominent technical challenges involved in calcium imaging with fluorescent indicators. Loading of dye-based small molecule indicators are most commonly delivered into cells by incubation with AM ester precursors. AM ester loading is a pharmacological technique in which the AM ester is an inactive pro-drug (AM esters are not sensitive to Ca^{2+}). As is usually the case in drug treatments, the process produces a desired activity together with a number of undesired side-effects (Fig. 8.2). For example, detection of signals from indicators compartmentalized in the endoplasmic reticulum can lead to overestimation of cytoplasmic calcium unless the signals are identified and excluded during image analysis. It is normal and expected for cytoplasmic fluorescence signals to be weak and diffuse under resting (non-stimulated) conditions. Typically, external AM ester concentrations of 1–10 μM are used, of which 10% or less actually loads into the cell, the remainder being removed by post-incubation washing. Nevertheless, intracellular indicator concentrations can be substantial (10–100 μM) and buffering of calcium transients, particularly by high-affinity (K_d = 150–500 nM) indicators, may require consideration when interpreting data. Other loading methods such as biolistic delivery, electroporation, and patch pipet infusion are generally more invasive and suited to loading of small cell populations in tissues. Response calibration is typically accomplished by extracellular addition of an ionophore such as ionomycin or 4-BrA23187 [17] to clamp intracellular Ca^{2+} to controlled extracellular concentrations. Since ionophore treatment is somewhat cytotoxic, it is preferable to perform the calibration as the last step in a Ca^{2+} imaging experiment. Even when calibrated responses are not required, ionophore treatment can be a useful positive control in cases where Ca^{2+} responses elicited by physiological stimulation are less than expected.

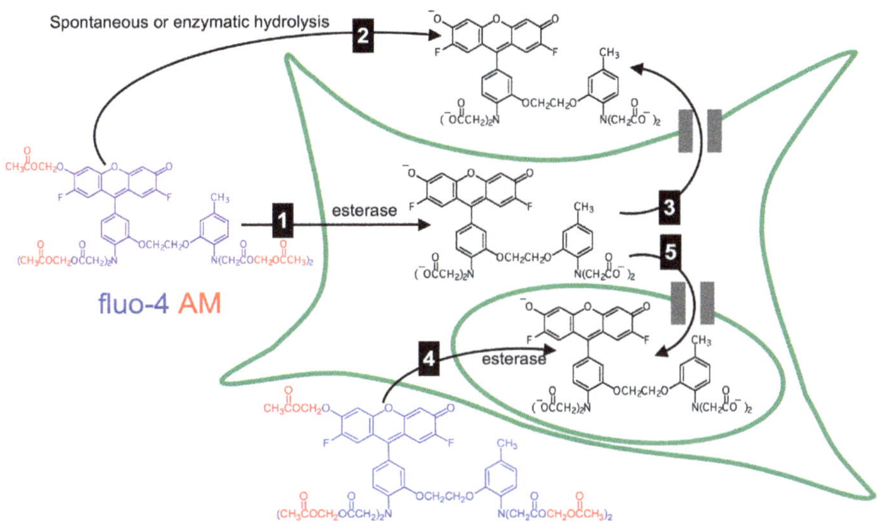

Figure 8.2 Schematic of AM ester loading of intracellular calcium indicators, exemplified by fluo-4. The objective is to deliver anionic, Ca^{2+}-sensitive fluo-4 into the cytoplasm (1). Competing processes that occur during incubation of cells with fluo-4 AM are as follows. (2) Spontaneous or enzymatic extracellular hydrolysis. The acetoxymethyl substituents indicated in red are decomposed to acetic acid and formaldehyde. The result of this process is a faint background of extracellular Ca^{2+}-bound fluo-4. (3) Extrusion of anionic fluo-4 via plasma membrane organic anion transporters. This process results in weak intracellular labeling in cells where transporter expression levels are high and adds to background signals derived from (2). (4) Compartmentalization of AM ester followed by hydrolysis to anionic fluo-4. (5) Compartmentalization via organic anion transporters. (4) and (5) result in fluorescence background signals in organelles with high internal Ca^{2+} levels such as the endoplasmic reticulum.

8.2 pH Indicators

In contrast to calcium, intracellular pH is generally highly regulated and not subject to fluctuation in mammalian cells. Instead, pH changes associated with endocytosis and exocytosis and pathological conditions such as cancer [18, 19] are the most usual subjects for imaging investigations. pH indicators are conceptually similar to calcium indicators. Instead of a calcium chelator, a protolytically

Table 8.2 Characteristics of pH indicators

Indicator	Ionizable substituent	Fluorescence change (+/−) on pH increase	pK_a	Notes
BCECF	hydroxyl	+	7.0	
SNARF	hydroxyl	+	7.5	Emission ratio 580/640 nm
pHrodo	amine	−	6.9	pK_a depends on coupled targeting group
LysoProbe	amine	−	5.7	
Ratiometric pHluorin	hydroxyl	−	7.2	Excitation ratio 395/475 nm
Superecliptic pHluorin	hydroxyl	+	7.2	

ionizable substituent is attached to the fluorophore. Protonation of this substituent is electronically transduced to the fluorophore, resulting in a coupled change in its fluorescence. The ionizable substituent is generally either a phenolic hydroxyl group or an amine (Table 8.2). The pK_a of the ionizable substituent is analogous to the K_d of a calcium indicator (Section 8.1) and determines the pH sensitivity range of the indicator.

BCECF and SNARF are generally used for imaging cytoplasmic pH and are loaded into cells in the form of acetoxymethyl ester derivatives [20], in the same way as calcium indicators (Fig. 8.2). BCECF typically provides much stronger fluorescence signals (~10 times under the same for the same conditions) than SNARF. The compensating advantages of SNARF are its capacity for ratio imaging (580/640 nm emission ratio with excitation at 543 nm) and its spectral compatibility with GFP-expressing cells. Indicators derived from rhodamine dyes such as pHrodo [19] and LysoProbe [21] are generally coupled to targeting groups such as inactivated bacteria (for phagocytic uptake), transferrin (for uptake via receptor-mediated endocytosis) or N-linked glycans (for lysosomal localization). In this form, they are added extracellularly and become increasingly fluorescent as they are translocated into phagosomes, endosomes and lysosomes. This fluorogenic response[c] provides enhanced contrast from extracellular background for visualization of small acidic intracellular compartments. Targeting groups + fluorophore conjugates are preferable to probes such as LysoSensor Green and quinacrine, for which both fluorescence intensity and intracellular distribution are dependent on protonation of the fluorophore, resulting in data interpretation ambiguities [22].

FP-based pH indicators exploit the fact that GFP itself is pH-sensitive, due to ionization of the phenol group of its 4-(*p*-hydroxybenzylidene)imidazolidin-5-one fluorophore. As with phenolic dyes such as BCECF and SNARF, the ionized phenolate form is more fluorescent and is optimally excited at longer wavelengths (~475 nm, compared to ~395 nm for the neutral phenol form). FPs engineered specifically for use as pH sensors are referred to as pHluorins and are of two types (Table 8.2). Ratiometric pHluorin exhibits increased 395 nm excitation with corresponding decreased 475 nm excitation as pH decreases [23]. It is primarily used for imaging endocytic processes with such as receptor internalization. Superecliptic pHluorin is essentially nonfluorescent at pH 6 and exhibits increasing fluorescence with increasing pH. It is primarily useful for monitoring exocytosis and protein secretion. Calibration of intracellular pH indicators is usually performed using ionophores in the same manner previously described for Ca^{2+} (Section 8.1). The K^+/H^+ ionophore nigericin, the Na^+/H^+ ionophore monensin or an equimolar nigericin–monensin mixture are typically used.

8.3 Membrane Potential Sensors

Membrane potential sensors and Ca^{2+} indicators are quite often used in combination. To the extent that Ca^{2+} influxes in excitable cells are the result of voltage-sensitive Ca^{2+} channel activation, Ca^{2+} indicators provide a surrogate measurement of membrane potential [6, 24]. Small molecule sensors of transmembrane electrical

---

[c]Probes with fluorogenic responses are initially nonfluorescent and become fluorescent upon interaction with their biomolecular targets.

potential are of two types: fast response probes and slow response probes [25]. The fluorescence of fast response probes tracks with membrane potential changes on the millisecond time scale whereas for slow probes, the temporal response is on the order of seconds. The two types of probes have reciprocal limitations. Slow response probes generate large fluorescence changes (~1%/mV) but cannot track rapid potential changes of physiological importance such as the action potential of neurons. The fluorescence response of fast probes is relatively insensitive—typically 10% fluorescence intensity change per 100 mV potential change. This gives rise to data with poor signal-to-noise characteristics, particularly when attempting to make parallel measurements on multiple cells. However, the ability to make such measurements is perhaps the most compelling advantage of the fluorescence microscopy approach over electrophysiological techniques. Consequently, there has been a long and continuing effort to achieve a merger of the response speed of fast probes with the potentiometric sensitivity of slow probes. Before summarizing progress toward this objective, it is useful to define the starting point. Slow probes, such as cationic carbocyanine (e.g., $DiOC_2(3)$, $DiIC_1(5)$) and anionic bisoxonols (e.g., $DiBAC_4(3)$) are electrophoretically translocated across the membrane in response to potential changes. The difference in the extracellular and intracellular solvent environments of the probe results in a change of its fluorescence quantum yield, producing a fluorescence change that is coupled, albeit indirectly, to the membrane potential. Indeed, the indirect nature of the response mechanism creates a liability for generating false-positive responses. When the plasma membrane potential is of primary interest, anionic bisoxonols are preferred to cationic carbocyanines, which are primarily sensitive to mitochondrial potential [26]. Carbocyanines are primarily used in experiments where the mitochondrial potential is the subject of interest or in cells lacking mitochondria (e.g., erythrocytes and bacteria). In fast response probes, the electrons of the probe molecule are redistributed under the direct influence of the external electric field (an effect known as electrochromism). The difference between moving electrons along the ~1 nm length of a dye molecule and moving the entire molecule (molecular weight >500) across the 40 nm width of the plasma membrane accounts for the difference in response speed. The most successful fast response probes are the aminonaphthylethenylpyridinium (ANEP) dyes [27, 28]. Their spectroscopic properties make them particularly suitable for two-photon excitation and second harmonic generation (SHG) microscopy [29]. As is generally the case with live cell probes, ANEP dyes are not physiologically inert [30] and should be added at the lowest practicable concentration to avoid artifacts. The response mechanism of ANEP dyes requires that labeling is confined to one half of the membrane bilayer, a condition that is hard to sustain over long periods of time. If both sides of the membrane are labeled, the voltage-sensitive fluorescence signals cancel out. Hybrid voltage sensors (hVOS) employ a membrane surface localized fluorophore or fluorescent protein that is quenched to a proximity-dependent extent by an electrophoretically translocated lipophilic anion such as dipicrylamine (DPA) in the membrane interior [31]. This requires simultaneously optimized labeling of cells and tissues with two probe species (fluorophore and quencher). This restriction is avoided by the most recently developed fast-response probes consisting of a cell surface-localized fluorophore connected to an aniline quencher by a "molecular wire" [32, 33]. These probes exhibit $\Delta F/F_0 > 20\%$ per 100 mV potential change with sub-millisecond response times. Unlike hVOS, molecular wire sensors do not affect neuronal excitability by adding to membrane capacitance.

Like their small molecule counterparts, FP-based genetically encoded voltage indicators (GEVI) have a history of extensive molecular engineering in pursuit of the objective of millisecond temporal responses with $\Delta F/F_0 > 20\%$ [6, 34]. At present, the most sensitive GEVI is ArcLight, a fusion of superecliptic pHluorin (A227D) with the voltage-sensitive domain of *C. intestinalis* voltage sensitive phosphatase [35]. The voltage-sensitive fluorescence response of ArcLight is quite strong ($\Delta F/F_0 =$ 30% per 100 mV) but relatively slow (>10 ms). The other major group of GEVIs are archaerhodopsins [36] which exhibit millisecond response times but have very low fluorescence quantum yields, requiring overexpression and high-intensity excitation for imaging. Calibration of potentiometric probes can be accomplished by imposing a transmembrane potential using valinomycin or gramicidin in conjunction with externally applied K^+ solutions [26]. The ultimate test of calibration veracity is quantitative agreement with electrophysiological measurements [32].

8.4 Reactive Oxygen Species (ROS)

The most ubiquitous probes for reactive oxygen species are chemically reduced fluorescein, rhodamine and cyanine [37–39] dyes (Fig. 8.3). In all cases, reduction renders the parent molecule nonfluorescent. Two general properties of these probes have a dominant impact on their applications. (i) Oxidation of the probe irreversibly regenerates the parent fluorophore, thereby generating a corresponding fluorescence signal. (ii) Oxidation can be accomplished by a wide variety of ROS, including some that may be artifactually generated. Photosensitized oxidation initiated by excitation light is particularly pertinent in fluorescence microscopy [39]. Because of (i), processes that generate transient fluxes of ROS cannot be tracked in real time in the same way that Ca^{2+} fluxes can be tracked using reversible fluorescent indicators such as fluo-4 and fura-2. If the chemical identity of the ROS is a question of interest, (ii) dictates that it cannot be directly deduced from fluorescence image data. In the face of these limitations, probes

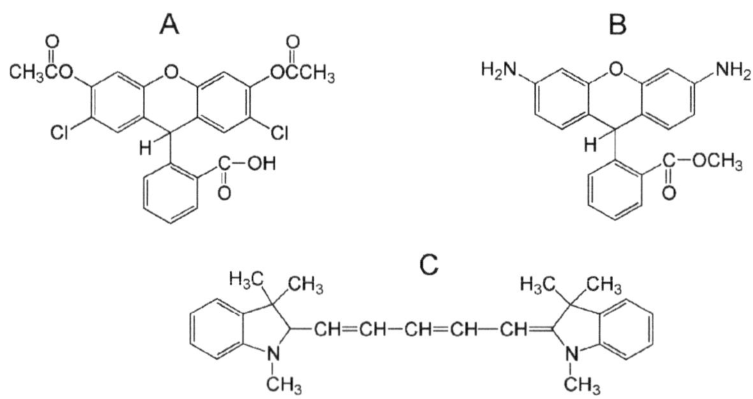

Figure 8.3 Structures of reduced dichlorofluorescein (A), rhodamine (B), and cyanine (C) dyes used for detection of reactive oxygen species.

with greater ROS specificity oxidant have been developed [40, 41] and reversible small-molecule ROS sensors are becoming available [42, 43].

As with other FP-based sensors, the primary advantage of reduction-oxidation sensitive green fluorescent proteins (roGFP) is the capacity to fuse them to specific molecular targets, decreasing interference from parallel redox reactions outside the scope of investigation [44]. These sensors are obtained by inserting an intramolecular disulfide bridge in the vicinity of the GFP fluorophore. Reduction of the disulfide bridge causes deprotonation of the phenol group of the fluorophore, resulting in a decrease in the 405 nm/475 nm fluorescence excitation ratio [45, 46]. Because disulfide reduction is reversible, roGFPs can provide real-time monitoring of intracellular redox status [47]. The mid-point redox potentials of roGFPs are functionally equivalent to the dissociation constants (K_d) of ion indicators (Section 8.1) and can be engineered for optimum response in oxidizing (e.g., endoplasmic reticulum) and reducing (e.g., mitochondria) environments [47]. The tripeptide glutathione is the primary determinant of intracellular redox potential. Sensitivity of roGFP to glutathione is enhanced by linking it to human glutaredoxin 1 (Grx1).

8.5 Enzyme Activity Indicators

Differences between the expression and activity levels of enzymes are often physiologically consequential, leading to the need to measure both quantities in parallel. While expression levels are readily assessed using immunofluorescence in fixed cells (Chapter 5) or fluorescent protein fusions (Chapter 6) in live cells, quantitative imaging of activity is a more challenging exercise, particularly within the constraints imposed by living cells or model organisms [48]. In relation to the most common approach, that of using substrate analog probes, these constraints are threefold. Firstly, turnover of the substrate analog probe must generate a fluorescence change that is detectable in situ. Secondly, unless the enzyme of interest resides on the cell surface, the substrate has to be delivered and maintained intracellularly at concentrations sufficient to saturate the enzyme. Thirdly, information on localization of enzymatic activity is predicated on the fluorescent product of enzymatic turnover not diffusing or being transported away from the site of generation. Note that, since most enzyme reactions are unidirectional, the substrate is consumed and must be continually refreshed in order to accomplish measurements over long time periods. Given these constraints, it is perhaps not surprising that imaging of intracellular enzyme activity has been a less widespread application than those described in earlier sections of this chapter. Nevertheless, in favorable circumstances, fluorogenic[d] enzyme substrates have been and continue to be successfully developed and applied. PED6, a fluorescent reporter designed for in situ detection of phospholipase activity in zebrafish, provides an instructive example [49]. The PED6 substrate is a phospholipid analog incorporating a fluorescent dye (BODIPY FL) and a quencher group (2,4-dinitroaniline) organized such that cleavage by phospholipase A2 separates the BODIPY dye from the quencher, resulting in increased fluorescence that is proportional to enzymatic activity. PED6 is ingested by zebrafish and is distributed throughout the body. However, because the background fluorescence from the quenched substrate is minimal, increases of fluorescence associated with localized enzyme activity are detectable in situ (Fig. 8.4). Imaging the activity of extracellular enzymes such as that of matrix metalloproteinases (MMP) is not impeded by the obstacle of maintaining saturating substrate concentrations. Consequently, this has been one of the most successful applications of enzyme activity imaging [48]. The substrates are generally extracellular matrix proteins such as collagen and elastin that are extensively labeled, thereby promoting dye–dye association accompanied by fluorescence quenching. Proteolysis of the substrate generates singly labeled, unquenched peptide fragments, indicated by increased fluorescence.

The most notable FP-based enzyme activity indicators are fluorogenic[e] protein kinase substrates [50–52]. This is both a reflection of the crucial role of kinases in regulation of gene expression and the

[d]Probes with fluorogenic responses are initially nonfluorescent and become fluorescent upon interaction with their biomolecular targets.

[e]Probes with fluorogenic responses are initially nonfluorescent and become fluorescent upon interaction with their biomolecular targets.

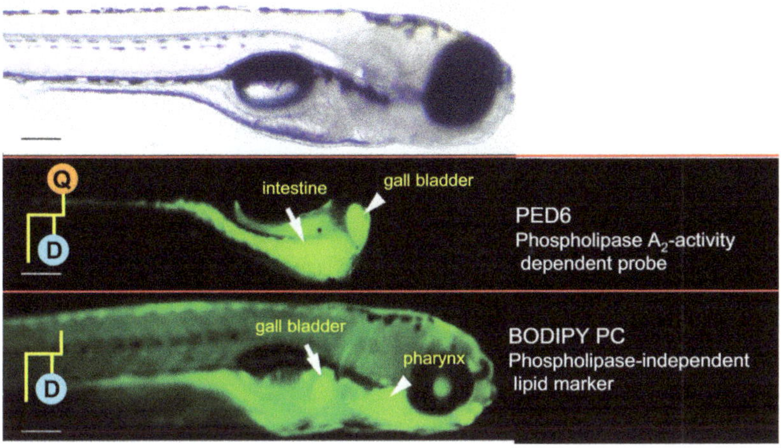

Figure 8.4 In vivo imaging of lipid metabolism in zebrafish using a fluorogenic substrate probe. The substrate (PED6) is a phospholipid analog incorporating a fluorescent dye (D) and a quencher group (Q) organized such that cleavage by phospholipase A2 separates the dye from the quencher, resulting in increased fluorescence that is proportional to enzymatic activity (middle panel). A probe lacking the quencher group (lower panel) produces an unresolved image of substrate and product distribution that provides no information as to enzyme activity. Reproduced with permission from Farber et al. [49].

fact that development of small-molecule kinase substrate analogs has proved to be largely impracticable. FP-based kinase substrates generally consist of a phosphorylatable peptide sequence and an adjacent phosphopeptide binding domain flanked by CFP and YFP (Fig. 8.5). Kinase-mediated phosphorylation of the substrate peptide drives interaction with the phosphopeptide binding domain with the resulting conformational change detected by CFP-to-YFP FRET. Sensors incorporating single, conformation-sensitive FPs instead of a FRET pair have also been described [51]. The principle alternatives to fluorogenic substrates are activity-based probes (ABP) [53]. These small-molecule probes react specifically and irreversibly with enzyme active sites. Consequently, they allow spatial localization of enzyme activity but cannot provide information on its time-dependent variation [50–52].

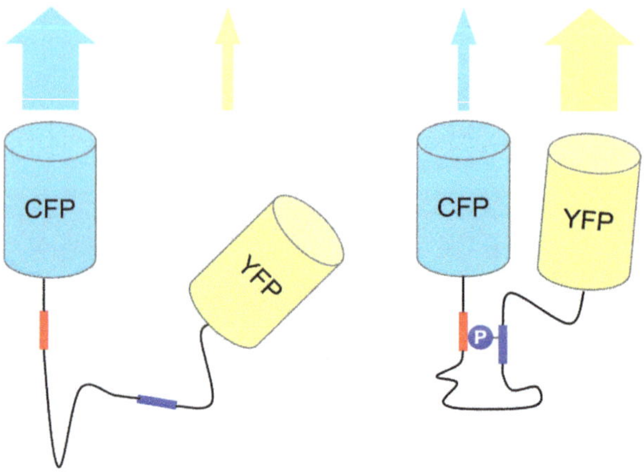

Figure 8.5 Schematic representation of a FP-based kinase sensor. CFP and YFP are linked by a peptide containing a phosphate-binding domain (red) and a substrate domain (blue). Intracellular kinase activity produces phosphorylation of the substrate domain, resulting in capture by the phosphate binding domain. The resulting conformational change is registered by proximity-dependent fluorescence resonance energy transfer (FRET) from CFP to YFP, detected as increased YFP fluorescence and decreased CFP fluorescence. The sensitivity of this detection mechanism is limited by two factors. (1) Direct excitation of YFP, resulting in a non-zero baseline signal and (2) the physical size of CFP and YFP which limits their mutual proximity.

8.6 Perspective

At the risk of stating the obvious, only a limited range of molecular targets has been considered in this chapter. Indicators for other important targets such as glutamate, glucose, cAMP, ATP, and inositol trisphosphate have been developed following the general strategies laid out above [6, 52, 54]. Indirect detection approaches can also will be implemented. For example, imaging the translocation of GFP-tagged glucose transporter GLUT1 to the cell surface provides a surrogate indication of glucose uptake. Another quite versatile indirect detection method is small molecule-induced GFP expression [55]. However, while such approaches can be useful adjuncts to the direct detection methods outlined above, they are inevitably inferior in terms of temporal and spatial precision.

References

1. Johnson, I. D., and Spence, M. T. Z. (2010). Indicators for Ca^{2+}, Mg^{2+}, Zn^{2+} and other metal ions, in *Molecular Probes Handbook*, 11th ed., pp. 829–882.
2. Paredes, R. M., Etzler, J. C., Watts, L. T., Zheng, W., and Lechleiter, J. D. (2008). Chemical calcium indicators. *Methods*, **46**, 143–151.
3. Oheim, M., van't Hoff, M., Feltz, A., Zamaleeva, A., Mallet, J. M., and Collot, M. (2014). New red-fluorescent calcium indicators for optogenetics, photoactivation and multi-color imaging. *Biochim. Biophys. Acta*, **1843**, 2284–2306.
4. Miyawaki, A., Llopis, J., Heim, R., McCaffery, J. M., Adams, J. A., Ikura, M., and Tsien, R. Y. (1997). Fluorescent indicators for Ca^{2+} based on green fluorescent proteins and calmodulin. *Nature*, **388**, 882–887.
5. Dana, H., Mohar, B., Sun, Y., Narayan, S., Gordus, A., Hasseman, J. P., Tsegaye, G., Holt, G. T., Hu, A., Walpita, D., Patel, R., Macklin, J. J., Bargmann, C. I., Ahrens, M. B., Schreiter, E. R., Jayaraman, V., Looger, L. L., Svoboda, K., and Kim, D. S. (2016). Sensitive red protein calcium indicators for imaging neural activity. *Elife*, **5**, e12727.
6. Lin, M. Z., and Schnitzer, M. J. (2016). Genetically encoded indicators of neuronal activity. *Nat. Neurosci.*, **19**, 1142–1153.
7. Suzuki, J., Kanemaru, K., Ishii, K., Ohkura, M., Okubo, Y., and Iino, M. (2014). Imaging intraorganellar Ca^{2+} at subcellular resolution using CEPIA. *Nat. Commun.*, **5**, 4153.
8. Dynes, J. L., Amcheslavsky, A., and Cahalan, M. D. (2016). Genetically targeted single-channel optical recording reveals multiple Orai1 gating states and oscillations in calcium influx. *Proc. Natl. Acad. Sci. U.S.A.*, **113**, 440–445.
9. Larsch J., Flavell, S. W., Liu, Q., Gordus, A., Albrecht, D. R., and Bargmann, C. I. (2015). A circuit for gradient climbing in *C. elegans* chemotaxis. *Cell Rep.*, **12**, 1748–1760.
10. Helassa, N., Zhang, X. H., Conte, I., Scaringi J., Esposito, E., Bradley, J., Carter, T., Ogden, D., Morad, M., and Török, K. (2015). Fast-response calmodulin-based fluorescent indicators reveal rapid intracellular calcium dynamics. *Sci. Rep.*, **5**, 15978.
11. Lock, J. T., Parker, I., and Smith, I. F. (2015). A comparison of fluorescent Ca^{2+} indicators for imaging local Ca^{2+} signals in cultured cells. *Cell Calcium*, **58**, 638–648.

12. Thestrup, T., Litzlbauer, J., Bartholomäus, I., Mues, M., Russo, L., Dana, H., Kovalchuk, Y., Liang, Y., Kalamakis, G., Laukat, Y., Becker, S., Witte, G., Geiger, A., Allen, T., Rome, L. C., Chen, T. W., Kim, D. S., Garaschuk, O., Griesinger, C., and Griesbeck, O. (2014). Optimized ratiometric calcium sensors for functional in vivo imaging of neurons and T lymphocytes. *Nat. Methods*, **11**, 175–182.
13. Hyrc, K. L., Minta, A., Escamilla, P. R., Chan, P. P., Meshik, X. A., and Goldberg, M. P. (2013). Synthesis and properties of Asante Calcium Red – a novel family of long excitation wavelength calcium indicators. *Cell Calcium*, **54**, 320–333.
14. Komatsubara, A. T., Matsuda, M., and Aoki, K. (2015). Quantitative analysis of recombination between YFP and CFP genes of FRET biosensors introduced by lentiviral or retroviral gene transfer. *Sci. Rep.*, **5**, 13283.
15. Hannanta-Anan, P., and Chow, B.Y. (2016). Optogenetic control of calcium oscillation waveform defines NFAT as an integrator of calcium load. *Cell Syst.*, **2**, 283–288.
16. Ambrosi, C. M., and Entcheva, E. (2014). Optogenetic control of cardiomyocytes via viral delivery. *Methods Mol. Biol.*, **1181**, 215–228.
17. Erdahl ,W. L., Chapman, C. J., Taylor, R. W., and Pfeiffer, D. R. (1994). Ca^{2+} transport properties of ionophores A23187, ionomycin, and 4-BrA23187 in a well defined model system. *Biophys. J.*, **66**, 1678–1693.
18. Anderson, M., Moshnikova, A., Engelman, D. M., Reshetnyak, Y. K., and Andreev, O. A. (2016). Probe for the measurement of cell surface pH in vivo and ex vivo. *Proc. Natl. Acad. Sci. U.S.A.*, **113**, 8177–8181.
19. Ogawa, M., Kosaka, N., Regino, C. A., Mitsunaga, M., Choyke, P. L., and Kobayashi, H. (2010). High sensitivity detection of cancer in vivo using a dual-controlled activation fluorescent imaging probe based on H-dimer formation and pH activation. *Mol. Biosyst.*, **6**, 888–893.
20. Nakata, E., Yukimachi, Y., Nazumi, Y., Uto, Y., Maezawa, H., Hashimoto, T., Okamoto Y., and Hori, H. (2010). A newly designed cell-permeable SNARF derivative as an effective intracellular pH indicator. *Chem. Commun.*, **46**, 3526–3528.
21. Yapici, N. B., Bi, Y., Li, P., Chen, X., Yan, X., Mandalapu1, S. R., Faucett, M., Jockusch, S., Ju, J., Gibson, K. M., Pavan, W. J., and Lanrong Bi, L. (2015). Highly stable and sensitive fluorescent probes (LysoProbes) for lysosomal labeling and tracking. *Sci. Rep.*, **5**, 8576.

22. Haggie, P. M., and Verkman, A. S. (2009). Unimpaired lysosomal acidification in respiratory epithelial cells in cystic fibrosis. *J. Biol. Chem.*, **284**, 7681–7686.
23. Mahon, M. J. (2011). pHluorin2: an enhanced, ratiometric, pH-sensitive green fluorescent protein. *Adv. Biosci. Biotechnol.*, **2**, 132–137.
24. Herron, T. J., Lee, P., and Jalife, J. (2012). Optical imaging of voltage and calcium in cardiac cells and tissues. *Circ. Res.*, **110**, 609–623.
25. Johnson, I. D., and Spence, M. T. Z. (2010). Probes for membrane potential, in *Molecular Probes Handbook*, 11th ed., pp. 923–936.
26. Klapperstück, T., Glanz, D., Klapperstück, M., and Wohlrab, J. (2009). Methodological aspects of measuring absolute values of membrane potential in human cells by flow cytometry. *Cytometry A*, **75**, 593–608.
27. Loew, L. M. (2015). Design and use of organic voltage sensitive dyes. *Adv. Exp. Med. Biol.*, **859**, 27–53.
28. Aseyev, N., Roshchin, M., Ierusalimsky, V. N., Balaban, P. M., and Nikitin, E. S. (2013). Biolistic delivery of voltage-sensitive dyes for fast recording of membrane potential changes in individual neurons in rat brain slices. *J. Neurosci. Methods*, **212**, 17–27.
29. Fisher, J. A., Salzberg, B. M., and Yodh, A. G. (2005). Near infrared two-photon excitation cross-sections of voltage-sensitive dyes. *J. Neurosci. Methods*, **148**, 94–102.
30. Mennerick, S., Chisari, M., Shu, H. J., Taylor, A., Vasek, M., Eisenman, L. N., and Zorumski, C. F. (2010). Diverse voltage-sensitive dyes modulate $GABA_A$ receptor function. *J. Neurosci.*, **30**, 2871–2879.
31. Ghitani, N., Bayguinov, P. O., Yihe, M., and Jackson, M. B. (2015). Single-trial imaging of spikes and synaptic potentials in single neurons in brain slices with genetically encoded hybrid voltage sensor. *J. Neurophysiol.*, **113**, 1249–1259.
32. Miller, E. W., Lin, J. Y., Frady, E. P., Steinbach, P. A., Kristan, W. B., and Tsien, R. Y. (2012). Optically monitoring voltage in neurons by photoinduced electron transfer through molecular wires. *Proc. Natl. Acad. Sci. U.S.A.*, **109**, 2114–2119.
33. Huang, Y. L., Walker, A. S., and Miller E. W. (2015). A photostable silicon rhodamine platform for optical voltage sensing. *J. Am. Chem. Soc.*, **137**, 10767–10776.
34. Storace, D, Rad, M. S., Kang, B., Cohen, L. B., Hughes, T., and Baker, B. J. (2016). Toward better genetically encoded sensors of membrane potential. *Trends Neurosci.*, **39**, 277–289.

35. Han, Z., Jin, L., Chen, F., Loturco, J. J., Cohen, L. B., Bondar, A., Lazar, J., and Pieribone, V. A. (2014). Mechanistic studies of the genetically encoded fluorescent protein voltage probe ArcLight. *PLoS One*, **9**, e113873.
36. Flytzanis, N. C., Bedbrook, C. N., Chiu, H., Engqvist, M. K., Xiao, C., Chan, K. Y., Sternberg, P. W., Arnold, F. H., and Gradinaru, V. (2014). Archaerhodopsin variants with enhanced voltage-sensitive fluorescence in mammalian and Caenorhabditis elegans neurons. *Nat. Commun.*, **15**, 4894.
37. Johnson, I. D., and Spence, M. T. Z. (2010). Probes for reactive oxygen species, including nitric oxide, in *Molecular Probes Handbook*, 11th ed., pp. 803–828.
38. Kundu, K., Knight, S. F., Willett, N., Lee, S., Taylor, W. R., and Murthy, N. (2009). Hydrocyanines: a class of fluorescent sensors that can image reactive oxygen species in cell culture, tissue, and in vivo. *Angew. Chem. Int. Ed. Engl.*, **48**, 299–303.
39. Chen, X., Zhong, Z., Xu, Z., Chen, L., and Wang, Y. (2010). 2',7'-Dichlorodihydrofluorescein as a fluorescent probe for reactive oxygen species measurement: forty years of application and controversy. *Free Radical Res.*, **44**, 587–604.
40. Setsukinai, K., Urano, Y., Kakinuma, K., Majima, H. J., and Nagano, T. (2003). Development of novel fluorescence probes that can reliably detect reactive oxygen species and distinguish specific species. *J. Biol. Chem.*, **278**, 3170–3175.
41. Robinson, K. M., Janes, M. S., and Beckman, J. S. (2008). The selective detection of mitochondrial superoxide by live cell imaging. *Nat. Protoc.*, **3**, 941–947.
42. Rayner, C. L., Gole, G. A., Bottle, S. E., and Barnett, N. L. (2014). Dynamic, in vivo, real-time detection of retinal oxidative status in a model of elevated intraocular pressure using a novel, reversibly responsive, profluorescent nitroxide probe. *Exp. Eye Res.*, **129**, 48–56.
43. Miller, E. W., Bian, S. X., and Chang, C. J. (2007). A fluorescent sensor for imaging reversible redox cycles in living cells. *J. Am. Chem. Soc.*, **129**, 3458–3459.
44. Pal, R., Basu, T. P., Li, S., Minard, C., and Rodney, G. G. (2013). Real-time imaging of NADPH oxidase activity in living cells using a novel fluorescent protein reporter. *PLoS One*, **8**, e63989.
45. Hanson, G. T., Aggeler, R., Oglesbee, D., Cannon, M., Capaldi, R. A., Tsien, R. Y., and Remington, S. J. (2004). Investigating mitochondrial redox

potential with redox-sensitive green fluorescent protein indicators. *J. Biol. Chem.*, **279**, 13044–13053.

46. Sugiura, K., Nagai, T., Nakano, M., Ichinose, H., Nakabayashi, T., Ohta, N., and Hisabori, T. (2015). Redox sensor proteins for highly sensitive direct imaging of intracellular redox state. *Biochem. Biophys. Res. Commun.*, **457**, 242–248.
47. van Lith, M., Tiwari, S., Pediani, J., Milligan, G., and Bulleid, N. J. (2011). Real-time monitoring of redox changes in the mammalian endoplasmic reticulum. *J. Cell Sci.*, **124**, 2349–2356.
48. Van Noorden, C. J. F. (2010). Imaging enzymes at work: metabolic mapping by enzyme histochemistry. *J. Histochem. Cytochem.*, **58**, 481–497.
49. Farber, S. A., Pack, M., Ho, S. Y., Johnson, I. D., Wagner, D. S., Dosch, R., Mullins, M. C., Hendrickson, H. S., Hendrickson, E. K., and Halpern, M. E. (2001). Genetic analysis of digestive physiology using fluorescent phospholipid reporters. *Science*, **292**, 1385–1388.
50. Ryu, H., Chung, M., Dobrzyński, M., Fey, D., Blum, Y., Lee, S. S., Peter, M., Kholodenko, B. N., Jeon, N. L., and Pertz, O. (2015). Frequency modulation of ERK activation dynamics rewires cell fate. *Mol. Syst. Biol.*, **11**, 838.
51. Bonnot, A., Guiot, E., Hepp, R., Cavellini, L., Tricoire, L., and Lambolez, B. (2014). Single-fluorophore biosensors based on conformation-sensitive GFP variants. *FASEB J.*, **28**, 1375–1385.
52. Miyawaki, A., and Niino, Y. (2015). Molecular spies for bioimaging—fluorescent protein-based probes. *Mol. Cell*, **58**, 632–643
53. Blum, G., Weimer, R. M., Edgington, L. E., Adams, W., and Bogyo, M. (2009). Comparative assessment of substrates and activity based probes as tools for non-invasive optical imaging of cysteine protease activity. *PLoS One*, **4**, e6374.
54. Germond, A., Fujita, H., Ichimura, T., and Watanabe, T. M. (2016). Design and development of genetically encoded fluorescent sensors to monitor intracellular chemical and physical parameters. *Biophys. Rev.*, **8**, 121–138.
55. Flavahan, W. A., Wu, Q., Hitomi, M., Rahim, N., Kim, Y., Sloan, A. E., Weil, R. J., Nakano, I., Sarkaria, J. N., Stringer, B. W., Day, B. W., Li, M., Lathia, J. D., Rich, J. N., and Hjelmeland, A. B. (2013). Brain tumor initiating cells adapt to restricted nutrition through preferential glucose uptake. *Nat. Neurosci.*, **16**, 1373–1382.

Chapter 9

New Approaches to Cancer Therapy Using in vivo Fluorescent-Protein Imaging

Robert M. Hoffman
Department of Surgery, University of California, San Diego, CA
AntiCancer Inc., 7917 Ostrow Street, San Diego, CA 92111, USA
all@anticancer.com

Fluorescent proteins have revolutionized biology by enabling what was formerly invisible to be seen clearly. Fluorescent proteins enable imaging of all aspects of cancer in living animals. This chapter highlights the potential therapeutic uses of imaging with fluorescent proteins. Introducing the GFP gene in malignant tissue enabled fluorescence-guided surgery to resect all tumor nodules labeled with GFP and increases survival of tumor-bearing animals. Double amino acid auxotrophs of *Salmonella typhimurium* labeled with GFP have been used to target and eradicate tumors and metastasis labeled with RFP in metastatic nude mouse models of human cancer. Fluorescence-guided ultraviolet C (UVC) irradiation inhibited the growth of murine melanoma expressing GFP in the ear of RFP transgenic mice. The efficacy was readily visible using noninvasive imaging. UVC irradiation also inhibited growth of 143B

Fundamentals of Fluorescence Imaging
Edited by Guy Cox

ISBN 978-981-4774-85-7 (Hardcover), 978-1-351-12940-4 (eBook)
www.jennystanford.com

human osteosarcoma cells, HT-1080 human fibrosarcoma cells, Lewis lung carcinoma (LLC), and XPA-1 human pancreatic cancer all labeled with fluorescent proteins. Primary patient tumors acquired fluorescent protein-expressing stroma from passage in GFP and RFP transgenic nude mice. The tumors, with very bright fluorescent stroma, were then orthotopically passaged to nontransgenic nude mice where it was possible to image the brightly fluorescent tumors non-invasively and longitudinally as they progressed in the nontransgenic nude mice. This non-invasive imageable tumorgraft model will be valuable to screen for effective treatment options for individual patients. Many more clinical applications of fluorescent proteins will be developed in the future.

9.1 The in vivo GFP Revolution

9.1.1 Properties of GFP for in vivo Cellular Imaging

The GFP gene was cloned from the bioluminescent jellyfish *Aequorea victoria* [1] and has revolutionized cellular imaging [2, 3]. The GFP gene encodes a 283-amino acid protein of MW of 27,000 kDa [4, 5] that requires no cofactors to fluoresce [6]. Bright mutants of the GFP gene have been generated [7–9] and have been humanized [10]. Red fluorescent protein (RFP) from the *Discosoma* coral has as the advantage of longer wavelength emission [11–14] (see Chapter 6).

Our laboratory was the first to use GFP in vivo for imaging, in this case, for cancer cells in mice [15] and many laboratories followed our lead [16–18]. Fine cellular details of cancer cells, expressing GFP, such as pseudopodial projections, even after extended periods of in vivo growth were observed by Naumov et al. [19] (Fig. 9.1).

Dual-color cells, with GFP in the nucleus and RFP in the cytoplasm, were developed to image cytoplasmic and nuclear dynamics of cells migrating in capillaries. The cytoplasm and nuclei of the cancer cells deformed sufficiently to fit the width of these vessels [20] (Fig. 9.2). The cellular dynamics of dual-color cancer cells could be observed in real time in larger blood vessels as they trafficked or adhered to the vessel surface. During extravasation, both cytoplasm and nuclei underwent deformation as shown by

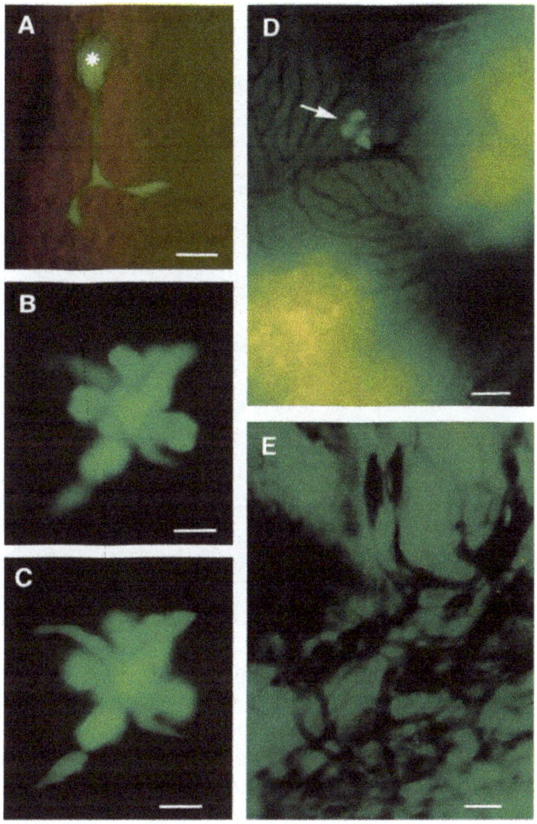

Figure 9.1 Morphological observations by IVVM: interactions of cells and metastases with liver microenvironment. (A) An extravasated CHO-K1-GFP cell at day 1, in the process of migration to the hepatocyte layer in subcapsular region. Cell body (*), slightly below the plane of focus, extends a long, bifurcated pseudopodial projection up towards the liver surface (viewed by epifluorescence plus transillumination). Scale bar: 25 mm. (B, C) Micrometastasis at day 4 observed by epifluorescence alone, at two different focal depths ("optical slicing"). (B) Viewed at a deeper plane of focus, center of micrometastasis is sharply defined but peripheral cells appear out of focus. Scale bar: 25 mm. (C) Viewed nearer to the liver surface, peripheral cells are more sharply defined than cells at the center, indicating that the peripheral cells had invaded the hepatocyte layer located just below the liver capsule. Scale bar: 25 mm. (D) Solitary tumor cells (→) present in normal liver tissue at day 11, in close proximity to two metastases (viewed by epifluorescence plus transillumination). Scale bar: 50 mm. (E) Highly distorted microvasculature within metastasis at day 11, viewed by epifluorescence alone. Blood vessels appear dark against GFP-fluorescent background of tumor tissue. Scale bar: 25 mm. Views A to E are photomicrographs taken with a 35 mm camera. Reproduced with permission from [19].

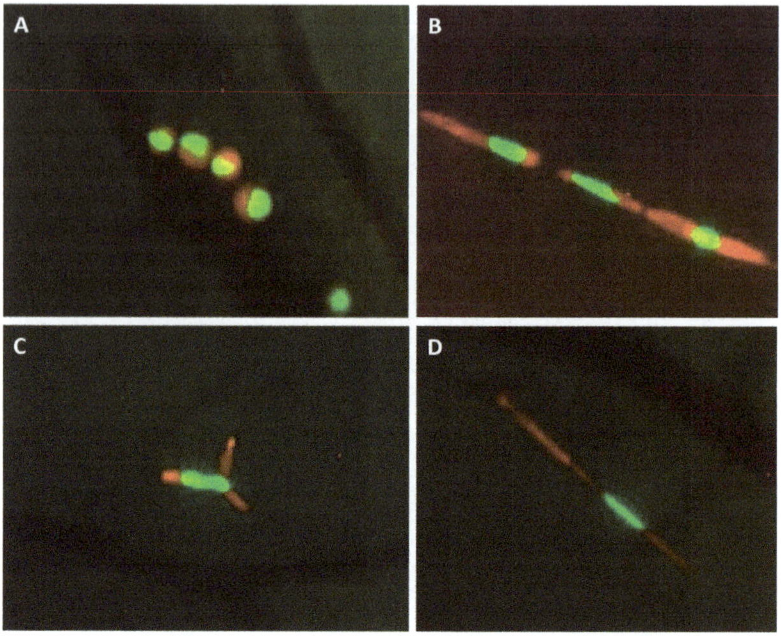

Figure 9.2 Classification of the deformation of HT-1080-dual-color cells in the vessels in the skin. (A) nondeformed cells are within a microvessel. The cells in the microvessel are round and the nuclei oval. The cells occupy the full diameter of the vessel. (B) the cells and nuclei are elongated to fit a capillary. (C) the cells are arrested at the capillary bifurcation. Because of the difference of the deformability between cytoplasm and nucleus, only the cytoplasm is bifurcated. The nucleus is also deformed but remains intact. (D) cytoplasmic fragmentation in very thin capillary. Scale bar: 50 μm [20]

time-lapse dual-color imaging [21]. Dual-color cancer cells were also imaged trafficking through lymphatic vessels until they entered the axillary lymph node [22].

Cancer-cell–stromal-cell interaction and cellular dynamics as well as tumor blood flow could be visualized by color-coded imaging [23]. We demonstrated using GFP and RFP imaging that stromal cells are necessary for metastasis. After splenic injection of cancer cells, splenocytes co-trafficked with the cancer cells to the liver metastasis of the cancer cells could not occur without the splenocytes [24] (Fig. 9.3A,B).

Pretreatment with cyclophosphamide before cancer-cell injection enhanced intravascular proliferation and extravasation of human HT-1080 GFP-RFP fibrosarcoma cells along with extravascular colony formation. In the nonpretreated mice, most cancer cells remained quiescent in vessels without extravasation. Cyclophosphamide did not directly affect the cancer cells because cyclophosphamide had been cleared at the time the cancer cells were injected. These results show an "opposite effect" of chemotherapy which enhances critical steps in malignancy rather than inhibiting them [25].

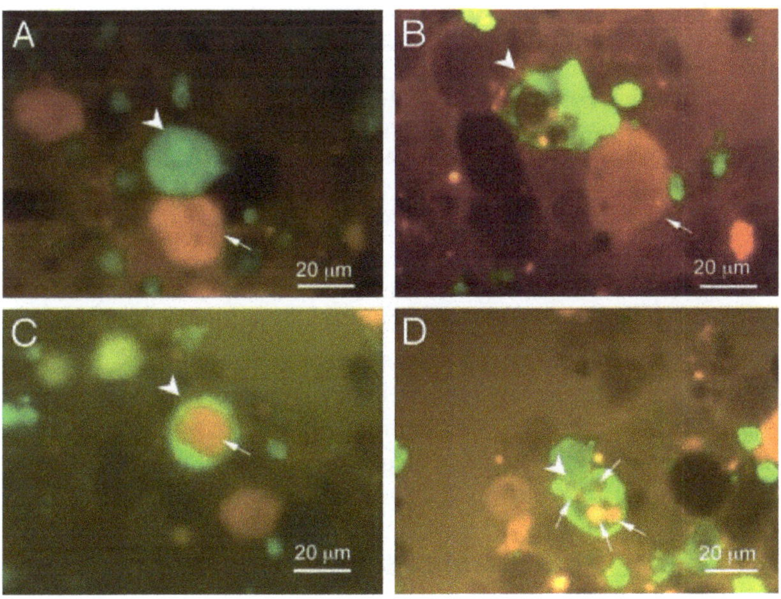

Figure 9.3A Visualization of host macrophage–tumor cell interaction in fresh tumor tissue. Images show host macrophages expressing GFP interacting with human PC-3-RFP prostate cancer cells on day 35 after orthotopic implantation of PC-3-RFP cells in the transgenic GFP nude mouse. (A) Host GFP macrophage (arrowhead) contacting RFP cancer cell (arrow). (B) GFP macrophage (arrowhead) engulfing RFP cancer cell (arrow). (C) RFP cancer cell (arrow) engulfed by GFP macrophage (arrowhead). (D) RFP cancer cell (arrows) digested by GFP macrophage (arrowhead). Scale bar: 20 μm [62].

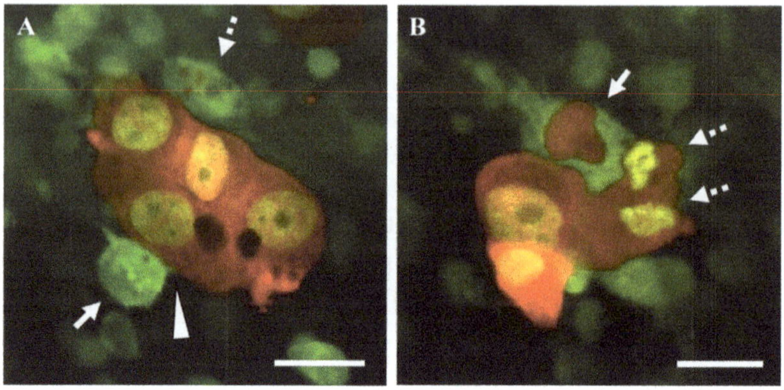

Figure 9.3B Visualization of host macrophage cancer cell interaction in live mice. (A) Host GFP macrophages were visualized contacting HT-1080-GFP-RFP cells (arrow, broken arrow). Macrophage processes can be seen extending into the HT-1080-GFP-RFP cell (arrowhead). Another macrophage (broken arrow) contained cytoplasmic fragments after digesting an HT-1080-GFP-RFP cell. (B) A macrophage (arrow), engulfing the RFP-expressing cytoplasm of an HT-1080 cell shown in contact with additional HT-1080-GFP-RFP cells (broken arrows). Scale bar: 20 μm [63].

A model to image single cancer-cell seeding in the lung in real time was developed. Dual-color HT-1080 human fibrosarcoma cells were injected in the tail vein of the mouse. The right chest wall was then opened in order to image seeding of and subsequent growth of cancer cells on the lung surface directly [26]. This model provides a direct visual target to inhibit cancer cell seeding on the lung.

9.2 Fluorescence-Guided Surgery

A herpes simplex virus (NV1066) vector that expresses the GFP gene was used to label metastatic lung tumor foci in a mouse model [6]. The fluorescently labeled tumors were detected by using a thoracoscopic endoscope system. Kishimoto et al. selectively labeled tumors with GFP using a telomerase-dependent adenovirus (OBP-401) that expresses the GFP gene only in cancer cells [27–29]. The

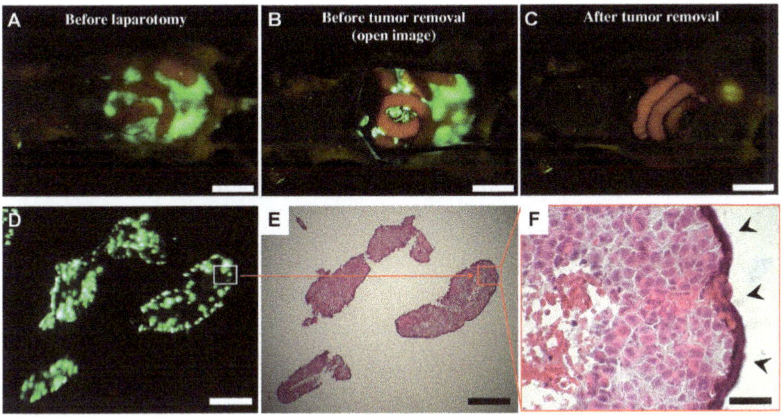

Figure 9.4 Fluorescence-guided surgical removal of peritoneal disseminated HCT-116 tumors after GFP labeling with OBP-401. Non-colored HCT-116 human colon cancer cells were injected into the abdominal space of nude mice. Ten days later, 1×10^8 PFU of OBP-401 were i.p. injected. (A) Disseminated nodules were efficiently labeled and noninvasively visualized by GFP expression 5 days after virus administration. (B) Under general anesthesia, laparotomy was performed to remove intra-abdominal disease under GFP-guided navigation. (C) Disseminated nodules visualized by GFP-guided navigation were removed (scale bars: A–C, 10 mm). (D) Frozen section of resected HCT-116 disseminated nodules with fluorescence detection (scale bar: 500 μm). (E) H&E section of HCT-116 disseminated nodules shown in D. The box outlines a region of D and E analyzed in F (scale bar: 500 μm). (F) Detail of the boxed region of D and E (scale bar: 50 μm) [27].

labeled tumors could then be resected under fluorescence-guided surgery (FGS). Recurrent cancer cells stably expressed GFP enabling their potential resection under FGS (Fig. 9.4).

Fluorescence-guided laparoscopy could detect fluorescent metastatic lesions smaller than 1 mm, which are not visible under bright-field when unlabeled [30, 31]. Fluorescence laparoscopy with an LED light source enabled real-time visualization of fluorescent protein-labeled tumor metastasis in the peritoneal cavity. An LED light source permitted simultaneous detection of metastasis of different fluorescent colors with minimal interfering autofluorescence from surrounding structures [32].

Bright light surgery (BLS) was compared to FGS in orthotopic mouse models of human pancreatic cancer expressing RFP. A more complete resection of pancreatic cancer was achieved using FGS compared with BLS resulting in significantly longer disease-free survival than BLS [33].

A modified neurological microscope was used in the clinic for FGS of glioma [34]. For FGS of ovarian cancer in the clinic, a real-time intraoperative multispectral fluorescence imaging system was used [35]. FGS holds much promise for surgery of today and tomorrow [31, 36].

9.3 Bacterial Therapy of Cancer

In the late 19th and early 20th century, Coley infected cancer patients with *Streptococcus pyrogenes* and in later years treated the patients with extracts of the bacteria, which became known as Coley's toxins. Coley had remarkable results with his toxins, but this treatment fell out of favor after Coley's death in 1936 [37, 38].

A strain of *S. typhimurium*, A1-R, expressing GFP has been developed which has greatly increased antitumor efficacy. *S. typhimurium* A1-R is auxotrophic for Leu Arg which prevents it from mounting a continuous infection in normal tissues. A1-R was able to eradicate primary and metastatic tumors in monotherapy in nude mouse models of prostate, breast, and pancreatic cancer, as well as sarcoma and glioma [39–45]. Tumors with a high degree of vascularity were more sensitive to A1-R, and vascular destruction appears to play a role in A1-R antitumor efficacy [46]. GFP imaging of the bacteria enables visualization of single bacteria in vivo interacting with cancer cells [47] (Fig. 9.5).

The barriers in tumors for standard therapy to be effective, such as hypoxia, acidic pH, disorganized vascular architecture, and dissemination, are in contrast to opportunities for bacteria to target cancer and suggest important clinical potential [48].

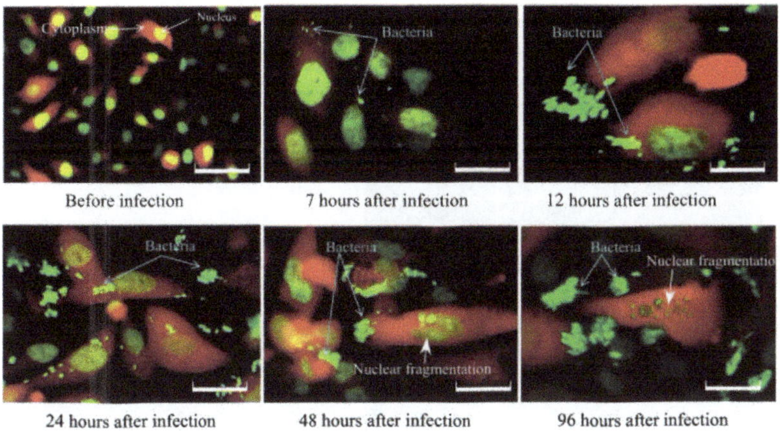

Figure 9.5 Intracellular growth of *S. typhimurium* A1. (A) Time course of GFP-labeled *S. typhimurium* A1 growing in GFP–RFP-labeled PC-3 human prostate cells in vitro. PC-3 human prostate tumor cells labeled with RFP in the cytoplasm and GFP in the nucleus by means of a fusion with histone H2B were grown in 24-well tissue culture plates to a density of $\sim 10^4$ cells per well. Bacteria were grown in LB and harvested at late-logarithmic phase, then diluted in cell culture medium and added to the tumor cells (1×10^5 cfu per well). After 1 h of incubation at 37°C, the cells were rinsed and cultured in medium containing gentamycin sulfate (20 μg/ml) to kill external but not internal bacteria. Interaction between bacteria and tumor cells was observed at the indicated time points under fluorescence microscopy magnification. scale bar: 156 μm for upper left panel; otherwise, scale bar: 78 μm [39].

9.4 Tumor Grafts from Patients Made Imageable in Fluorescent-Protein Expressing Transgenic Nude Mice

Pancreatic-cancer-patient tumor specimens were initially established subcutaneously in NOD/SCID mice immediately after surgery [49, 50]. The patient tumors were then harvested from NOD/SCID mice and passaged orthotopically in transgenic nude mice ubiquitously expressing RFP. The primary patient tumors acquired RFP-expressing stroma. The RFP-expressing stroma included cancer-associated fibroblasts (CAFs) and tumor-associated macrophages

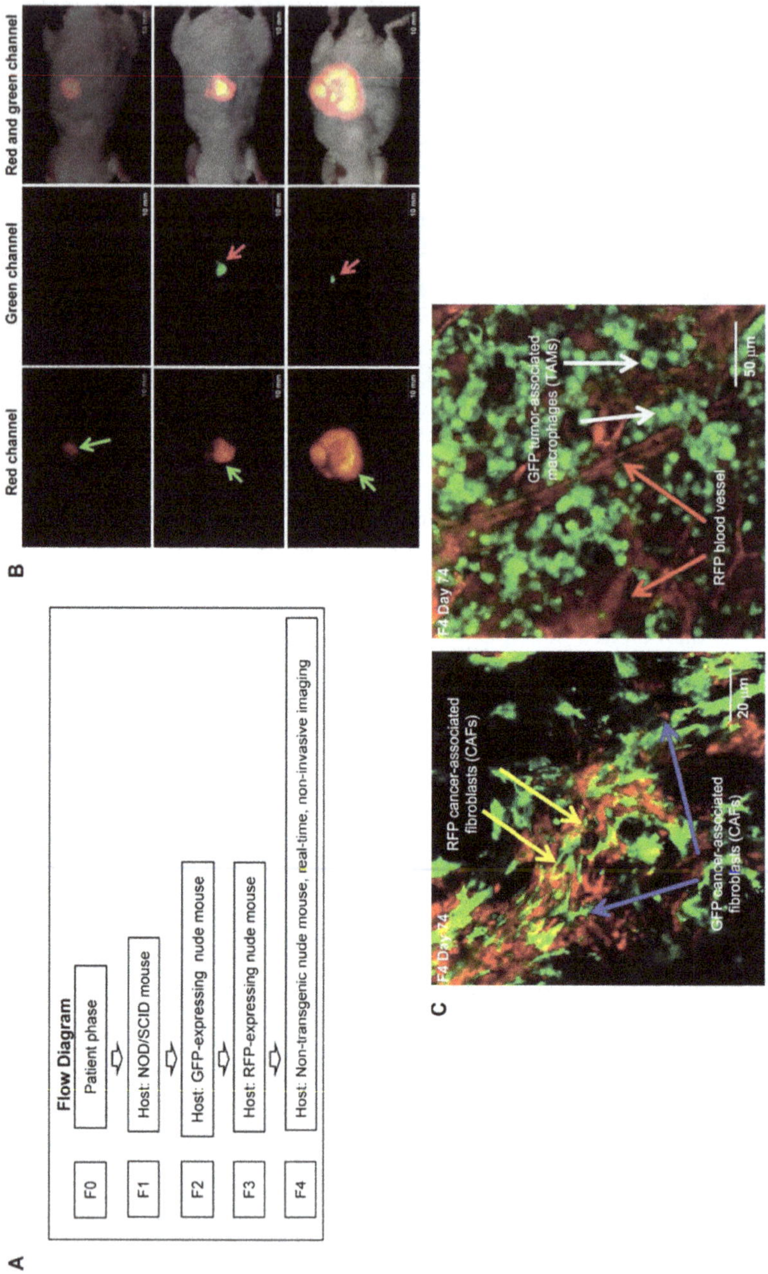
A
Flow Diagram
F0
Patient phase
F1
Host: NOD/SCID mouse
F2
Host: GFP-expressing nude mouse
F3
Host: RFP-expressing nude mouse
F4
Host: Non-transgenic nude mouse, real-time, non-invasive imaging
B
Red channel
Green channel
Red and green channel
C
F4 Day 74
RFP cancer-associated fibroblasts (CAFs)
GFP cancer-associated fibroblasts (CAFs)
20 μm
F4 Day 74
GFP tumor-associated macrophages (TAMs)
RFP blood vessel
50 μm

Figure 9.6 Non-invasive imaging of fluorescent tumor from (a) patient with pancreatic cancer growing orthotopically in nude mice. (A) Flow diagram of the experimental protocol. (B) Whole-body non-invasive imaging of human pancreatic cancer orthotopic tumorgraft in nontransgenic nude mice. Mice were non-invasively imaged at day 21 (upper row), day 30 (middle row), and day 74 (lower row). The tumors in the nontransgenic nude mice are in the F4 passage after F1, in NOD/SCID mice after patient surgery; F2, in transgenic green fluorescent protein (GFP)-expressing nude mice; and F3 in transgenic red fluorescent protein (RFP)-expressing nude mice. The tumor acquired GFP and RFP stroma in the F2 and F3 passages, respectively. Green arrows indicate tumor with RFP stroma. Red arrows indicate tumor with GFP stroma. Images were taken with the Olympus OV100 Small Animal Imaging System. (C) Image of human pancreatic cancer tumor tissue resected from the F4 passage with RFP and GFP stroma. Images were taken with an FV1000 confocal laser microscope. Left panel: RFP-expressing and GFP-expressing cancer-associated fibroblast cells (CAFs). Right panel: RFP-expressing blood vessels and GFP-expressing tumor-associated macrophages (TAMs). Yellow arrows indicate RFP-expressing CAFs. Blue arrows indicate GFP-expressing CAFs. White arrows indicate GFP-expressing TAMs. Red arrows indicate RFP-expressing blood vessels [53].

(TAMs). Further passage to transgenic nude mice ubiquitously expressing GFP resulted in tumors that acquired GFP stroma in addition to their RFP stroma, including CAFs and TAMs as well as blood vessels. The RFP stroma persisted in the tumors growing in the GFP mice [51].

To image metastasis by patient tumors in the fluorescent-protein expressing mouse models, the primary patient tumors acquired GFP-expressing stroma in GFP-expressing nude mice. Subsequent liver metastases, and disseminated peritoneal metastases maintained the stroma from the primary tumor, and possibly recruited additional GFP-expressing stroma, resulting in their very bright fluorescence [52].

In order to make a non-invasive imaging nude mouse model of patient tumors, with very bright GFP and RFP stroma, they were then orthotopically passaged to nontransgenic nude mice. It was possible to image the brightly fluorescent tumors non-invasively longitudinally as they progressed in the nontransgenic nude mice. This non-invasive imageable tumorgraft model will be valuable to screen for effective treatment options for individual patients with pancreatic cancer, as well as for the discovery of improved agents for this treatment-resistant disease [53] (Fig. 9.6).

9.5 Stroma Therapy

A cancer cell/stromal interaction in the tumor microenvironment (TME) was developed. GFP–expressing mice were transplanted in the foot pad with dual-color cancer cells labeled with GFP in the nucleus and in the cytoplasm. Non-invasively imaging visualized mitotic and apoptotic cancer cells, stromal cells interacting with the cancer cells, tumor vasculature, and tumor blood flow as well as drug response of both cancer and stromal cells was possible in the intact live animal in real time [23].

Harvested GFP-expressing macrophages from GFP mice with subcutaneous BxPC3-RFP tumor were defined as "tumor-educated macrophages." Macrophages were also harvested from transgenic GFP mice without tumors and identified as "naïve macrophages." The tumor-educated macrophages strongly promoted tumor progression compared to naïve macrophages. Tumor-educated macrophages can be a target for stromal therapy [54].

A powerful TME imaging model using the GFP transgenic mice and non-colored colon cancer cells which metastasize to the liver was developed [55]. As the liver metastasis progresses, brightly GFP-fluorescent CAFs, much brighter than the liver parenchymal cells are recruited by the liver metastasis. CAFs recruitment to the liver is readily imaged in live mice making them visual targets for stromal therapy [55].

9.6 Fluorescence-Guided UV Therapy

Fluorescence-guided ultraviolet C (UVC) irradiation inhibited the growth of murine melanoma expressing GFP in the ear of RFP mice using a non-invasive ear tumor imaging model developed previously [14]. UVC damaged blood vessels in the tumor which may be important for growth inhibition. Thus, UVC irradiation has a direct effect on melanoma growth as well as an anti-angiogenesis effect. These results suggest that fluorescence-guided UVC irradiation is a promising therapeutic strategy for melanoma [56]. UVC irradiation also suppressed RFP-expressing HT1080 human fibrosarcoma (HT1080), 143B human osteosarcoma (143B), Lewis lung carcinoma (LLC), and XPA-1 human pancreas cancer cell (XPA-1) growth in nude mice in a model of minimal residual cancer (MRC). No apparent side effects of UVC exposure were observed [57].

9.7 Conclusion

Fluorescent proteins have revolutionized biology by enabling what was formerly invisible to be seen clearly. Fluorescent proteins enable imaging of all aspects of cancer in living animals. This chapter highlights the potential therapeutic uses of imaging with fluorescent protein. Introducing the GFP gene in malignant tissue enabled fluorescence-guided surgery to resect all tumor nodules labeled

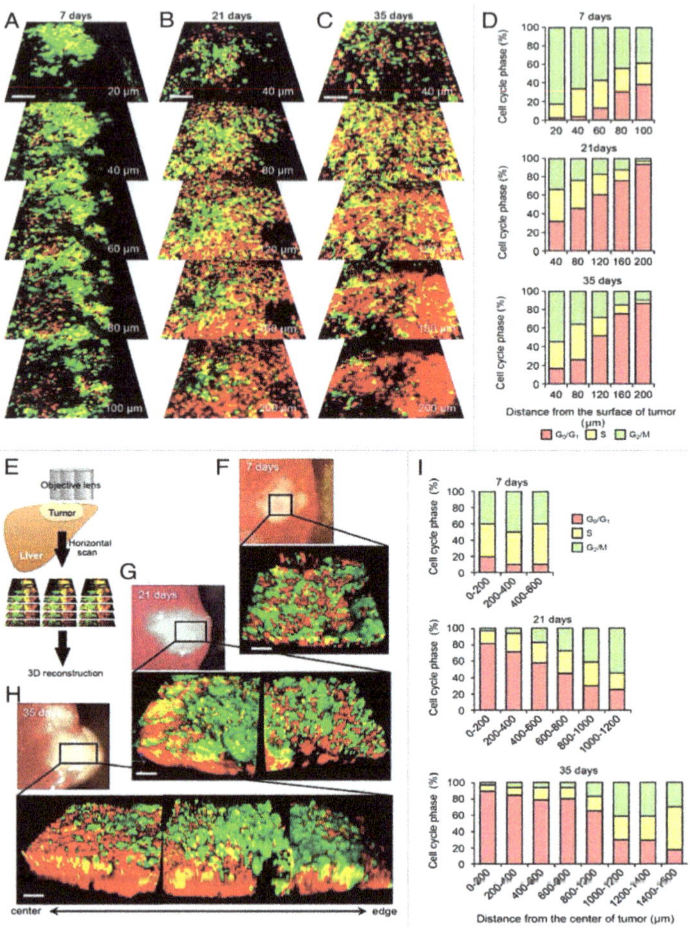

Figure 9.7 Cell cycle phase distribution of cancer cells at the tumor surface and center. (A–C) FUCCI-expressing MKN45 cells were implanted directly in the liver of nude mice and imaged at 7 d (A), 21 d (B), or 35 d (C). (D) Histograms show the cell cycle distribution in the tumor at 7 d (top), 21 d (middle), and 35 d (lower) after implantation. (E) Schematic diagram of in vivo CLSM imaging of different sized tumors. Tumors were scanned from the center to the edge. The scanned images were then three-dimensionally reconstructed. (F–H) Representative 3D reconstruction images of a nascent tumor at 7 d after cancer cell implantation (F), 21 d (G) and 35 d (H) after implantation. (I) Histogram shows the distribution of FUCCI-expressing cells at different distances from the center. The number of cells in each cell cycle phase were assessed by counting the number of cells of each color at the indicated time points and depth. The percentages of cells in the G_2/M, S, and G_0/G_1 phases of the cell cycle are shown (E and I). Data are means (each group for $n = 5$). Scale bars: 100 μm [59].

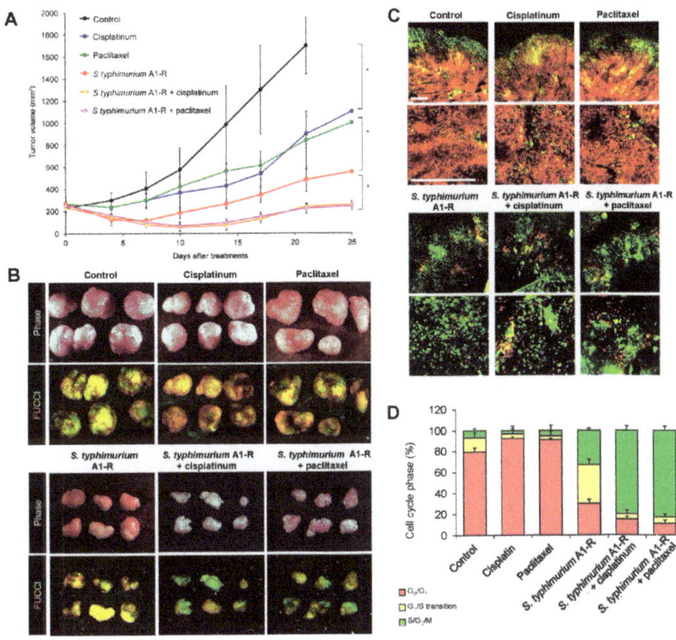

Figure 9.8 *S. typhimurium* A1-R-decoyed tumors became sensitive to chemotherapy. FUCCI-expressing MKN45 cells (5 × 106 cells/mouse) were injected subcutaneously into the left flank of nude mouse. When the tumors reached approximately 8 mm in diameter (tumor volume, 300 mm^3), mice were administered iv *S. typhimuium* A1-R alone, or in combination with cisplatinum (4 mg/kg ip) or paclitaxel (5 mg/kg ip) for 5 cycles every 3 d. (A) Growth curves of tumors derived from FUCCI-expressing MKN45 cells after treatment with chemotherapy, *S. typhimuium* A1-R or the combination of *S. typhimuium* A1-R and chemotherapy. The difference between control and cisplatinum-treated, $P < 0.01$; the difference between control and paclitaxel-treated, $P < 0.05$; the difference between control and *S. typhimuium* A1-R, $P < 0.05$; the difference between control and the combination of *S. typhimuium* A1-R and cisplatinum, $P < 0.01$; the difference between control and the combination of *S. typhimuium* A1-R and paclitaxel, $P < 0.01$. (B) Macroscopic photographs of FUCCI-expressing tumors, untreated control, *S. typhimuium* A1-R-treated, cisplatinum-treated, paclitaxel-treated, or treated with the combination of *S. typhimuium* A1-R and either cisplatinum or paclitaxel (right). Scale bars: 10 mm. (C) Representative images of cross-sections of FUCCI-expressing MKN45 subcutaneous tumors, untreated control, *S. typhimuium* A1-R-treated, cisplatinum-treated, paclitaxel-treated, or treated with the combination of *S. typhimuium* A1-R and either cisplainum or paclitaxel. (D) Histogram shows cell cycle phase of FUCCI-expressing MKN45 subcutaneous tumors, including untreated control, *S. typhimuium* A1-R-treated, cisplatinum-treated, paclitaxel-treated, or the combination of *S. typhimuium* A1-R and either cisplatinum or paclitaxel. Scale bars: 500 mm [60].

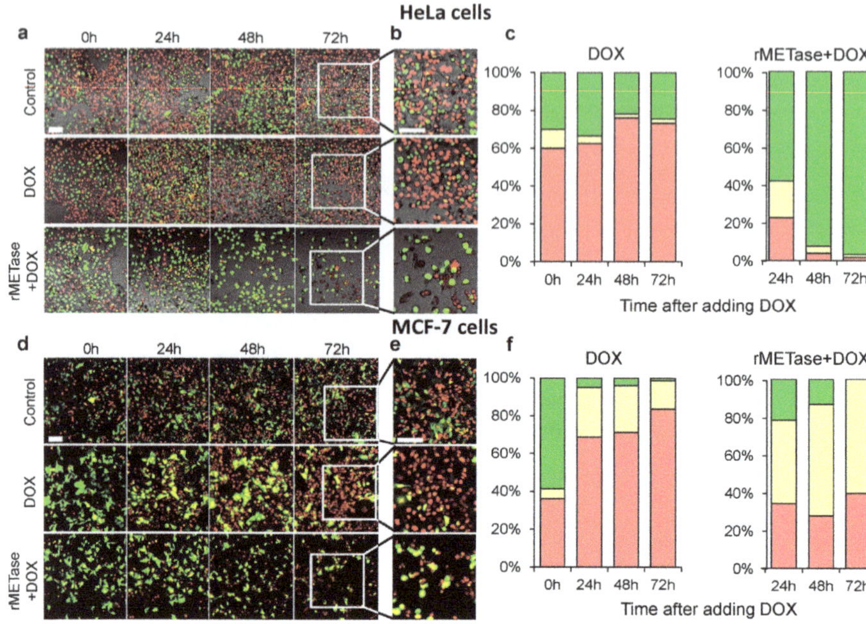

Figure 9.9 FUCCI cell cycle analysis during chemotherapy with and without rMETase. After overnight culture, HeLa cells (a, b, and c) and MCF-7 cells (d, e, and f) were treated with 0.25 unit/ml rMETase for 48 hours and, then were treated with 0.5 μg/ml doxorubicin (HeLa cells) or 2.5 μg/ml doxorubicin (MCF-7) for 72 hours. For conventional chemotherapy, after culture for 48 hours, HeLa cells (a, b, and c) and MCF-7 cells (d, e, and f) were treated with 0.5 μg/ml DOX (HeLa cells) or 2.5 μg/ml DOX (MCF-7) for 72 hours. (a, b, d, e) Representative images acquired with the FV-1000 are shown. (c, f) Histograms show the percentages of cells G_1(red), early-S (yellow), or late-S/G_2/M (green). Cells at each cell cycle phase were quantitatively assessed by counting the number of cells with each color. $N =$ 5 experiments were analyzed. Scale bars: 50 μm [61].

with GFP and increases survival of tumor-bearing animals. Double amino acid auxotrophs of *Salmonella typhimurium* labeled with GFP have been used to target and eradicate tumors and metastasis labeled with RFP in metastatic nude mouse models of human cancer. Fluorescence-guided ultraviolet C (UVC) irradiation inhibited the growth of murine melanoma expressing GFP in the ear of RFP transgenic mice. The efficacy was readily visible using noninvasive imaging. UVC irradiation also inhibited growth of 143B human osteosarcoma cells, HT-1080 human fibrosarcoma cells, Lewis lung carcinoma (LLC), and XPA-1 human pancreatic cancer. Primary patient tumors acquired fluorescent protein-expressing stroma from passage in GFP and RFP transgenic nude mice. The tumors, with very bright fluorescent stroma, were then orthotopically passaged to nontransgenic nude mice where it was possible to image the brightly fluorescent tumors noninvasively longitudinally as they progressed in the nontransgenic nude mice. This non-invasive imageable tumorgraft model will be valuable to screen for effective treatment options for individual patients. Many more applications of fluorescent proteins will be developed in the future.

Previous use of cell cycle imaging where the cells change fluorescent color in real time with cell cycle phase change: red for G_1/G_0, green/yellow for $S/G_2/M$ in the fluorescent ubiquitination-based cell cycle indicator (FUCCI) [58]. FUCCI has shown only the surface cells of a tumor are dividing with the majority quiescent [59] (Fig. 9.7). FUCCI has also shown that the tumor-targeting bacteria, *Salmonella typhimurium* A1-R, described above, can decoy the quiescent cells to cycle and become chemosensitive [60] (Fig. 9.8) and that methionine deprivation using recombinant methioninase (rMETase) can trap cancer cells in S/G_2, where they become chemosensitive [61] (Fig. 9.9).

References

1. Prasher, D. C., Eckenrode, V. K., Ward, W. W., Prendergast, F. G., and Cormier, M. J. (1992). Primary structure of the *Aequorea victoria* green fluorescent protein. *Gene,* **111**, 229–233.
2. Chalfie, M., Tu, Y., Euskirchen, G., Ward, W. W., and Prasher, D. C. (1994). Green fluorescent protein as a marker for gene expression. *Science,* **263**, 802–805.
3. Cheng, L., Fu, J., Tsukamoto, A., and Hawley, R. G. (1996). Use of green fluorescent protein variants to monitor gene transfer and expression in mammalian cells. *Nat. Biotechnol.*, **14**, 606–609.
4. Cody, C. W., Prasher, D. C., Westler, W. M., Prendergast, F. G., and Ward, W. W. (1993). Chemical structure of the hexapeptide chromophore of the Aequorea green fluorescent protein. *Biochemistry,* **32**, 1212–1218.

5. Yang, F., Moss, L. G., and Phillips, G. N., Jr. (1996). The molecular structure of green fluorescent protein. *Nat. Biotechnol.*, **14**, 1246–1251.
6. Morin, J., and Hastings, J. (1971). Energy transfer in a bioluminescent system. *J. Cell Physiol.*, **77**, 313–318.
7. Cormack, B., Valdivia, R., and Falkow, S. (1996). FACS-optimized mutants of the green fluorescent protein (GFP). *Gene*, **173**, 33–38.
8. Crameri, A., Whitehorn, E. A., Tate, E., and Stemmer, W. P. (1996). Improved green fluorescent protein by molecular evolution using DNA shuffling. *Nat. Biotechnol.*, **14**, 315–319.
9. Delagrave, S., Hawtin, R. E., Silva, C. M, Yang, M. M., and Youvan, D. C. (1995). Red-shifted excitation mutants of the green fluorescent protein. *Biotechnology (N.Y.)*, **13**, 151–154.
10. Heim, R., Cubitt, A. B., and Tsien, R. Y. (1995). Improved green fluorescence. *Nature*, **373**, 663–664.
11. Zolotukhin, S., Potter, M., Hauswirth, W. W., Guy, J., and Muzyczka, N. (1996). A 'humanized' green fluorescent protein cDNA adapted for high-level expression in mammalian cells. *J. Virol.*, **70**, 4646–4654.
12. Gross, L. A., Baird, G. S., Hoffman, R. C., Baldridge, K. K., and Tsien, R. Y. (2000). The structure of the chromophore within DsRed, a red fluorescent protein from coral. *Proc. Natl. Acad. Sci. U.S.A.*, **97**, 11990–11995.
13. Fradkov, A. F., Chen, Y., Ding, L., Barsova, E. V., Matz, M. V., and Lukyanov, S. A. (2000). Novel fluorescent protein from Discosoma coral and its mutants possesses a unique far-red fluorescence. *FEBS Lett.*, **479**, 127–130.
14. Yamamoto, N., Jiang, P., Yang, M., Xu, M., Yamauchi, K., Tsuchiya, H., Tomita, K., Wahl, G. M., Moossa, A. R., and Hoffman, R. M. (2004). Cellular dynamics visualized in live cells *in vitro* and *in vivo* by differential dual-color nuclear cytoplasmic fluorescent-protein expression. *Cancer Res.*, **64**, 4251–4256.
15. Chishima, T., Miyagi, Y., Wang, X., Yamaoka, H., Shimada, H., Moossa, A. R., and Hoffman, R. M. (1997). Cancer invasion and micrometastasis visualized in live tissue by green fluorescent protein expression. *Cancer Res.*, **57**, 2042–2047.
16. Wang, W., Wyckoff, J. B., Frohlich, V. C., Oleynikov, Y., Hüttelmaier, S., Zavadil, J., et al. (2002). Single cell behavior in metastatic primary mammary tumors correlated with gene expression patterns revealed by molecular profiling. *Cancer Res.*, **62**, 6278–6288.

17. Goswami, S., Sahai, E., Wyckoff, J. B., Cammer, M., Cox, D., Pixley, F. J., et al. (2005). Macrophages promote the invasion of breast carcinoma cells via a colony-stimulating factor- 1/epidermal growth factor paracrine loop. *Cancer Res.*, **65**, 5278–5283.
18. Mook, O. R. F., Van Marle, J. V., Vreeling-Sindelarova, H., et al. (2003). Visualization of early events in tumor formation of eGFP transfected rat colon cancer cells in liver. *Hepatology*, **38**, 295–304.
19. Naumov, G. N., Wilson, S. M., MacDonald, I. C., Schmidt, E. E., Morris, V. L., Groom, A. C., Hoffman, R. M., and Chambers, A. F. (1999). Cellular expression of green fluorescent protein, coupled with high-resolution in vivo videomicroscopy, to monitor steps in tumor metastasis. *J. Cell Sci.*, **112**, 1835–1842.
20. Yamauchi, K., Yang, M., Jiang, P., Yamamoto, N., Xu, M., Amoh, Y., Tsuji, K., Bouvet, M., Tsuchiya, H., Tomita, K., Moossa, A. R., and Hoffman, R. M. (2005). Real-time *in vivo* dual-color imaging of intracapillary cancer cell and nucleus deformation and migration. *Cancer Res.*, **65**, 4246–4252.
21. Yamauchi, K., Yang, M., Jiang, P., Xu, M., Yamamoto, N., Tsuchiya, H., Tomita, K., Moossa, A. R., Bouvet, M., and Hoffman, R. M. (2006). Development of real-time subcellular dynamic multicolor imaging of cancer-cell trafficking in live mice with a variable-magnification whole-mouse imaging system. *Cancer Res.*, **66**, 4208–4214.
22. Hayashi, K., Jiang, P., Yamauchi, K., Yamamoto, N., Tsuchiya, H., Tomita, K., Tomita, K., Moossa, A. R., Bouvet, M., and Hoffman, R. M. (2007). Real-time imaging of tumor-cell shedding and trafficking in lymphatic channels. *Cancer Res.*, **67**, 8223–8228.
23. Yang, M., Jiang, P., and Hoffman, R. M. (2007). Whole-body subcellular multicolor imaging of tumor-host interaction and drug response in real time. *Cancer Res.*, **67**, 5195–5200.
24. Bouvet, M., Tsuji, K., Yang, M., Jiang, P., Moossa, A. R., and Hoffman, R. M. (2006). *In vivo* color-coded imaging of the interaction of colon cancer cells and splenocytes in the formation of liver metastases. *Cancer Res.*, **66**, 11293–11297.
25. Yamauchi, K., Yang, M., Hayashi, K., Jiang, P., Yamamoto, N., Tsuchiya, H., Tomita, K., Moossa, A. R., Bouvet, M., and Hoffman, R. M. (2008). Induction of cancer metastasis by cyclophosphamide pretreatment of host mice: an opposite effect of chemotherapy. *Cancer Res.*, **68**, 516–520.

26. Kimura, H., Hayashi, K., Yamauchi, K., Yamamoto, N., Tsuchiya, H., Tomita, K., Kishimoto, H., Bouvet, M., and Hoffman, R. M. (2010). Real-time imaging of single cancer-cell dynamics of lung metastasis. *J. Cell. Biochem.*, **109**, 58–64.
27. Kishimoto, H., Zhao, M., Hayashi, K., Urata, Y., Tanaka, N., Fujiwara, T., Penman, S., and Hoffman, R. M. (2009). *In vivo* internal tumor illumination by telomerase-dependent adenoviral GFP for precise surgical navigation. *Proc. Natl. Acad. Sci. U.S.A.*, **106**, 14514–14517.
28. Kishimoto, H., Urata, Y., Tanaka, N., Fujiwara, T., and Hoffman, R. M. (2009). Selective metastatic tumor labeling with green fluorescent protein and killing by systemic administration of telomerase-dependent adenoviruses. *Mol. Cancer Ther.*, **8**, 3001–3008.
29. Kishimoto, H., Aki, R., Urata, Y., Bouvet, M., Momiyama, M., Tanaka, N., Fujiwara, T., and Hoffman, R. M. (2011). Tumor-selective adenoviral-mediated GFP genetic labeling of human cancer in the live mouse reports future recurrence after resection. *Cell Cycle*, **10**, 2737–2741.
30. Tran Cao, H. S., Kaushal, S., Menen, R. S., Metildi, C. A., Lee, C., Snyder, C. S., Talamini, M. A., Hoffman, R. M., and Bouvet, M. (2011). Submillimeter-resolution fluorescence laparoscopy of pancreatic cancer in a carcinomatosis mouse model visualizes metastases not seen with standard laparoscopy. *J. Laparoendosc. Adv. Surg. Tech. A*, **21**, 485–489.
31. Bouvet, M., and Hoffman, R. M. (2011). Glowing tumors make for better detection and resection. *Sci. Transl. Med.*, **3**, 110sf10.
32. Metildi, C. A., Kaushal, S., Lee, C., Hardamon, C. R., Snyder, C. S., Luiken, G. A., Talamini, M. A., Hoffman, R. M., and Bouvet, M. (2012). An LED light source and novel fluorophore combinations improve fluorescence laparoscopic detection of metastatic pancreatic cancer in orthotopic mouse models. *J. Am. Coll. Surg.*, **214**, 997–1007.
33. Metildi, C. A., Kaushal, S., Hardamon, C. R., Snyder, C. S., Pu, M., Messer, K. S., Talamini, M. A., Hoffman, R. M., and Bouvet, M. (2012). Fluorescence-guided surgery allows for more complete resection of pancreatic cancer, resulting in longer disease-free survival compared with standard surgery in orthotopic mouse models. *J. Am. Coll. Surg.*, **215**, 126–135.
34. Stummer, W., Pichlmeier, U., Meinel, T., Wiestler, O. D., Zanella, F., Reulen, H. J., and ALA-Glioma Study Group (2006). Fluorescence-guided surgery with 5-aminolevulinic acid for resection of malignant glioma: A randomised controlled multicentre phase III trial. *Lancet Oncol.*, **7**, 392–401.

35. Urano, Y., Sakabe, M., Kosaka, N., Ogawa, M., Mitsunaga, M., Asanuma, D., Kamiya, M., Young, M. R., Nagano, T., Choyke, P. L., and Kobayashi, H. (2011). Rapid cancer detection by topically spraying a ϒ-glutamyltranspeptidase–activated, fluorescent probe. *Sci. Transl. Med.*, **3**, 110ra119.

36. Bouvet, M., Metildi, C. A., and Hoffman, R. M. Glowing tumors for surgery of today and tomorrow. *J. Assoc. Coll. Surg.*, in press.

37. Nauts, H. C., Fowler, G. A., and Bogatko, F. H. (1953). A review of the influence of bacterial infection and of bacterial products (Coley's toxins) on malignant tumors in man; a critical analysis of 30 inoperable cases treated by Coley's mixed toxins, in which diagnosis was confirmed by microscopic examination selected for special study. *Acta Med. Scand. Suppl.*, **145**, 1–102.

38. Hoffman, R. M. (2012). The preclinical discovery of bacterial therapy for the treatment of metastatic cancer with unique advantages. *Expert Opin. Drug Discov.*, **7**, 73–83.

39. Zhao, M., Yang, M., Li, X.-M., Jiang, P., Baranov, E., Li, S., Xu, M., and Hoffman, R. M. (2005). Tumor-targeting bacterial therapy with amino acid auxotrophs of GFP-expressing *Salmonella typhimurium*. *Proc. Natl. Acad. Sci. U.S.A.*, **102**, 755–760.

40. Zhao, M., Yang, M., Ma, H., Li, X., Tan, X., Li, S., Yang, Z., and Hoffman, R. M. (2006). Targeted therapy with a *Salmonella typhimurium* leucine-arginine auxotroph cures orthotopic human breast tumors in nude mice. *Cancer Res.*, **66**, 7647–7652.

41. Zhao, M., Geller, J., Ma, H., Yang, M., Penman, S., and Hoffman, R. M. (2007). Monotherapy with a tumor-targeting mutant of *Salmonella typhimurium* cures orthotopic metastatic mouse models of human prostate cancer. *Proc. Natl. Acad. Sci. U.S.A.*, **104**, 10170–10174.

42. Nagakura, C., Hayashi, K., Zhao, M., Yamauchi, K., Yamamoto, N., Tsuchiya, H., Tomita, K., Bouvet, M., and Hoffman, R. M. (2009). Efficacy of a genetically-modified *Salmonella typhimurium* in an orthotopic human pancreatic cancer in nude mice. *Anticancer Res.*, **29**, 1873–1878.

43. Hayashi, K., Zhao, M., Yamauchi, K., Yamamoto, N., Tsuchiya, H., Tomita, K., Kishimoto, H., Bouvet, M., and Hoffman, R. M. (2009). Systemic targeting of primary bone tumor and lung metastasis of high-grade osteosarcoma in nude mice with a tumor-selective strain of *Salmonella typhimurium*. *Cell Cycle*, **8**, 870–875.

44. Yam, C., Zhao, M., Hayashi, K., Ma, H., Kishimoto, H., McElroy, M., Bouvet, M., and Hoffman, R. M. (2010). Monotherapy with a tumor-targeting mutant of *S. typhimurium* inhibits liver metastasis in a mouse model of pancreatic cancer. *J. Surg. Res.*, **164**, 248–255.

45. Kimura, H., Zhang, L., Zhao, M., Hayashi, K., Tsuchiya, H., Tomita, K., Bouvet, M., Wessels, J., and Hoffman, R. M. (2010). Targeted therapy of spinal cord glioma with a genetically-modified *Salmonella typhimurium*. *Cell Prolif.*, **43**, 41–48.

46. Liu, F., Zhang, L., Hoffman, R. M., and Zhao, M. (2010). Vessel destruction by tumor-targeting *Salmonella typhimurium* A1-R is enhanced by high tumor vascularity. *Cell Cycle*, **9**, 4518–4524.

47. Hoffman, R. M., and Zhao, M. (2006). Whole-body imaging of bacterial infection and antibiotic response. *Nat. Protoc.*, **1**(6), 2988–2994.

48. Forbes, N. S. (2010). Engineering the perfect (bacterial) cancer therapy. *Nat. Rev. Cancer*, **10**, 785–794.

49. Kim, M. P., Evans, D. B., Wang, H., Abbruzzese, J. L., Fleming, J. B., and Gallick, G. E. (2009). Generation of orthotopic and heterotopic human pancreatic cancer xenografts in immunodeficient mice. *Nat. Protoc.*, **4**, 1670–1680.

50. Kim, M. P., Truty, M. J., Choi, W., Kang, Y., Chopin-Lally, X., Gallick, G. E., Wang, H., McConkey, D. J., Hwang, R., Logsdon, C., Abbruzzesse, J., and Fleming, J. B. (2011). Molecular profiling of direct xenograft tumors established from human pancreatic adenocarcinoma after neoadjuvant therapy. *Ann. Surg. Oncol.*, **19**, S395–S403.

51. Suetsugu, A., Katz, M., Fleming, J., Truty, M., Thomas, R., Moriwaki, H., Bouvet, M., Saji, S., and Hoffman, R. M. (2012). Multi-color palette of fluorescent proteins for imaging the tumor microenvironment of orthotopic tumorgraft mouse models of clinical pancreatic cancer specimens. *J. Cell. Biochem.*, **113**, 2290–2295.

52. Suetsugu, A., Katz, M., Fleming, J., Truty, M., Thomas, R., Saji, S., Moriwaki, H., Bouvet, M., and Hoffman, R. M. (2012). Imageable fluorescent metastasis resulting in transgenic GFP mice orthotopically implanted with human-patient primary pancreatic cancer specimens. *Anticancer Res.*, **32**, 1175–1180.

53. Suetsugu, A., Katz, M., Fleming, J., Truty, M., Thomas, R., Saji, S., Moriwaki, H., Bouvet, M., and Hoffman, R. M. (2012). Non-invasive fluorescent-protein imaging of orthotopic pancreatic-cancer-patient tumorgraft progression in nude mice. *Anticancer Res.*, **32**, 3063–3068.

54. Menen, R. S., Hassanein, M. K., Momiyama, M., Suetsugu, A., Moossa, A. R., Hoffman, R. M., and Bouvet, M. (2012). Tumor-educated macrophages promote tumor growth and peritoneal metastasis in an orthotopic nude mouse model of human pancreatic cancer. *In Vivo*, **26**, 565–569.

55. Suetsugu, A., Osawa, Y., Nagaki, M., Saji, S., Moriwaki, H., Bouvet, M., and Hoffman, R. M. (2011). Imaging the recruitment of cancer-associated fibroblasts by liver-metastatic colon cancer. *J. Cell. Biochem.*, **112**, 949–953.

56. Tsai, M.-H., Aki, R., Amoh, Y., Hoffman, R. M., Katsuoka, K., Kimura, H., Lee, C., and Chang, C.-H. (2010). GFP-fluorescence-guided UVC irradiation inhibits melanoma growth and angiogenesis in nude mice. *Anticancer Res.*, **30**, 3291–3294.

57. Kimura, H., Lee, C., Hayashi, K., Yamauchi, K., Yamamoto, N., Tsuchiya, H., Tomita, K., Bouvet, M., and Hoffman, R. M. (2010). UV light killing efficacy of fluorescent protein-expressing cancer cells *in vitro* and *in vivo*. *J. Cell. Biochem.*, **110**, 1439–1446.

58. Sakaue-Sawano, A., Kurokawa, H., Morimura, T., Hanyu, A., Hama, H., Osawa, H., Kashiwagi, S., Fukami, K., Miyata, T., Miyoshi, H., Imamura, T., Ogawa, M., Masai, H., and Miyawaki, A. (2008). Visualizing spatiotemporal dynamics of multicellular cell-cycle progression. *Cell*, **132**, 487–498.

59. Yano, S., Li, S., Han, Q., Tan, Y., Bouvet, M., Fujiwara, T., and Hoffman, R. M. (2014). Selective methioninase-induced trap of cancer cells in S/G_2 phase visualized by FUCCI imaging confers chemosensitivity. *Oncotarget*, **5**, 8729–8736.

60. Yano, S., Zhang, Y., Zhao, M., Hiroshima, Y., Miwa, S., Uehara, F., Kishimoto, H., Tazawa, H., Bouvet, M., Fujiwara, T., and Hoffman, R. M. (2014). Tumor-targeting *Salmonella typhimurium* A1-R decoys quiescent cancer cells to cycle as visualized by FUCCI imaging and become sensitive to chemotherapy. *Cell Cycle*, **13**(24), 3958–3963.

61. Yano, S., Li, S., Han, Q., Tan, Y., Bouvet, M., Fujiwara, T., and Hoffman, R. M. (2014). Selective methioninase-induced trap of cancer cells in S/G_2 phase visualized by FUCCI imaging confers chemosensitivity. *Oncotarget*, **5**, 8729–8736.

62. Yang, M., Li, L., Jiang, P., Moossa, A. R., Penman, S., and Hoffman, R. M. (2003). Dual-color fluorescence imaging distinguishes tumor cells from induced host angiogenic vessels and stromal cells. *Proc. Natl. Acad. Sci. U.S.A.*, **100**, 14259–14262.

63. Yamauchi, K., Tome, Y., Yamamoto, N., Hayashi, K., Kimura, H., Tsuchiya, H., Tomita, K., Bouvet, M., and Hoffman, R. M. (2012). Color-coded real-time subcellular fluorescence imaging of the interaction between cancer and host cells in live mice. *Anticancer Res.*, **32**, 39–43.

Chapter 10

Fluorescence Lifetime Imaging in Living Cells

Chittanon Buranachai,[a,c] John P. Eichorst,[a] Kei Wei Teng,[a] and Robert M. Clegg[a,b]

[a] *Center for Biophysics and Computational Biology, University of Illinois, 600 South Mathews Avenue, Urbana, Illinois 61801, USA*
[b] *Department of Physics, University of Illinois, 1110 West Green Street, Urbana, IL 61801, USA*
[c] *Department of Physics and Center of Excellence for Trace Analysis and Biosensor, Faculty of Science, Prince of Songkla University, 15 Karnjanavanich Road, Hat Yai, Songkhla 90110, Thailand*
chittanon.buranachai@gmail.com

Fluorescence lifetime-resolved imaging (FLI) records fluorescence lifetimes in an image and the spectroscopic information related to the lifetime of the fluorescence decay is displayed at every pixel of the image. FLI measurements supplement normal intensity fluorescence imaging measurements; the intensity at every pixel is also acquired during a FLI experiment. Fluorescence microscopy has become a common measurement in every field of cellular biology. Advanced, sophisticated techniques and instrumentation are commercially available, which reveal detailed morphological information with high spatial resolution. The lifetime-resolved measurement of FLI combines the spatial resolution of a fluorescence

---

Fundamentals of Fluorescence Imaging
Edited by Guy Cox

ISBN 978-981-4774-85-7 (Hardcover), 978-1-351-12940-4 (eBook)
www.jennystanford.com

image with the temporal dimension of a nanosecond fluorescence lifetime-resolved spectroscopic measurement.

10.1 Introduction

Standard non-imaging fluorescence lifetime measurements are nowadays easily carried out in a cuvette. This has been done for many years, and commercial equipment is available. An extensive literature on applications of fluorescence lifetime measurements in biology exists, and nanosecond time-resolved analysis of fluorescence emission has yielded a wealth of critical information about biological molecules and biological systems. Until recently, lifetime measurements were available only in cuvette-based (single-channel) experiments, where no spatial information is available. The purpose of fluorescence lifetime-resolved imaging (FLI) is to provide this lifetime information in an imaging environment. The majority of publications dealing with FLI have been performed in optical microscopes; therefore, the technique is often referred to as FLIM, where the "M" stands for "microscopy." However, lifetime-resolved imaging has been carried out on macroscopic objects and in endoscopes, as well as in microscopes. Therefore, in order to stress the broad applications of FLI, we use the acronym FLI. The combination of lifetime, intensity, and spatial resolution into a single fluorescence image provides unique, novel opportunities for the experimenter. The more sophisticated information content of time-resolved fluorescence measurements extends significantly the capability of the investigator to reveal molecular scale information about complex biological samples. The aim of this chapter is to acquaint the biologist with the wealth of information offered by time-resolved fluorescence imaging, to describe how the experiments are carried out, and to explain the basis of the different methods of analysis.

Why would a biologist want to do FLI? The rate of fluorescence decay of a fluorophore is a characteristic of the specific fluorophore and is sensitive to its molecular environment. This sensitivity to the molecular environment is one of the major reasons for making fluorescence measurements. Often the fluorophore can be considered a "spy," reporting to the experimenter detailed information about the local environment of the fluorophore. In the case of FLI, the location of the fluorophore in a biological sample is also available. For instance, changes in pH, temperature, ion concentrations, or the presence of a fluorescence quencher, such as O_2, significantly affect fluorescence lifetimes [1–3]. FLI can differentiate and contrast fluorescent species within an image from their dissimilar lifetimes, even when their emission spectra overlap [4, 5] thus, FLI is an effective way to reveal heterogeneous environments within an imaged sample. Because the lifetime of a fluorophore is independent of its concentration, FLI can distinguish changes in concentration from changes in quantum yield, QY. There are two ways to define the fluorescence quantum yield: from lifetimes, $QY \equiv QY_\tau = \tau_{meas}/\tau_0$, where $\tau_{meas} \equiv$ measured fluorescence lifetime, and $\tau_0 \equiv$ natural radiative lifetime (see Section 2.1); from intensity, $QY \equiv QY_{inten} =$ (# photons emitted) / (# photons absorbed). It is true for any single fluorescent molecule $QY_{inten} = QY_\tau$; however, it is often difficult, or impossible to interpret intensity measurements. A decrease in the intensity could be due either to a decrease in the concentration of fluorophores, or to a quenching effect. Similarly, intensity based Förster resonance energy transfer (FRET) experiments are common in biological microscopy (Chapter 12), but intensity based experiments are often difficult to interpret because controls to determine concentrations of donor and acceptor molecules are difficult, and sometimes impossible, to carry out. FLI circumvents this difficulty. Lifetimes are independent of concentration, and each component (donor alone, and donor–acceptor pairs) can easily be distinguished by their disparate lifetimes. Lifetime measurements of FRET efficiency are also highly reliable because no assumption must be made about the percentage of labeling of the donor/acceptor pair, as in the case of intensity based measurements [6, 7].

We have selected a few publications and a few results from our laboratory to illustrate some of the applications of FLI. Such a short bibliography does not do justice to the many excellent FLI studies. These applications are only selected as examples. We apologize to the extended community of FLI investigators; this short chapter is only meant as an introduction to FLI methods, and there is only room for a few examples.

10.2 Background

10.2.1 What Is the Meaning of the Fluorescence Lifetime?

The fluorescence lifetime of a fluorophore is the mean duration that the molecule spends in the electronically excited state before exiting the excited state through one of the available pathways (Fig. 10.1). The reciprocal of the measured dwell time τ in the excited state is the sum over all the deexcitation rates ($1/\tau = \sum_i k_i$). As a result, the fluorescence lifetime is dependent on which relaxation processes are present. A single molecular species in a defined molecular environment will decay as a single exponential; that is, the time dependence of a single fluorescence species is proportional to $\exp(-t/\tau)$, where τ is given above. The same molecular species of

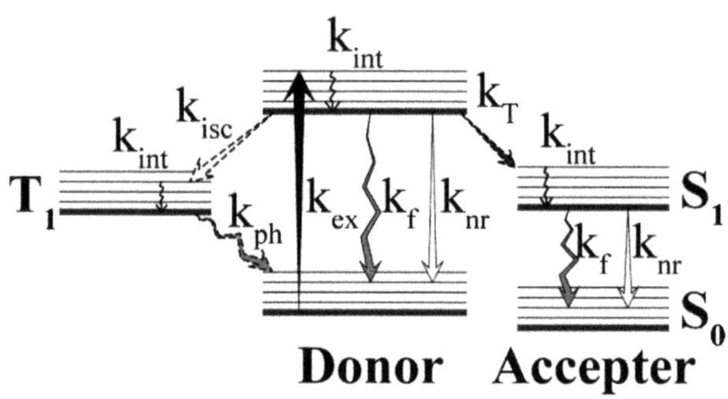

Figure 10.1 The simplified Jablonski energy diagram showing multiple relaxation pathways to exit the excited state. The relaxation pathways shown in the figure have rates of (i) fluorescence emission ($k_f \cong 10^9 s^{-1}$), (ii) nonradiative transition (k_{nr} varies over a wide range depending on the type of nonradiative processes), (iii) radiative energy transfer to a proximate FRET acceptor molecule ($k_T \cong 10^9 s^{-1}$ depends on the intermolecular distance), (iv) inter-system crossing from the first excited singlet state S_1 to the first excited triplet state T_1 ($k_{isc} \cong 10^6 s^{-1}$), and (v) the internal conversion ($k_{int} \cong 10^{12} s^{-1} - 10^{13} s^{-1}$). The process of absorption of a photon takes place extremely fast time scale ($k_{ex} \cong 10^{15} s^{-1}$) and phosphorescence occurs in a much slower time scale ($k_{ex} \cong 10^6 s^{-1} - 1\ s^{-1}$). The sum of the relaxation rates from the S_1 state is the inverse of the detected fluorescence lifetime ($1/\tau_F = \sum_i k_i$). Therefore, the measured fluorescence lifetime is a measure of the rates of the pathways.

fluorophore will have different lifetimes depending on its molecular surroundings, and this sensitivity is the major reason for measuring fluorescence lifetimes. All kinetic pathways out of the excited state compete in a first-order kinetic fashion; therefore, by measuring the change in the rate of the fluorescence decay, we can monitor changes in the rate (or presence) of any of the other (nonfluorescence) pathways (for instance, FRET or dynamic quenching).

This kinetic aspect of fluorescence lifetimes, together with the spectral distribution and polarization of the emission, is the basis of all fluorescence measurements, and is directly available to the experimenter when measuring the fluorescence lifetime, rather than the indirect effect of lifetime changes on the intensity. Fluorescence intensities measure only the steady state of emission. The most straightforward means to reveal the presence of molecular dynamic processes affecting a fluorescence signal, or to detect changes in these processes, is to measure the dynamics of the fluorescence decay. The principal incentive for making FLI measurements is to extend the information conveyed by the normal fluorescence intensity at every pixel of a fluorescence image with the mechanistic insight embodied in the dynamics of fluorescence.

10.2.2 The FLI Measurement: Basics of the Time Domain and Frequency Domain

Fluorescence lifetimes are typically measured in two ways (Fig. 10.2): in the time domain or the frequency domain. The time domain is a direct measurement of the time decay process; data points are acquired at specific times following the excitation of the sample, usually with a very short pulse of light. The frequency domain carries out the measurement by repetitively modulating the light continuously at a very high frequency, and detecting the frequency dispersion of the signal. Both measurement modes can be carried out in scanning mode (such as a scanning confocal microscope, where a focused excitation beam is scanned across the sample, collecting the image points sequentially) or in wide-field mode (such as in a normal wide-field fluorescence microscope, where all locations of the imaged sample are excited and detected simultaneously).

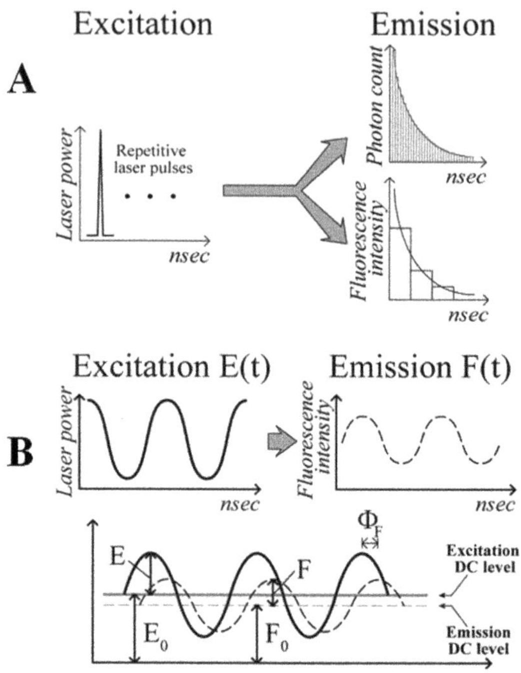

Figure 10.2 Lifetime measurement in time domain (A) and in frequency domain (B). In the time-domain, a series of repetitive excitation light pulses excites the sample. An exponential decay function (or summation of several exponential functions) is fitted to either (i) the histogram of the number of photons detected at different times following the light pulse, as shown in the top right panel of (A), or (ii) the integrated (average) signal within a few pre selected time bins, as shown in the bottom right panel of (A). The first type of acquisition is represented by single point scanning FLIM using TCSPC. The second type is usually carried out with wide-field excitation using a fast time-gated image intensifier. In the frequency-domain, (B), a repetitive modulated form of excitation light (the sinusoidal form is shown in (B) for simplicity) is used to excite the sample, and the modulated fluorescence emission is usually acquired in the homodyne mode for wide-field excitation, or in the heterodyne mode for point scanning. Lifetimes are calculated from the phase shift ϕ_F and modulation ratio M, defined in (B) and Eqs. 10.1 and 10.2. With kind permission from Springer Science+Business Media.

10.2.3 Time Domain

Figure 10.2A is a schematic of a time domain experiment. In the scanning mode, the data acquisition is carried out point by point at every location of a sample by scanning the sample with a focused light beam that is repetitively pulsed (the pulse widths are in the range of a few hundreds of femtoseconds to a few nanoseconds [2, 8, 9]). The dynamic fluorescence response at every pixel is averaged over many pulses. The image acquisition is performed confocally, either in a normal confocal microscope [10], or using two photon excitation with scanning [8, 9, 11]. The emitted photons are counted at selected time intervals (time bins) following each light pulse. Histograms of the averages of photon counts per time bin represent the decay of fluorescence. This method has recently become more accessible (and commercially available) due to the advent of reliable, high repetition (10–100 MHz) pulsed lasers with very short pulses (e.g., Ti–sapphire lasers with tens to hundreds of femtosecond pulses) and very fast data acquisition (picosecond resolution) that can handle high repetition rates [9, 12]. The use of parallel detectors (multi-anode) in the time domain with TCSPC, which require the development of the technology to acquire the data simultaneously from the multiple detectors, is also an attractive alternative [13]. The data analysis for the TCSPC modes of data acquisition is usually carried out by iterative nonlinear regression techniques, fitting the histogram of the number of photons arriving at the detector at different time delays to exponential functions.

Other methods in time-domain FLI have been developed. By acquiring data in wide-field mode integrated over just a few pre-selected time intervals, the data acquisition and analysis can be considerably speeded up [14, 15]. All pixels of the image are collected simultaneously by carrying out the data acquisition in a wide-field data acquisition mode, whereby all locations of the sample are illuminated concurrently. This was historically the first method of time domain acquisition [16–20]. For fluorescence, the time intervals and starting time for recording the image is controlled by actuating a fast image intensifier following every excitation light pulse. The intensifier operates as an "image light switch," opening for pre-selected time intervals. For fluorescence measurements the light switch operates in fractions of nanoseconds to a few nanoseconds; phosphorescence lifetime imaging has also been developed, where the "light switch" operates from microseconds to seconds [16]. The start of each acquisition interval is delayed over a time range of a fraction of a nanosecond to several nanoseconds (microseconds to seconds for phosphorescence). The output of the image intensifier is integrated (at every pixel over many pulses) on a CCD camera. The sequentially delayed averaged images are transferred to a computer for a pixel-by-pixel analysis of the time dependence. Since in the wide-field mode only a few time intervals are acquired, the data can be analyzed by simply taking ratios of images taken at different time intervals (see Data Analysis below). Because of the limited number of time intervals (often just two or three), this method is not amenable for the analysis of multi-exponential signals; however, often in lifetime imaging an exact representation of the multi-exponential decay is not sought, but one is interested in changes in an "average" lifetime throughout the image. This wide-field time domain method using only a few time intervals is amenable for measuring lifetime-resolved fluorescence during a rapid kinetic process. Recent developments on the instrumentation, such as the segmented fast-gated image intensifier allow data acquisition at different time delays to be taken in a single shot further improving the speed of data acquisition [21].

In all time domain measurements, one averages many pulses to acquire sufficient photons. An important consideration of the TCSPC time domain measurement is that the time between the excitation pulses must be long enough so that there is no interference between the consecutive decays (sometimes corrections can be made if this is not possible). For highly accurate TCSPC data this usually means that fewer than one in 100 pulses should generate a detected photon originating from any location (to avoid pulse pile-up; that is, one must be sure that the photon detected is a result of the last excitation pulse). However, because such high accuracy, which is necessary for separating multiple lifetimes with high statistical significance, is often not the premium goal when imaging, accuracy of the lifetimes can be somewhat sacrificed by raising the repetition rate. When not photon-counting, and integrating an analog time-domain detector signal over only a few pre-determined time intervals (such as discussed in the last paragraph), this limitation on photon counting is not important. Then, a repetition period only about 5 times the expected time decay is sufficient. This is not an issue for the frequency domain measurement.

10.2.4 Frequency Domain

The excitation light is modulated repetitively at a high frequency, usually between 10 and 100 MHz, in order for the frequency dependent amplitude and phase of the nanosecond fluorescence response to be sensitive to the frequency of repetition, ω (see Fig. 10.2B and Eqs. 10.1 and 10.2). The waveform of the repetitive excitation is often sinusoidal, but it can be any repetitive shape that can be decomposed into multiple harmonics in a Fourier series, and each harmonic component is treated separately. In this review, for simplicity, we focus on using a single harmonic sinusoidal excitation waveform. The general case, using Fourier analysis of a general repetitive waveform, can be found in the literature [22]. The basics of frequency domain measurements are not as familiar to many biological researchers as the time mode, but the basic idea is the same as in many frequency domain experiments; for instance, frequency domain measurements are the basis of many radio communication and radar techniques, as well as laboratory measurements, such as dielectric dispersion. Because the frequency domain method of FLI is becoming more widespread, we will discuss this method in more detail.

A schematic of the frequency domain measurement using a single harmonic sine wave excitation light is shown in Fig. 10.2B. The waveform of the emission is the time convolution of the sinusoidal excitation profile and the time response to a delta function excitation pulse (which is an exponential decay function or a sum of exponentials). The fluorescence emission is thereby modulated at the same frequency. At very low frequencies, $\omega \ll 1/\tau_{\text{meas}}$, the depth of modulation (the frequency dependent amplitude divided by the time-average signal) and the phase of the sinusoidal modulation are nearly identical to the modulation and phase of the excitation light. At higher frequencies, the depth of modulation of the fluorescence is reduced by a factor M (M is the ratio of the depth of modulation of the fluorescence to the depth of modulation of the excitation light) and the sine wave is phase-shifted (delayed) by a phase angle ϕ_F, relative to the excitation light. The lifetime information is contained in ϕ_F and M (Fig. 10.2B and Eqs. 10.1 and 10.2). The phase and modulation are determined at every individual pixel of the image. Thorough derivations of Eqs. 10.1 and 10.2 are available in the literature [22–24], and extensive references to earlier literature are given therein. For a single lifetime component, the phase and modulation ratio can be expressed as

$$\phi_F = \tan^{-1}(\omega\tau_\phi) \tag{10.1}$$

$$M = \frac{b/B}{a/A} = \frac{1}{\sqrt{1+(\omega\tau_M)^2}}. \tag{10.2}$$

According to Eqs. 10.1 and 10.2, a "phase lifetime" (τ_ϕ) and a "modulation lifetime" (τ_M) can be determined, and these are equal for a single lifetime component. If more than one lifetime component is present, τ_ϕ and τ_M can be defined by Eqs. 10.1 and 10.2; however, this is just a formal mathematical transformation, and they do not refer to lifetimes of any single component. For multiple lifetime components, the phase and modulation ratio are expressed by complex mathematical expressions of the amplitudes and phases of all the contributing fluorescence lifetime components. The phase and the modulation ratio of the fluorescence signal at the repetition frequency are the raw data of a frequency domain experiment.

Multiple lifetimes can be determined by making measurements at several frequencies [25–27], as in single-channel cuvette experiments. If need be, this can be carried out in FLI [28]; however, the goal of imaging experiments is often different from cuvette experiments. Biological samples are multifaceted, especially if one is dealing with live systems. The acquisition of multiple frequencies is time consuming, and often the interesting biological processes are dynamic and rapidly changing. In such cases, when the data acquisition must be performed in real time, e.g., up to a video rates (25 frames per second) [29, 30], frequency domain measurements at one frequency are often preferred. Valuable information can be obtained by carrying out FLI measurements at a single frequency, as we will show.

10.2.5 The Frequency Domain Heterodyne/Homodyne Measurement

Accurate measurements of the phase shift and modulation ratio of the fluorescence are the key for accurate estimations of the fluorescence lifetimes in the frequency domain. To avoid complications working with very high frequencies (hundreds of MHz), the frequency analysis of the emitted fluorescence signal is not carried out at the high frequency. Rather, the amplification of the detector is modulated at a frequency either very close to the excitation modulation frequency (the heterodyne method) or the same as the excitation modulation frequency (the homodyne method).

In the heterodyne method the difference frequency signal (oscillating at the frequency difference between the light modulation and the detector modulation) is measured; this frequency difference is usually chosen to be in the range between a few hundred Hz to a few thousand Hz. The phase shift and extent of demodulation of the fluorescence relative to the excitation is preserved in the slowly varying heterodyne signal, which is easy to acquire at the lower frequency. For imaging, the heterodyne method has been applied exclusively in the scanning mode [9]. This is similar to the way a radio operates.

In the homodyne method, the high frequency signal is reduced to a DC signal and the phase shift and modulation ratio is obtained by acquiring the DC signal at different phase offsets between the detector amplification and the excitation light [7]. The examples shown in Figs. 10.7, 10.8, 10.9, and 10.10 in this chapter have been acquired with the homodyne frequency-domain method. Figure 10.3 explains the homodyne measurement in more detail. The time average of the multiplication of two high-frequency waveforms, the emission signal $F(t)$ and the modulation of the detector amplification $D(t)$, results in a DC component. This is achieved by using a high-frequency modulated (or fast-gated) image intensifier as the primary detector. The gain of the intensifier is repetitively modulated at the same frequency of the excitation modulation. The averaged amplitude of this multiplication process depends on the phase difference between the two high frequency signals. The DC homodyne signal, at every phase offset, is integrated over time on

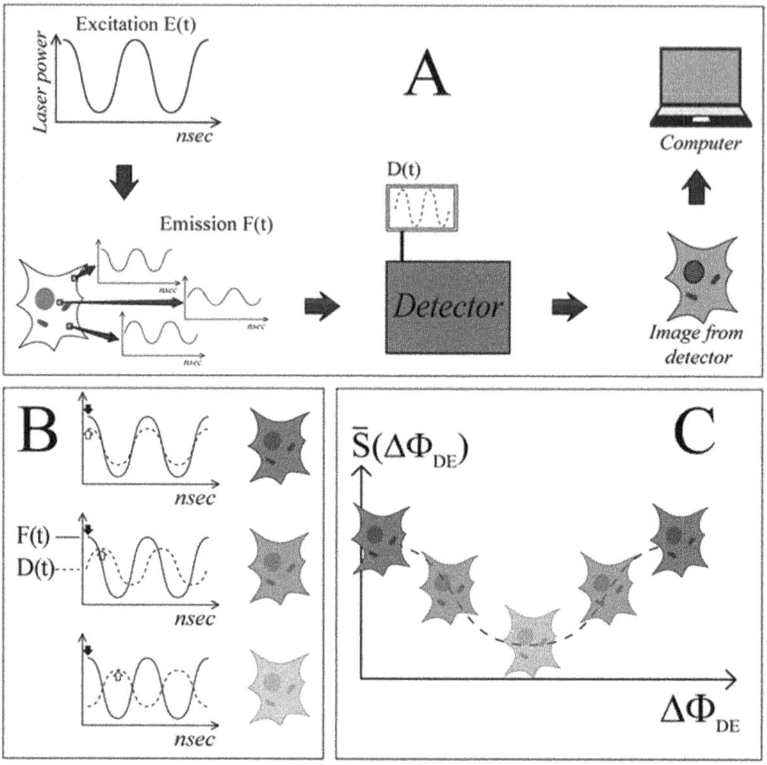

Figure 10.3 (A) Schematic explaining the frequency-domain lifetime measurement using the homodyne method. The amplification gain of the PMT, $D(t)$, is modulated at the same high frequency as that of the excitation $E(t)$; the emission $F(t)$ is modulated at the same high frequency, but delayed from $E(t)$. (B) The high-frequency-modulated fluorescence signal, $F(t)$ (solid line), which is detected on the intensifier, is multiplied by the modulated amplification gain of the intensifier $D(t)$ (dashed line). The output signal resulting from this multiplication, $\bar{S}(\Delta\phi_{\mathrm{DE}})$, is a DC signal, which varies as the phase difference between $E(t)$ and $D(t)$, $\Delta\phi_{\mathrm{DE}}$, is varied. (C) Integrated DC component $\bar{S}(\Delta\phi_{\mathrm{DE}})$ as a function of phase difference $\Delta\phi_{\mathrm{DE}}$. With kind permission from Springer Science+Business Media.

the CCD camera to increase the signal-to-noise ratio. Sequential images at different phase offsets are acquired with a computer (Fig. 10.3A). The phase difference between the modulation of the detector and the excitation light, $\Delta\phi_{\mathrm{DE}} = \phi_{\mathrm{D}} - \phi_{\mathrm{E}}$, is varied over one

2π period, usually in equal steps. The series of data points taken at different phase offsets $\Delta\phi_{\mathrm{DE}}$ form a sinusoidal signal (Fig. 10.3B,C), as in Eq. 10.3.

$$S(\Delta\phi_{\mathrm{DE}}) = \langle F(t) \cdot D(t) \rangle_{Av} = S_0 \left(1 + \frac{M}{2} \cos(\Delta\phi_{\mathrm{DE}} - \phi_{\mathrm{F}})\right) \quad (10.3)$$

$S(\Delta\phi_{\mathrm{DE}})$ is the measured signal at the phase offset $\Delta\phi_{\mathrm{DE}}$ mentioned in the last paragraph, and S_0 is the average DC signal. Just as in the heterodyne mode, the phase delay, ϕ_{F}, and the modulation ratio, M, of the fluorescence at the high frequency (Fig. 10.2B) are preserved in the integrated homodyne (or heterodyne) signal. After straightforward corrections for instrumental offsets, these parameters can then be used to estimate the lifetimes, as in Eqs. 10.1 and 10.2. The data acquisition for the homodyne frequency domain FLI measurement is generally performed in wide-field excitation, where the fluorescence emission is collected simultaneously from every point in the sample. This is partly due to the fact that it is difficult to carry out the heterodyne measurement in the wide-field mode. More discussions about the instrument setup will be presented in the next section.

10.3 Materials and Methods

10.3.1 FLI Instrumentation Setups and Data Acquisition

FLI was introduced in the early 1990s. Technical developments in instrumentation have contributed greatly to the applications of FLI [24]. These include single-beam confocal scanning [10, 31], two-photon excitation [32], streak cameras [33, 34], spectrally resolved FLI [35, 36], and fast data acquisition hardware [23]. Optical sectioning and three-dimensional lifetime imaging [37–40], anisotropy FLIM [41], low-cost, compact, and user-friendly FLI instrumentation [42], lifetime imaging in real time and at video rate have all been developed [7, 23, 30].

The basic details of FLI instrumentation setup in the time domain and the frequency domain are presented in this section (and in Fig. 10.4). We also give an example of some improvements in FLI

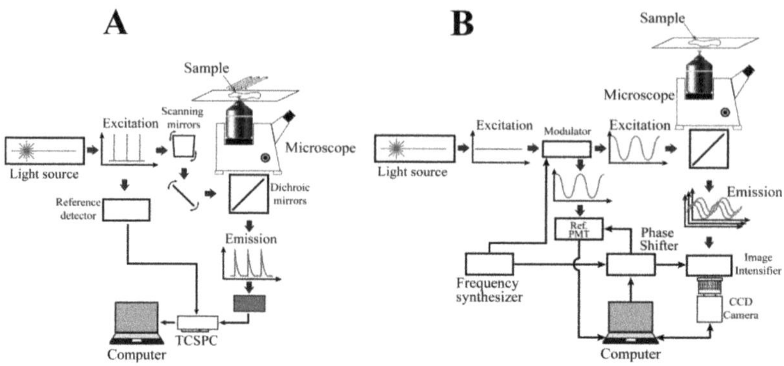

Figure 10.4 Examples of FLI instrumentation. (A) Single-beam scanning confocal and multiphoton excitation time domain and (B) wide-field frequency domain. In (A), a series of very short laser pulses excites the sample. The scanning mirrors control the location of the focused excitation spot on the sample. The arrival time of each emitted photon is recorded by a photon counting module (TCSPC module) following the detector (PMT). The signal from the TCSPC is transferred to the computer, where a histogram of the number of photons acquired at different photon arrival times is built up. This process is repeated at every location of the excitation spot. By scanning the excitation focused spot, an image is built up. (B) The excitation light is modulated at high frequency (10–100 MHz) by an AOM or Pockels cell, and the modulated excitation light fills the whole field of observation in the sample, in wide-field fashion, so that every point in the field of view is excited simultaneously. The image is focused onto an image intensifier, which records the modulated fluorescence emission at every pixel in parallel; the output of the image intensifier is focused onto a CCD camera, where the image is integrated for every phase offset between the excitation and intensifier modulations (see Fig. 10.3B). The phase shift of the fluorescence ϕ_F and modulation ratio M is then determined in the computer (Fig. 10.3C) at every pixel separately. The fluorescence lifetime image is constructed using the known ϕ_F and M and Eqs. 10.1 and 10.2.

instrumentation, developed in our lab, designed especially for live cell imaging where both speed and confocality are important.

10.3.2 Light Source and Modulator

Light sources for the time domain are typically fast-pulsed lasers with pulse widths in a range of hundreds of femtoseconds to a few nanoseconds. The lasers often have repetition frequencies up to several tens of MHz. Tunable Ti–sapphire lasers are used for two-photon excitation time domain lifetime measurements [9, 43]. Q-switched diode pumped solid state (DPSS) lasers or picosecond pulsed diode lasers [44–46] have been used in wide-field and scanning confocal time domain lifetime measurement [43, 47]. Low-cost light-emitting diodes (LEDs) have been shown to be applicable to frequency domain lifetime measurement as well [48–50]. In a single-beam scanning time domain measurement an adequate number of photons per pixel must be collected to ensure reliable data analysis. For example, 2000 photons (total) yield a reasonable fit for single decay fitting, but many more photons must be collected in order to fit multiple decays reliably [9, 24, 51, 52]. For weaker signals, long data acquisition times and/or a high laser power are required. However, too high excitation power causes photobleaching of the sample. Two-photon excitation reduces the overall out-of-focus photobleaching of the sample; however, two-photon excitation can increase the rate of photobleaching at the focal point when the technique is used in biological samples [53]. Ti–sapphire lasers produce a pulse every 12.5 ns, which is too rapid for lifetimes longer than a few nanoseconds; in that case, a pulse picker is used to reduce the pulse frequency.

Frequency domain FLI requires excitation light modulated repetitively at a high frequency (10–200 MHz) [23, 54]. Continuous-wave (CW) lasers or an arc lamp can be modulated with a Pockels cell or acousto-optical modulator (AOM). Details about choosing a laser source suitable for fluorescent proteins can be found at the Nikon Microscopy U website (http://www.microscopyu.com/articles/livecellimaging/fpimaging.html). An AOM is inexpensive and easy to set up [55], but it operates at given frequencies and requires extensive thermal equilibration to prevent phase drift [23]. A Pockel's cell can be driven at any frequency and can produce close to 100% modulation [30, 56]. The system must be calibrated with data taken from a sample with known lifetimes. For example, at 488 nm excitation, fluorescein diluted in 0.1 N NaOH is known to have a single lifetime of 4.1 ns [57]. More complex forms of excitation signal, such as a square wave or a series of nonsinusoidal pulses can be used [54]. Strong modulation of the excitation light (great depth of modulation) is preferred for making reliable measurements.

As mentioned above, too high excitation power can cause photobleaching of the sample during data acquisition, in both the time and frequency domain. In general, in the time domain mode, photobleaching does not distort the form of the time-dependent fluorescence decay of a single decay process, because the fluorescence decay (in nanoseconds) is much faster than the time for significant photobleaching (usually greater than milliseconds to minutes); the signal only becomes less intense. However, extensive photobleaching during a homodyne frequency domain measurement will distort the signal (usually the signal is averaged at each phase setting for a time longer than 10 ms). Thus, photobleaching can lead to errors in the fitted phase and modulation values. This distortion can be largely corrected during homodyne data acquisition by first increasing the phase delay from zero through 2π, and then decreasing the phase to the original value, and summing the signals at the same phase delays of the two sequences [58]. It is also possible to take this into account during the fitting procedure; if this is done using a nonlinear iterative regression, the time of analysis becomes very long. However, by using Chebyshev fitting methods [59], which avoid carrying out iterative fitting; very rapid data analysis (in real time) is possible. Therefore, for both the time and frequency domain modes, photobleaching does not pose a problem with rapid analysis, provided the right algorithms are used.

10.3.3 Microscope

Any research grade microscope can be used in frequency or time domain FLI. In the wide-field excitation, all points in the field of view are excited at the same time. In single-beam scanning FLI (either time domain or frequency domain), galvano mirrors are used to raster scan the excitation beam over a sample (i.e., a scanning confocal arrangement). Commercial equipment from several firms has become recently available [60].[a]

---

[a] **Becker & Hickl** http://www.becker-hickl.de; **Carl Zeiss** http://www.zeiss.com; **Hamamatsu Photonics** http://www.hamamatsu.com; **Lambert Instruments** http://www.lambert-instruments.com; **Leica Microsystems** http://www.leica-microsystems.com; **Nikon** http://www.nikon.com; **Olympus** http://www.olympus.com; **PerkinElmer Life; Analytical Sciences** http://las.perkinelmer.com; **Pico-Quant** http://www.picoquant.com; **TauTec** http://www.tautec.com

10.3.4 Detectors

Single-beam scanning FLI uses either a photo-multiplier tube (PMT) or an avalanche photodiode (APD) as a detector. PMTs are widely used as single-channel detectors in the frequency domain because of their short response times, and the amplification gain of a PMT can easily be modulated at high frequencies, which is a key requirement for the frequency domain measurement. However, PMTs have relatively low quantum efficiency (5–10% in a general PMT and less than 40% for high quality PMT) and they perform poorly at very long wavelengths in the near-IR range. Nevertheless, PMTs are suitable because they are widely commercially available and relatively inexpensive. Single-photon counting APDs are commonly used in photon counting time domain measurements. They are sensitive and have high quantum efficiencies (more than 80%) over a wide wavelength range in the visible region. However, APDs perform well only at lower repetition frequencies, due to their long dead time between repetitive measurements (time of recovery following each avalanche).

Time-correlated single-photon counting (TCSPC) provides the highest sensitivity in the single-beam scanning mode. The TCSPC records the number of photons arriving at the detector at certain times following the excitation pulse. Details on the mechanism of TSCPC are available in the literature [61, 62].

The wide-field FLI arrangement requires a high-frequency-modulated, or fast-gated, image intensifier and a CCD camera. The fluorescence image is focused onto the cathode of the image intensifier, which emits electrons. The two-dimensional microchannel plate stage of the intensifier amplifies this signal in an electron cascade at every channel individually (typically 10^4 channels); the electrons exiting the MCP are accelerated through a high voltage onto the intensifier phosphor screen, and the emission of the phosphor screen is focused onto the face of a CCD camera. Therefore, fluorescence emission from wide-field excitation is measured at all pixels in parallel, simultaneously. The signal is summed (averaged) in the pixel wells of the CCD camera for a set time. The averaged data stored in the pixel wells of the CCD camera are then individually, serially read out to a computer.

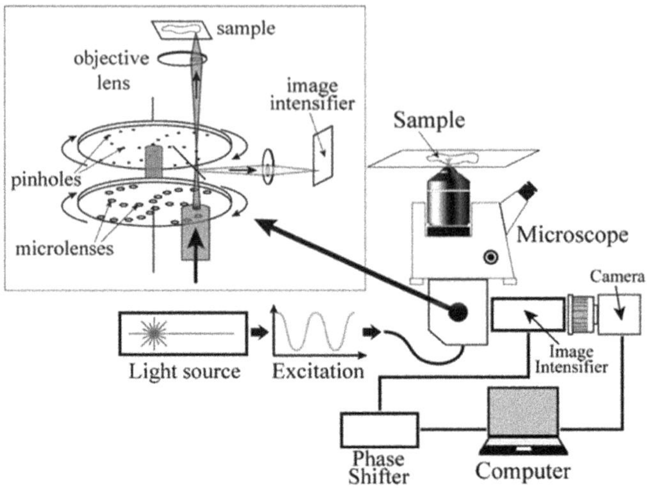

Figure 10.5 Spinning disk fluorescence lifetime confocal microscopy in the frequency domain. The dichroic mirror appearing beneath the microscope in Fig. 10.4B is replaced by a spinning disk confocal module (model CSU-10, Yokogawa Corp. of America, GA). The excitation beam illuminates a region on the lenslet disk (inset), covering approximately 1000 microlenses; each microlens focuses the beam passing the dichroic mirror onto a corresponding pinhole on the pinhole disk. The lenslet disk (Yokogawa design [162]) in front of the pinhole disk increases the excitation light to about 40–60% of the input intensity, instead of 1% without the microlenses [64]. The pinhole disk is designed to be at the optical conjugate plane to the microscope's objective focal plane [63]; the excitation beams are refocused by the tube lens and the objective lens onto the sample. Fluorescence emitted from the sample is collected and focused onto the same set of pinholes. As a result, each pinhole is used twice—for excitation and emission. The in-focus fluorescence reflects off the dichroic mirror and is focused onto the image intensifier. The position of the illuminated microlenses and the pinholes constantly change while the disks are spinning resulting in the raster scanning of the excitation points on the sample, and the scanned points are then focused onto the image intensifier. On the image intensifier, each pixel "sees" the fluorescence signal intermittently during the scanning. The time for excitation beam to sweep the diffraction limited spot approximately the time the excitation passes a microlens. This time is ~100 μs [163], and is much longer than the period of the modulation (10 ns at 100 MHz); therefore, several complete periods of the signal arrive at the image intensifier during that time. As a result, the data acquisition can be performed in the same fashion as done in the case of a conventional wide-field FLI setup. With kind permission from Springer Science+Business Media.

For the time domain wide-field measurement, the intensifier is set to have a fixed gate width and is gated on after a certain time delay after the arrival of an excitation pulse, as described in Section 2.2.1 [14]. In the homodyne wide-field frequency domain measurement, the amplification gain is modulated at the chosen frequency, usually at the cathode of the image intensifier, and the DC signal is averaged on the CCD array at every phase delay setting; then the averaged image of the CCD is read out to the computer for analysis and display.

10.3.5 An Improvement in the Wide-Field Frequency Domain: The Implementation of the Spinning Disk Microscope to Provide Confocal FLI Images

The conventional wide-field FLI setup suffers from the same out-of-focus background fluorescence in a thick sample as any normal non-confocal fluorescence microscope. In order to limit the out-of-focus background fluorescence without sacrificing the speed of data acquisition, as is the case for a single-beam scanning confocal setup, we have incorporated a spinning disk confocal head (model CSU-10, Yokogawa Corp. of America, GA) into a wide-field frequency domain FLI instrument (manuscript in preparation); see Section 5.5. With the spinning rate of 30 Hz, 360 confocal images are collected. For more technical details of the CSU-10, refer to Kawamura et al. [63]. Thus, rapid dynamic studies can be made in vivo [64], which is not be possible with a single-beam confocal scanning microscope. A schematic of the spinning disk confocal FLI instrument is shown in Fig. 10.5. The spinning disk confocal has also recently been incorporated into wide-field FLI setup in the time domain [65]. A practical advantage showing the benefit of spinning disk confocal FLI will be shown in Section 5.5.

10.4 Data Analysis

This section provides a short discussion of data acquisition and data analysis for the time and frequency domain lifetime measurements. A more detailed discussion for the frequency domain lifetime measurement is given, because the frequency mode is not as familiar to many readers. Reviews regarding both modes of measurement can be found in the literature [24, 30, 47, 54, 66]. Often 10^5–10^6 pixels must be fit individually if high spatial resolution is desired; therefore, it is important to streamline the analysis algorithms whenever possible. The signal-to-noise ratios of the frequency and time domain measurements for certain instrumentation configurations have been analyzed and compared [9, 67, 68].

10.4.1 Time Domain

If the lifetime measurement is carried out in a single-beam time domain scanning mode, the fluorescence signal at each pixel is a sum of exponentially decaying fluorescence components (Section 2.2.1). Basically, a statistical fit of the time-resolved data to a sum (often only one) of weighted exponentially decaying components with different lifetimes is carried out at every separate pixel. The derived parameters are the lifetimes τ_i and the fractional population a_i of each molecular species $(F(t) = \sum_i^{a_i \cdot \exp(-t/\tau_i)})$. An extensive literature is available describing different methods of analysis of multi-exponential signals, and these have been applied extensively for the measurement of cuvette-based fluorescence lifetimes [61]. Often, the fluorescence signal must be deconvolved from the shape and duration of the excitation light, and algorithms for this deconvolution are readily available [69–71]. For many time points, especially when fitting 10^4–10^5 separate image locations (pixels), these methods of data analysis take very long times. However, for high signal-to-noise data this method of analysis can fit multi-component lifetime decays with high accuracy [61, 72]. Recently the scanning time domain method has been significantly improved and streamlined for moderately rapid kinetic measurements (http://www.becker-hickl.de/).

The wide-field time domain mode produces images taken at selected, gated time intervals following the excitation pulse [47, 73, 74]. When only a few time intervals are selected, the analysis of these data at every pixel is often carried out by taking ratios of the consecutive time intervals; for instance, for a single exponentially decaying signal, this ratio will be $\exp(-\Delta t/\tau)$, and the logarithm of this ratio is simply $(\Delta t/\tau)$, from which the lifetime can be determined. More on analysis on the data taken under the wide-field excitation is published elsewhere [14].

10.4.2 Frequency Domain

Data analysis in the frequency domain is usually carried out by Fourier analysis, and sometimes by direct nonlinear regression (minimizing the variance between the data and the fit iteratively). The regression analysis is equivalent to the Fourier analysis, but digital Fourier analysis is much faster [54, 58, 75]. As mentioned above, multiple frequency measurements can also be carried out in imaging mode; however, most FLI measurements in the frequency domain have been carried out at a single frequency. In general, the analysis of frequency domain data at every pixel is similar to the analysis of single-channel measurements. Since Fourier analyses are rapid, especially when only calculating the fundamental frequency parameters, the time for analyzing the frequency domain data is usually shorter than for the time domain. It is not essential to use a fast Fourier transform algorithm if only the phase and modulation of the fundamental repetitive frequency component is needed. In this case, only three Fourier components (the phase, the amplitude of the modulation depth, and the amplitude of the average: DC fluorescence) are required. This speeds up the analysis considerably [22, 30, 54]. For multiple frequency measurements, maximum entropy methods for fitting the data have been found highly accurate [76, 77], and a lifetime moments analysis has been used to estimate errors [78]. Global analysis on FLI data [79] and an extensive discussion of errors in lifetime values within and between lifetime images, the importance of careful lifetime calibration and the analysis on the minimum resolvable lifetime differences in the frequency domain lifetime measurement have also been reported [80].

10.4.3 Rapid, Informative Analysis and Display Methods are Needed for FLI Data: The Polar Plot Analysis in the Frequency Domain

We use the following example to emphasize that analysis and display methods different from conventional single-channel analyses are often convenient when working with image data.

A convenient and rapid method for presenting and analyzing frequency domain FLI data is the "polar plot" [24, 81–84]. The polar plot is a simple mathematical representation of the measured phase (ϕ_F of Eq. 10.1) and modulation (M of Eq. 10.2) values as vectors in an x–y plot ($x = M\cos(\phi_F)$ and $y = M\sin(\phi_F)$), as shown in Fig. 10.6 below. It is called a "polar plot" because the phase shift and modulation values of each time-resolved data point are represented as a vector in polar coordinates. $M\sin(\phi_F)$ is the out-of-phase fraction of the fluorescence signal, and $M\cos(\phi_F)$ is the in-phase fraction. Since the complex exponential is defined as $M\exp(i\phi_F) = M[\cos(\phi_F) + i\sin(\phi_F)]$, this is equivalent to plotting the polar representation of a complex number. This is a common way to represent many different frequency domain measurements, such as dielectric dispersion data [85–87], and has a long history. The "polar plot" is a convenient way to compress the raw data (that is, the phase and modulation values from many pixels) of FLI experiments on a single, model-independent plot. The features of the plot often show distinctive characteristics of the spatial and temporal changes of the fluorescence data, and this way of representation provides a very convenient and informative overview of the large number of data points.

If the fluorescent sample has only a single lifetime component, the measured phase and modulation parameters must lie on a semicircle (dashed line, Fig. 10.6) on the polar plot; the semicircle has a radius of 0.5, and the circle center lies at $(x, y) = (0.5, 0)$. For a single fluorescence component, the position of the x–y data point on the semicircle is determined only by the fluorescence lifetime and the frequency of the modulation. A clustering of the polar plot points lying on the semicircle is proof that the signal has only one distinguishable fluorescence component. The value of the lifetime can be read directly from the location on the semicircle

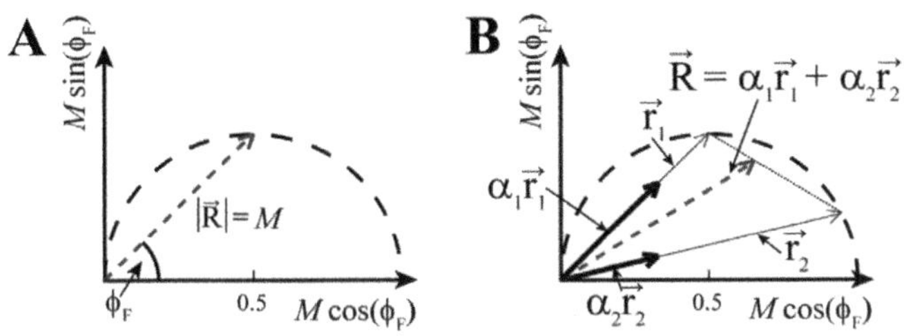

Figure 10.6 The polar plot representation. (A) For a single lifetime and (B) for a combination of two lifetimes. The ends of the dashed vectors, $\vec{R}$, mark the points on the polar plot calculated from the measured phase shift and modulation ratio. The end points of $\vec{r}_1$ and $\vec{r}_2$ designate the points on the semicircle corresponding to the separate first and second lifetime components, respectively. Note that in 6B the measured point resides on a straight line connecting the two points on the semicircle corresponding to the lifetime values of the two components. The lengths $\vec{r}_1$ and $\vec{r}_2$ contributing to the measured vector ($\vec{R}$) are $\alpha_1\vec{r}_1$ and $\alpha_2\vec{r}_2$. The coefficients α_i are the fractional fluorescence intensities, defined as $\alpha_n = a_\text{n} \cdot \tau_\text{n}/\sum_i a_\text{i} \cdot \tau_\text{i}$, where a_i and τ_i are the fractional population and fluorescence lifetime of species i^{th}, respectively. $\sum_i \alpha_\text{i} = 1$. Any combination of the two lifetime components (intersections of the two long shaded vectors with the semicircle) will correspond to a point along this line.

(knowing the frequency of modulation). A signal with multiple lifetime components must lie inside of the semicircle (Fig. 10.6B and Fig. 10.7). Well-characterized, diverse situations are recognizable simply by observing the distribution of the plotted points (as seen in the examples shown in Fig. 10.7 and Fig. 10.8). Of course, given the same information, an equivalent analysis can be made algebraically without the polar plot. But this simple polar plot representation of the data provides a very convenient and informative overview of the data, as well as a convenient platform for further analysis.

There are several advantages of the polar plot that are especially valuable for image data:

(1) Only the raw data, the phase ϕ_F and modulation M values, are used for forming the plot—this is model independent. These parameters are acquired directly from the data acquisition—there is no fitting procedure necessary. So we can observe directly on the plot how the measured points move in time in

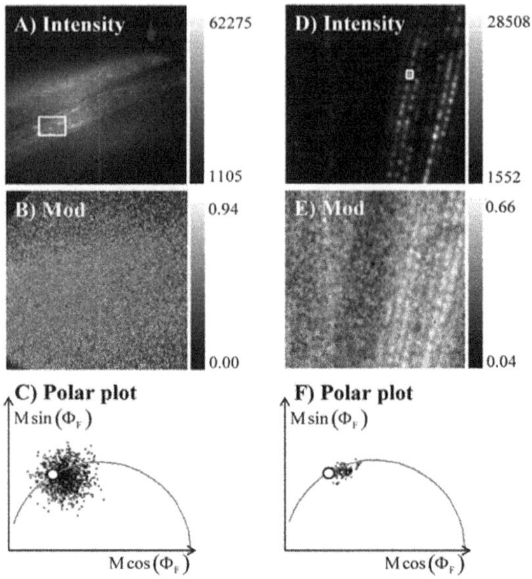

Figure 10.7 A sample data set taken from a *C. elegans* carrying PAT-4-yfp protein fusion (see text), taken with a conventional wide-field frequency domain FLI (A–C; left), and with the new setup with spinning disk confocal unit (D–F; right). Both instrumentation configurations show clear intensity images of muscle structure (A and D). The repeating pattern of bright spots is from the dense bodies. The M lines are less visible. Regions of interest (open rectangles) containing muscle structure are shown in Figs. 10.7A and 10.7D. The FLI data from these images produce a cluster of points on the polar plot (B and E, respectively). The polar plot shows that the measured lifetime of YFP attached to the PAT-4 protein in the transgenic worm clusters close to the reported value of approximately 2.9 ns of 10C YFP in vivo [5] marked on the polar plot with the open circle. There is the possibility of having more than one lifetime component, because the point cluster is centered slightly inside the semicircle, but the major component will have a lifetime close to the reported value according to the discussion made in Fig. 10.6B. The muscle structure is obscured in the demodulation image (7C). This is probably due to contributions to the lifetime-resolved image of the out of focus fluorescence in these images. On the other hand, the demodulation image from the data set taken using the spinning disk confocal FLI arrangement shows the same distinct muscle structures seen in the intensity images. The confocal FLI measurement obtained from the spinning disk confocal unit greatly improves the quality of the data. Even without the confocal mode, the visibility of the muscle structure in the FLI images can be improved by using only pixels for the FLI analysis with intensities above a set threshold (not shown). With kind permission from Springer Science+Business Media.

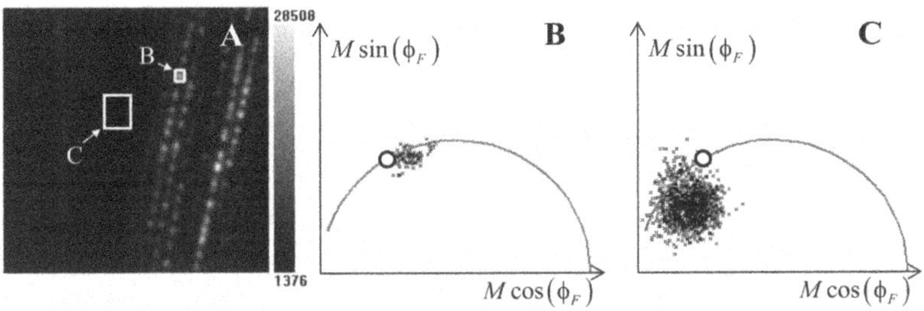

Figure 10.8 FLI data acquired with the spinning disc confocal microscope (Fig. 10.8A). A boxed region of interest containing a dense body (labeled B) produces a cluster of points near the reported 2.9 ns lifetime on the semicircle (Fig. 10.8B). A selected region outside the dense bodies (labeled C) produces a cluster of points centered at a different position. Due to the better out-of-focus background rejection, the spinning disk confocal instrumentation distinguishes the region of interest containing fluorescence proteins from the region outside. This figure demonstrates the benefit of the polar plot analysis to rapidly distinguish different lifetime information.

response to changes in the measured lifetimes, and how the polar plot points are distributed over the image. That is, we can watch how the clustering of points changes and progresses as kinetic experiments progress.

(2) For every frequency, each lifetime component *when measured separately* (i.e., when measuring only a single fluorescence lifetime component) occupies a distinct location lying on the semicircle.

(3) If only two lifetime components (or lifetime pools with a distribution similar to a two-dimensional Gaussian) are present, the data must lie on a straight line. In the case of two components, this line intersects the universal semicircle at the two positions, which are defined at any particular frequency by the separate values of the underlying lifetimes (Fig. 10.6B). This is diagnostic for two times. This is very useful for some experiments such as FRET where some of the donor molecules do not have an acceptor within the distance for energy transfer [24, 81].

(4) Any combination of lifetimes and intensities can easily be simulated and plotted on the polar plot for comparison to the data.

(5) If one of the lifetimes is known, it is often possible rapidly to gain information about the other lifetime as well as the fractional components.

This short introduction to the polar plot emphasizes an important point when making fluorescence lifetime imaging measurements in the microscope. In addition to new ways for carrying out the measurements, innovative rapid methods to analyze and display the data are also needed, which relate specifically to the imaging environment, and these analyses may be different than usually carried out in single channel experiments. It is essential to combine conveniently the photophysical data effectively with the spatial resolution of fluorescence imaging [88, 89] and the recognition of specific morphological features combined with lifetime information [90–92]. In a fluorescence image, especially in a biological cell, the environment of a fluorophore is not nearly as well controlled as in a cuvette-type experiment. One has very little control over the intercellular conditions in biological cells, and cells are likely individually different in many respects, even within one image. The spatial distributions of lifetime-resolved signals can furnish valuable information that extends intensity data, even without sophisticated fitting procedures.

10.4.4 Phase Suppression

Within live cells, the dynamics of ion transport and the kinetics of enzymes can take place on timescales of seconds. As discussed previously, lifetime images can be acquired at high speed (near video rates) but only with sufficient signal-to-noise, which is not often available when studying live cells. Furthermore, regardless of the brightness of the sample or the sensitivity of the imaging system, the temporal resolution of FLIM is also degraded because multiple exposures of the sample are necessary to generate a single lifetime image.

A technique referred to as phase suppression has recently been applied to study the kinetics of enzymes within live cells expressing fluorescent proteins. Each lifetime component contributes to the homodyne waveform a separate sinusoidal constituent with a distinct phase. If two phase resolved images are taken with phases differing by 180 degrees at phase settings such that the two sinusoid amplitudes are zero, and if these two images are subtracted, the resulting image will not show any fluorescence intensity from the specifically suppressed lifetime component. Other lifetime components that are faster or slower than the suppressed component will appear in the suppressed image with positive or negative values. Which components are positive or negative will depend on whether the lower phase image is subtracted from the greater phase image, or vice versa. Detailed descriptions and references for this method can be found in the literature [22, 24, 54, 93]. This is a very rapid, model independent, nonfitting way to identify locations in an image with different lifetimes. By scanning the phases (keeping the phase difference equal to 180 degrees) locations with different lifetimes can be identified rapidly and easily. This is particularly valuable when measuring FRET-FLIM in live biological cells. An example is given in Fig. 10.9 and discussed in Section 5.6.

10.4.5 Spectral FLIM

Spectral information assists in the separation of multiple lifetime components. By inserting a spectrograph with continuous wavelength resolution or a set of narrowband filters into the emission path of a homodyne, frequency domain FLIM apparatus, a fluorescence spectrum can be obtained at every pixel of a FLIM image. If the spectrum of each fluorescence species has been previously recorded in the FLIM instrument, the spectral components can be numerically unmixed at every pixel of each separate phase image of a homodyne data acquisition, resulting in phase images of the separate fluorescence species. This allows the FLIM analysis to be carried out on the signal from each fluorescence species individually. This helps resolve signal crosstalking from fluorescence emanating from different species with different lifetimes, which can interfere with an accurate analysis of multiple lifetime processes. This extends the number of different dyes that can be measured simultaneously in a FLIM measurement. Spectrally resolved FLIM (spectral FLIM) also provides a means to measure more accurately the lifetime of a dim fluorescence component in the presence of another fluorescence component with a much higher intensity (lifetimes from components with 2% of the total intensity can be determined accurately). A more reliable separation of the donor and acceptor fluorescence signals are possible for FRET measurements; for instance, contributions from acceptor fluorescence can be completely removed from the FLIM images, allowing more accurate FRET efficiencies to be determined from the donor lifetime. By combining the polar plot analysis with Spectral-FLIM data, the spectral dispersion of the acceptor signal can be used to derive the donor lifetime—and thereby the FRET efficiency—without iterative fitting. Details for spectral FLIM measurements and analyses, for both time and frequency domain FLIM, can be found in the more recent FLIM literature (many references can be found in Kusumi et al. [35], Chen et al. [94], and Chen and Clegg [95].

10.5 FLI Studies on Biological Samples

Lifetime-resolved fluorescence imaging in live cells and tissues was first presented in the early 1990s. The first studies were performed in wide-field excitation mode, either in frequency domain or in time domain measurements. Early studies were used to monitor intercellular components, such as in vivo Ca^{2+} concentration [96], pH [97], the lateral distribution of seminolipids [98], chloroplast fluorescence [99], cell nuclei and cell wall constituents [100, 101], and free and protein-bound state of NADH [3]. In the same period of time, the benefit of using FLI to visualize FRET over an image was realized [102, 103]. Lifetime imaging is not limited to nanosecond, or subnanosecond, lifetimes. Microsecond to second luminescence decay involving images of phosphorescence in a microscope was also reported and presented by Marriott et al. [16].

10.5.1 Fluorescent Proteins in Transgenic Cells and FLI

In vivo FLI has benefited greatly from the incorporation of the green fluorescent protein (GFP) gene into a host genome [104] to form protein hybrids. This has the same advantage for FLI as for normal fluorescence intensity imaging: eliminating the external labeling of organic fluorescent dyes and allowing more specific labeling. Fluorescence lifetimes and photo-physical properties of fluorescent proteins have been characterized [105–114]. It is much easier to interpret FRET studies if the donor has only one lifetime, even though average lifetimes can be used. The original wild type GFP and GFP variants exhibit complex decays (multi-exponential) from their excited states [115, 116] limiting the reliability of lifetime measurements. Fluorescent proteins better suited for FLI and FLI-based FRET studies have been obtained by random and site-directed point mutations [112] (for a nice and concise review of the development of monomeric fluorescent proteins, see Patterson [117] and Chapter 6).

Because the instrumentation and analysis are continually being improved and FLI instruments are now commercially available, FLI in live cells and tissues, which are endogenously labeled with fluorescent proteins, is an active area of study [43, 118, 119], and also in plants [120]. With an increasing interest in real time dynamics of protein-protein and protein-nucleic acids interaction in vivo, FLI will continue to contribute greatly to quantitative investigations of a broad variety of biological problems, especially when coupled with fluorescent proteins.

10.5.2 FLI-Based FRET Imaging

Because the fluorescence lifetime of a probe is in general independent of its concentration (unless the concentration variation changes levels of molecular association between the fluorophore and a dynamic quencher), FLI-based FRET imaging is often the most reliable method for measuring FRET efficiency in vivo. If one is interested in investigating macromolecular association, the interacting partners can be labeled exogenously with fluorophores (donor and acceptor chromophores, usually organic dye molecules), or endogenously with fluorescent proteins, to produce suitable FRET pairs. The FRET efficiency, E, is calculated from the decrease in donor's fluorescence lifetime in the presence of the acceptor; $E = 1 - \tau_{DA}/\tau_{D}$, where the subscripts refer to the fluorescence lifetime in the presence of the donor alone (τ_D), or in the presence of the donor-acceptor pair (τ_{DA}). Usually lifetime of the donor alone is known. Then the FRET efficiency can be determined simply by determining τ_{DA}. If there is more than one lifetime, the "average" lifetime is often used, although each lifetime component will be related to its own FRET efficiency. The accuracy of the lifetime determination depends on the signal-to-noise of the measurement, and whether one is trying to separate multi-component signals. Photon-counting methods, as mentioned above, have the best noise characteristics for very low signals, but if the fluorescence is very weak, the lengthy time required for time correlated single-photon counting (TCSPC) can become quite long. Exciting the samples in the scanning confocal/two-photon mode greatly minimizes out of focus background fluorescence present in thicker samples, and the longer wavelength for two-photon excitation penetrates deeper into macroscopic biological samples, such as tissue [32, 121, 122]. However, the scanning method is limited to studies of slower kinetics due to the slower acquisition speed. For measuring fast kinetic processes, or using thin samples where the out of focus background fluorescence is minor, wide-field excitation, both in the frequency domain and in the time domain, can be utilized to increase the data acquisition rate.

There is a large literature and many reviews on FLI-based FRET studies [7, 8, 43, 119, 123]. As examples, FLI has been used to monitor binding and activation states of signaling proteins [124], protein oligomerization [125], protein co-localization [126], nuclear protein-protein interactions [127], nuclear protein–DNA interaction [128], to characterize amyloid-beta plaques [129], conformational changes of proteins [130], and the interaction between a neuronal protein with a potassium channel [131].

10.5.3 FLI Applications with Environmental Probes and Sensors

FLI has been used to probe the molecular environment of the fluorescent proteins. FLI reveals the difference in refractive index and solvent viscosity surrounding GFP [132, 133] or surrounding charged environment [134]. The fluorescence lifetimes of GFP, both the anionic and the neutral conformers, show a strong dependence on pH [135], and can be used to measure the pH. In live cells, FLI can distinguish GFP-tagged proteins located at different discrete locations [5, 136]. Organic fluorescent probes are currently available for use with FLI for determining the concentration, or presence, of Ca^{2+} [137, 138], O_2 [139], Cl^- [140], and pH [97, 122]). An alternative way of using the endogenous fluorescent proteins to detect ions in vivo is either to utilize the intrinsic sensitivity of fluorescent proteins to ions under consideration, as shown in case of sensors of Cl^- [141] and pH [142], or to incorporate a FRET pair fluorescent protein onto a molecule that changes its conformation upon binding of the analyte. Examples of these chimeras include a fluorescent protein-based sensor for detecting Ca^{2+} [143], Zn^{2+} [144], or large molecules such as second messengers and ligand for nuclear receptors [145, 146]. FLI can then be used to measure the lifetime-based FRET efficiency as we discussed. However, as of this writing, chimeras such as the Ca^{2+} sensor Cameleon, are used usually in intensity base sensing. An extensive review on the topic, including different type of fluorescent protein based ion sensors, is available [147].

10.5.4 Contrast Enhancement and Simultaneous Detection

As previously mentioned, FLI provides another dimension of contrast besides the spectral difference seen in fluorescence intensity images. It has been shown in cases of organic dyes that FLI yields better contrast [148–150] and allows simultaneous detection of multiple fluorophores with different lifetimes [151, 152]. This is beneficial for simultaneous detection of fluorescent proteins with large spectral overlap such as GFP and YFP. It has been shown that up to three different colors can be detected in the same time [5]. Furthermore, nonradiative energy transfer between two FRET pairs with closely overlapping spectra, such as GFP and YFP, is much easier to detect when it is performed with FLI compared with the normal intensity-based FRET imaging [153].

10.5.5 The Advantage of Confocal Measurements

An example of a FLI study with live cells from our lab is from a continuing collaborative effort with Benjamin D Williams, at UIUC [154]. The main focus is to study the interactions between PAT (Paralyzed and Arrested elongation at Two-fold) proteins essential for muscle assembly in the body wall muscle cells of *C. elegans*. The PAT phenotype is found in the worms without the *pat* genes present resulting in the arrested in development when the worm reaches the two-fold state during embryogenesis [155]. The monomeric CFP (the W7 variant with A206K mutation) is used as the FRET donor and the monomeric YFP (the 10C variant with A206K mutation) is used as the FRET acceptor. The preliminary example data sets presented in this section are taken from transgenic worms carrying the PAT-4–mYFP fusion protein.

Figure 10.7 contrasts two FLI data sets taken with (i) a conventional wide-field FLI microscope (7A–D) and (ii) with the same microscope in conjunction with the spinning disk confocal arrangement (7E–H).

The better quality of lifetime data taken with the spinning disk confocal FLI does not compromise the acquisition speed. The intensity is decreased somewhat, so this may lead to increased averaging times for very low fluorescence intensities. Using the spinning disk FLI, FLI measurements can follow biological processes in live cell in real time.

10.5.6 Phase Suppression of Biosensor for Locating Activity of MT1-MMP

The kinetics of a biosensor specific for the protease MT1-MMP has recently been examined with phase suppression in live cancer cells from the line HT-1080. MT1-MMP is primarily responsible for the degradation of collagen in cell's local environment. The MT1-MMP biosensor is a membrane-tethered tandem construct containing the mOrange2 and mCherry fluorescent proteins in addition to a substrate specific for MT1-MMP. When the biosensor is initially expressed, FRET takes place between the mOrange2 and mCherry fluorescent proteins upon excitation of the mOrange2. A nearby and

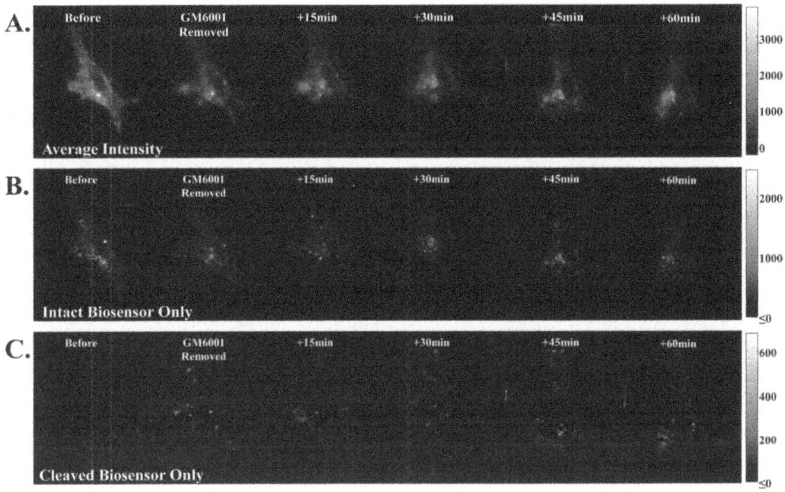

Figure 10.9 Phase Suppression Applied to Study Live Cells: (A) This HT-1080 cell was transfected with the mOrange2/mCherry variant of the MT1-MMP biosensor and imaged using phase suppression following the removal of the MMP inhibitor GM6001.Images of the average intensity determined from the four images selected for the phase suppression are presented. (B) Phase suppression was applied here to remove intensity from the cleaved biosensor. As a result, these images in time indicate the progression of the intact biosensor both before and after GM6001 was removed. (C) A set of detector phase angles were applied to remove intensity from the intact biosensor. As shown, the intensity from the cleaved biosensor advanced along the edge of the cell during the observed time.

active MT1-MMP enzyme can, however, cleave the substrate and consequently eliminate any possibility for FRET.

In the experiment presented in Fig. 10.9, the HT-1080 cells were pre-treated with an MMP inhibitor (GM6001) prior to imaging in order to prevent the MT1-MMP biosensor from being cleaved. The inhibitor was removed during the imaging so that the activation of MT1-MMP could be monitored. Two detector phases were chosen to suppress the intensity from the cleaved biosensor (Fig. 10.9B) and separately, two other detector phases were used to suppress the intensity from the intact biosensor (Fig. 10.9C). When compared to intensity images presented in Fig. 10.9A, the intensity from the cleaved biosensor emerged along the edge of the cell following the removal of GM6001. Interestingly, intensity from the cleaved biosensor progressively appeared in the protrusion that formed in the lower left portion of the cell. In parallel, the intensity from the intact biosensor was consistently reduced in area and constrained to regions in the inner part of the cell. In this example, phase suppression not only spatially mapped the activity of the biosensor with a minimum of two exposures, but also provided large contrast in intensity between regions of the cell containing different configurations of the biosensor (cleaved or intact).

10.5.7 Application of Polar Plot and FLIM with the Photosenstizer PPIX with ALA

Photodynamic therapy is a non-invasive way of removing tumor cells using reactive oxygen species (ROS) generated by excited intracellular photosensitizers. Protoporphyrin IX (PPIX) is a photosensitizer commonly studied due to its long fluorescence lifetime that allows it to generate significant amount of ROS in the excited state. When the precursor for PPIX called 5-aminolevulinic acids (ALA) are added exogenously to the tissue, malignant cells show significantly more accumulation of PPIX than the healthy cells [156, 157]. By exploiting this metabolic abnormality, the malignant cells will be selectively destroyed over the healthy cells after the tissue is irradiated with light. Since the monomer form of PPIX has a long fluorescence lifetime, it has adequate time to undergo intersystem crossing into the triplet state, where it can react with nearby triplet oxygen to form singlet oxygen. However, PPIX is known to aggregate heavily in solution, and in cells [158, 159]. PPIX aggregate on the other hand has a shorter lifetime and is not as efficient at generating ROS, and therefore less desirable for PDT.

The reactive oxygen species has a short diffusion distance before returning to triplet state. Therefore, the site of PPIX monomer accumulation is where the cell will receive the most oxidative damage. To predict which cellular organelle will receive the most damage during PDT, the site of monomer and aggregate accumulation needs to be located in vivo. Since monomer and aggregate emission spectra overlaps, conventional fluorescence microscope is not able to differentiate the two forms in cells. An

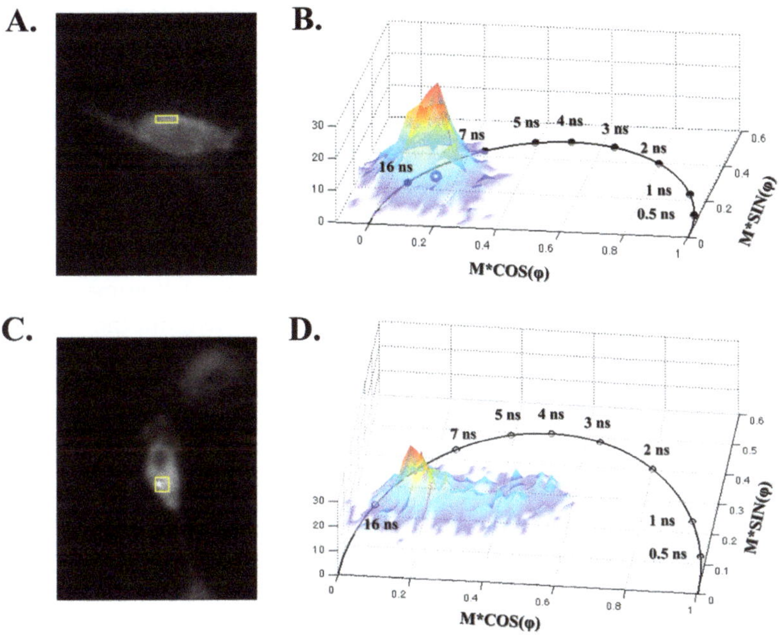

Figure 10.10 Polar plot analysis of PPIX lifetime data measured with FLIM: (A) Fluorescence intensity of PPIX measured in HT1080 cell incubated with ALA. The white square in the intensity image includes an area of plasma membrane that was analyzed with the polar plot. (B) Polar plot analysis of the enclosed region that showed a distribution of lifetime centered near 16 ns, the average of the distribution was also plotted (circle). (C) The intensity image showed a spot with high fluorescence intensity, and the region containing the bright spot enclosed by the white square was analyzed. (D) The polar plot analysis of the region enclosed by white square showed polar coordinates that spread across a line that connects the polar coordinate of 16 ns and a shorter lifetime between 1 and 2 ns.

effective way to distinguish between monomer and aggregated form of PPIX in cells is by using fluorescence lifetimes. The monomer form of PPIX is known to have a long lifetime (16–18 ns), while the aggregated form has a shorter lifetime (<6 ns) [160, 161]. Lifetime measurement that records a series of intensity images at various phase delays were performed on HT1080 cells, and selected cells were shown here as examples (Fig. 10.10A,C). Phase and modulation data in every pixels in the region enclosed by the white square in the intensity image were plotted in terms of polar coordinate previously defined. In the plasma membrane region of the cells, polar coordinate that shows lifetime close to 16 ns was detected (Fig. 10.10B). A few of the HT1080 cells showed bright spots that showed a mixture of components that consist of a long and a short lifetime approximately 16 ns and 1.5 ns respectively (Fig. 10.10D). As the polar plot revealed, the bright spot in the cell possibly contains both the PPIX monomer and aggregate. By combining polar plot with FLIM, multi-component lifetime in the cells due to aggregate formation can be resolved without the need to perform multi-frequency measurement that would otherwise cause significant photobleaching.

Acknowledgments

The spinning disk confocal head was loaned by courtesy of Prof. Akira Chiba, UIUC. We thank Ben Williams, Glen Redford, Christoph Gohlke, Oliver Holub, Peter Schneider, Tom Jovin, Enrico Gratton, Sophie Breusegem, and the growing community of FLI investigators with whom we have interacted over the years.

References

1. Valeur, B. (2002). *Molecular Fluorescence*. Wiley-VCH, New York.
2. Lakowicz, J. R. (1999). *Principles of Fluorescence Spectroscopy*, 2nd ed. Kluwer Academic/Plenum Publishers, New York.
3. Lakowicz, J. R., Szmacinski, H., Nowaczyk, K., and Johnson, M. L. (1992). Fluorescence lifetime imaging of free and protein-bound NADH. *Proc. Natl. Acad. Sci. U.S.A.*, **89**, 1271–1275.
4. Lakowicz, J. R., and Szmacinski, H. (1996). Imaging applications of time-resolved fluorescence spectroscopy. In: Wang, X. F., and Herman, B., eds. *Fluorescence Imaging Spectroscopy and Microscopy*. John Wiley & Sons, Inc., New York, pp. 273–311.
5. Pepperkok, R., Squire, A., Geley, S., and Bastiaens, P. I. H. (1999). Simultaneous detection of multiple green fluorescent proteins in live cells by fluorescence lifetime imaging microscopy. *Curr. Biol.*, **9**, 269–272.

6. Peter, M., and Ameer-Beg, S. M. (2004). Imaging molecular interactions by multiphoton FLIM. *Biol. Cell*, **96**, 231–236.
7. Clegg, R. M., Holub, O., and Gohlke, C. (2003). Fluorescence lifetime-resolved imaging: measuring lifetimes in an image. *Methods Enzymol.*, **360**, 509–542.
8. Cubeddu, R., Comelli, D., D'Andrea, C., Taroni, P., and Valentini, G. (2002). Time-resolved fluorescence imaging in biology and medicine. *J. Phys. D: Appl. Phys.*, **35**, R61–R76.
9. Gratton, E., Breusegem, S., Sutin, J., Ruan, Q., and Barry, N. P. (2003). Fluorescence lifetime imaging for the two-photon microscope: Time-domain and frequency-domain methods. *J. Biomed. Opt.*, **8**, 381–390.
10. Buurman, J. M., Knutson, J. R., Ross, J. B. A., Turner, B. W., and Brand, L. (1992). Fluorescence lifetime imaging using a confocal laser scanning microscope. *Scanning*, **14**, 155–159.
11. Sytsma, V., and Grauwd, G. (1998). Time-gated fluorescence lifetime imaging and microvolume spectroscopy using two-photon excitation. *J. Microsc.*, **191**, 39–51.
12. Chen, Y., and Periasamy, A. (2004). Two-photon FLIM-FRET microscopy for protein localization. In: Periasamy, A., So, P. T. C., eds. *Proc. SPIE; Multiphoton Microscopy in the Biomedical Sciences IV*. SPIE, Bellingham, pp. 431–439.
13. McLoskey, D., Birch, D. J. S., Sanderson, A., Suhling, K., Welch, E., and Hicks, P. J. (1996). Multiplexed single-photon counting. I. A time-correlated fluorescence lifetime camera. *Rev. Sci. Instrum.*, **67**, 2228–2237.
14. Agronskaia, A. V., Tertoolen, L., and Gerritsen, H. C. (2003). High frame rate fluorescence lifetime imaging. *J. Phys. D: Appl. Phys.*, **36**, 1655–1662.
15. Requejo-Isidro, J., McGinty, J., Munro, I., et al. (2004). High-speed wide-field time-gated endoscopic fluorescence-lifetime imaging. *Opt. Lett.*, **29**, 2249–2251.
16. Marriott, G., Clegg, R. M., Arndt-Jovin, D. J., and Jovin, T. M. (1991). Time resolved imaging microscopy. Phosphorescence and delayed fluorescence imaging. *Biophys. J.*, **60**, 1374–1387.
17. Cubeddu, R., Taroni, P., Valentini, G., and Canti, G. (1991). Use of time-gated fluorescence imaging for diagnosis in biomedicine. *J. Photochem. Photobiol., B*, **12**, 109–113.

18. Cubeddu, R., Taroni, P., and Valentini, G. (1993). Time-gated imaging system for tumor diagnosis. *Opt. Eng.*, **32**, 320–325.
19. Gerritsen, H. C., Vroom, J. M., and de Grauw, C. J. (1999). Combining two-photon excitation with fluorescence lifetime imaging. *IEEE Eng. Med. Biol. Mag.*, **18**, 31–36.
20. Dowling, K., Dayel, M. J., Lever, M. J., French, P. M. W., Hares, J. D., and Dymoke-Bradshaw, A. K. L. (1998). Fluorescence lifetime imaging with picosecond resolution for biomedical applications. *Opt. Lett.*, **23**, 810–812.
21. Elson, D. S., Munro, I., Requejo-Isidro, J., et al. (2004). Real-time time-domain fluorescence lifetime imaging including single-shot acquisition with a segmented optical image intensifier. *New J. Phys.*, **6**, 1–13.
22. Clegg, R. M., Schneider, P. C., and Jovin, T. M. (1996). Fluorescence lifetime-resolved imaging microscopy. In: Verga Scheggi, A. M., Martellucci, S., Chester, A. N., and Pratesi, R., eds. *Biomedical Optical Instrumentation and Laser-Assisted Biotechnology*. Kluwer Academic Publishers, Dordrecht, Boston, London, pp. 143–156.
23. Schneider, P. C., and Clegg, R. M. (1997). Rapid acquisition, analysis, and display of fluorescence lifetime-resolved images for real-time applications. *Rev. Sci. Instrum.*, **68**, 4107–4119.
24. Perisamy, A., and Clegg, R. M., eds. (2010). *FLIM Microscopy in Biology and Medicine*. Chapman and Hall/CRC.
25. Weber, G. (1981). Resolution of the fluorescence lifetimes in a heterogeneous system by phase and modulation measurements. *J. Phys. Chem.*, **85**, 949–953.
26. Gratton, E., and Limkeman, M. (1983). A continuously variable frequency cross-correlation phase fluorometer with picosecond resolution. *Biophys. J.*, **44**, 315–324.
27. Jameson, D. M., and Weber, G. (1981). Resolution of the pH-dependent heterogeneous fluorescence decay of tryptophan by phase and modulation measurements. *J. Phys. Chem.*, **85**, 953–958.
28. Squire, A., Verveer, P. J., and Bastiaens, P. I. H. (2000). Multiple frequency fluorescence lifetime imaging microscopy. *J. Microsc.*, **197**, 136–149.
29. Holub, O., Seufferheld, M., Gohlke, C., and Govindjee Clegg, R. M. (2000). Fluorescence lifetime-resolved imaging (FLI) in real-time - a new technique in photosynthetic research. *Photosynthetica*, **38**, 581–599.
30. Redford, G. I., and Clegg, R. M. (2005). Real-time fluorescence lifetime imaging and FRET using fast gated image intensifiers. In: Periasamy,

A., and Day, R. N., eds. *Molecular Imaging: FRET Microscopy and Spectroscopy*. Oxford University Press, New York, pp. 193–226.

31. Ghiggino, K. P., Harris, M. R., and Spizzirri, P. G. (1992). Fluorescence lifetime measurements using a novel fiber-optic laser scanning confocal microscope. *Rev. Sci. Instrum.*, **63**, 2999–3002.
32. Piston, D. W., Sandison, D. R., and Webb, W. W. (1992). Time-resolved fluorescence imaging and background rejection by two-photon excitation in laser scanning microscopy. *Proc. SPIE*, **1604**, 379–389.
33. Minami, T., and Hirayama, S. (1990). High quality fluorescence decay curves and lifetime imaging using an elliptical scan streak camera. *J. Photochem. Photobiol., A*, **53**, 11–21.
34. Kusumi, A., Tsuji, A., Murata, M., et al. (1991). Development of a streak-camera-based time-resolved microscope fluorimeter and its application to studies of membrane fusion in single cells. *Biochemistry*, **30**, 6517–6527.
35. Hanley, Q. S., Arndt-Jovin, D. J., and Jovin, T. M. (2002). Spectrally resolved fluorescence lifetime imaging microscopy. *Appl. Spectrosc.*, **56**, 155–166.
36. Nair, D. K., Jose, M., Kuner, T., Zuschratter, W., and Hartig, R. (2006). FRET-FLIM at nanometer spectral resolution from living cells. *Opt. Express*, **14**, 12217–12229.
37. Straub, M., and Hell, S. W. (1998). Fluorescence lifetime three-dimensional microscopy with picosecond precision using a multifocal multiphoton microscope. *Appl. Phys. Lett.*, **73**, 1769–1771.
38. Squire, A., and Bastiaens, P. I. H. (1999). Three dimensional image restoration in fluorescence lifetime imaging microscopy. *J. Microsc.*, **193**, 36–49.
39. Cole, M. J., Siegel, J., Webb, S. E. D., Jones, R., Dowling, K., and French, P. M. W. (2000). Whole-field optically sectioned fluorescence lifetime imaging. *Opt. Lett.*, **25**, 1361–1363.
40. Hanley, Q. S., Lidke, K. A., Heintzmann, R., Arndt-Jovin, D. J., and Jovin, T. M. (2005). Fluorescence lifetime imaging in an optically sectioning programmable array microscope (PAM). *Cytometry, Part A*, **67A**, 112–118.
41. Lidke, D. S., Nagy, P., Barisas, B. G., et al. (2003). Imaging molecular interactions in cells by dynamic and static fluorescence anisotropy (rFLIM and emFRET). *Biochem. Soc. Trans.*, **31**, 1020–1027.

42. Esposito, A., Oggierc, T., Gerritsenb, H. C., Lustenbergerc, F., and Woutersa, F. S. (2005). All-solid-state lock-in imaging for wide-field fluorescence lifetime sensing. *Opt. Express*, **13**, 9812–9821.
43. van Munster, E. B., and Gadella, T. W. J. (2005). Fluorescence lifetime imaging microscopy (FLIM). *Adv. Biochem. Eng./Biotechnol.*, **95**, 143–175.
44. Jones, R., Dowling, K., Cole, M. J., et al. (1999). Fluorescence lifetime imaging using a diode-pumped all-solid-state-laser system. *Electron. Lett.*, **35**, 256–258.
45. Elson, D. S., Siegel, J., Webb, S. E. D., et al. (2002). Fluorescence lifetime system for microscopy and multiwell plate imaging with a blue picosecond diode laser. *Opt. Lett.*, **27**, 1409–1411.
46. Kress, M., Meier, T., Steiner, R., et al. (2003). Time-resolved microspectrofluorometry and fluorescence lifetime imaging of photosensitizers using picosecond pulsed diode lasers in laser scanning microscopes. *J. Biomed. Opt.*, **8**, 26–32.
47. Suhling, K., French, P. M. W., and Phillips, D. (2005). Time-resolved fluorescence microscopy. *Photochem. Photobiol. Sci.*, **4**, 13–22.
48. Herman, P., Maliwal, B. P., Lin, H. J., and Lakowicz, J. R. (2001). Frequency-domain fluorscence microscopy with the LED as a light source. *J. Microsc.*, **203**, 176–181.
49. van der Oord, C. J. R., Stoop, K. W. J., and van Geest, L. K. (2001). Fluorescence lifetime attachment LIFA. *Proc. SPIE*, **4252**, 115–118.
50. Esposito, A., Gerritsen, H. C., Oggier, T., Lustenberger, F., and Wouters, F. S. (2006). Innovating lifetime microscopy: a compact and simple tool for life sciences, screening, and diagnostics. *J. Biomed. Opt.*, **11**, 034016.
51. Livesey, A. K., and Brochon, J. C. (1987). Recovering the distribution of decay constants in pulse-fluorimetry using maximum entropy. *Biophys. J.*, **52**, 693–706.
52. Brochon, J. C. (1994). Maximum entropy method of data analysis in time-resolved spectroscopy. In: Johnson, M. L., and Brand, L., eds. *Numerical Computer Methods Part B, Methods in Enzymology*, pp. 262–311.
53. Patterson, G. H., and Piston, D. W. (2000). Photobleaching in two-photon excitation microscopy. *Biophys. J.*, **78**, 2159–2162.
54. Clegg, R. M., and Schneider, P. C. (1996). Fluorescence lifetime-resolved imaging microscopy: a general description of the lifetime-resolved

imaging measurements. In: Slavik, J., ed. *Fluorescence Microscopy and Fluorescent Probes*. Plenum Press, New York, pp. 15–33.

55. Piston, D. W., Marriott, G., Radivoyevich, T., Clegg, R. M., Jovin, T. M., and Gratton, E. (1989). Wide-band acousto-optic light modulator for frequency domain fluorometry and phosphorimetry. *Rev. Sci. Inst.*, **60**, 2596–2600.
56. Redford, G. I. (2005). *Fast Fluorescence Lifetime Imaging Using a Full-field Homodyne System with Applications in Biology*. The University of Illinois at Urbana-Champaign, Urbana.
57. Sjöback, R., Nygren, J., and Kubista, M. (1995). Absorption and fluorescence properties of fluorescein. *Spectrochim. Acta, Part A*, **51**, L7–L21.
58. Clegg, R. M., Gadella, T. W. J., and Jovin, T. M. (1994). Lifetime-resolved fluorescence imaging. *Proc. SPIE*, **2137**, 105–118.
59. Malachowski, G., and Ashcroft, R. (1986). The Chebyshev transformation linearises the fitting of exponential, gaussian and binding function data. In: *Proceedings of the Australian Society of Biophysics, 10 Annual Meeting*, Canberra, Australia, p. 19.
60. Sage, L. (2005). Making the most of fluorescence: fluorescence lifetime imaging microscopy and spectral imaging complement conventional techniques. *Scientist*, **19**, 29.
61. Cundall, R. B., and Dale, R. E., eds. (1983). *Time-Resolved Fluorescence Spectroscopy in Biochemistry and Biology*. Plenum, New York.
62. Becker, W. (2005). *Advanced Time-Correlated Single Photon Counting Techniques*, Springer, Berlin, Heidelberg.
63. Kawamura, S., Negishi, H., Otsuki, S., and Tomosada, N. (2002). Confocal laser microscope scanner and CCD camera. *Yokogawa Tech. Rep. Engl. Ed.*, **33**, 17–33.
64. Inoue, S., and Inoue, T. (2002). Direct-view high-speed confocal scanner: the CSU-10. *Methods Cell Biol.*, **70**, 87.
65. Grant, D. M., Elson, D. S., Schimpf, D., et al. (2005). Optically sectioned fluorescence lifetime imaging using a Nipkow disk microscope and a tunable ultrafast continuum excitation source. *Opt. Lett.*, **30**, 3353–3355.
66. Morgan, C. G., and Mitchell, A. C. (1996). Fluorescence lifetime imaging: an emerging technique in fluorescence microscopy. *Chromosome Res.*, **4**, 261–263.

67. Philip, J., and Carlsson, K. (2003). Theoretical investigation of the signal-to-noise ratio in fluorescence lifetime imaging. *J. Opt. Soc. Am. A*, **20**, 368–379.

68. Pelet, S., Previte, M. J. R., Kim, D., Kim, K. H., Su, T.-T. J., and So, P. T. C. (2006). Frequency domain lifetime and spectral imaging microscopy. *Microsc. Res. Tech.*, **69**, 861–874.

69. Grinvald, A., and Steinberg, I. Z. (1974). On the analysis of fluorescence decay kinetics by the method of least-squares. *Anal. Biochem.*, **59**, 583–593.

70. O'Connor, D. V., Ware, W. R., and Andre, J. C. (1979). Deconvolution of fluorescence decay curves. A critical comparison of techniques. *J. Phys. Chem.*, **83**, 1333–1343.

71. Knutson, J. R., Beechem, J. M., and Brand, L. (1983). Simultaneous analysis of multiple fluorescence decay curves: a global approach. *Chem. Phys. Lett.*, **102**, 501–507.

72. Kollner, M., and Wolfrum, J. (1992). How many photons are necessary for fluorescence-lifetime measurements. *Chem. Phys. Lett.*, **200**, 199–204.

73. Dowling, K., Hyde, S. C. W., Dainty, J. C., French, P. M. W., and Hares, J. D. (1997). 2-D fluorescence lifetime imaging using a time-gated image intensifier. *Opt. Commun.*, **135**, 27–31.

74. Gerritsen, H. C., Asselbergs, N. A. H., Agronskaia, A. V., and Van Sark, W. G. J. H. M. (2002). Fluorescence lifetime imaging in scanning microscopes: acquisition speed, photon economy and lifetime resolution. *J. Microsc.*, **206**, 218–224.

75. Hamming, R. W. (1973). *Numerical Methods for Scientists and Engineers*, 2nd ed. Dover, New York.

76. Brochon, J.-C., Livesey, A. K., Pouget, J., and Valeur, B. (1990). Data analysis in frequency-domain fluorometry by the maximum entropy method — recovery of fluorescence lifetime distributions. *Chem. Phys. Lett.*, **174**, 517–522.

77. Shaver, J. M., and McGown, L. B. (1996). Maximum entropy method for frequency-domain fluorescence lifetime analysis. 2. Timing, mismatched intensity, and reference lifetime errors. *Anal. Chem.*, **68**, 611–620.

78. Esposito, A., Gerritsen, H. C., and Wouters, F. S. (2005). Fluorescence lifetime heterogeneity resolution in the frequency domain by lifetime moments analysis. *Biophys. J.*, **89**, 4286–4299.

79. Verveer, P. J., Squire, A., and Bastiaens, P. I. H. (2000). Global analysis of fluorescence lifetime imaging microscopy data. *Biophys. J.*, **78**, 2127–2137.

80. Hanley, Q. S., Subramaniam, V., Arndt-Jovin, D. J., and Jovin, T. M. (2001). Fluorescence lifetime imaging: multi-point calibration, minimum resolvable differences, and artifact suppresion. *Cytometry*, **43**, 248–260.

81. Redford, G. I., and Clegg, R. M. (2005). Polar plot representation for frequency-domain analysis of fluorescence lifetimes. *J. Fluoresc.*, **15**, 805–815.

82. Clayton, A. H. A., Hanley, Q. S., and Verveer, P. J. (2004). Graphical representation and multicomponent analysis of single-frequency fluorescence lifetime imaging microscopy data. *J. Microsc.*, **213**, 1–5.

83. Hanley, Q. S., and Clayton, A. H. A. (2005). AB-plot assisted determination of fluorophore mixtures in a fluorescence lifetime microscope using spectra or quenchers. *J. Microsc.*, **218**, 62–67.

84. Digman, M. A., Caiolfa, V. R., Zamai, M., and Gratton, E. (2008). The phasor approach to fluorescence lifetime imaging analysis. *Biophys. J.*, **94**, L14–L16.

85. Cole, K. S., and Cole, R. H. (1941). Dispersion and absorption in dielectrics. *J. Chem. Phys.*, **9**, 341.

86. von Hippel, A. R. (1954). *Dielectrics and Waves*. Wiley, New York.

87. Hill, N. E., Vaughan, W. E., Price, A. H., and Davies, M. (1969). *Dielectric Properties and Molecular Behavior*. van Nostrand Reinhold Company, New York.

88. Verveer, P. J., Squire, A., and Bastiaens, P. I. H. (2001). Improved spatial discrimination of protein reaction states in cells by global analysis and deconvolution of fluorescence lifetime imaging microscopy data. *J. Microsc.*, **202**, 451–456.

89. Verveer, P. J., and Bastiaens, P. I. H. (2003). Evaluation of global analysis algorithms for single frequency fluorescence lifetime imaging microscopy data. *J. Microsc.*, **209**, 1–7.

90. Buranachai, C., Kamiyama, D., Chiba, A., Williams, B. D., and Clegg, R. M. (2008). Rapid frequency-domain FLIM spinning disk confocal microscope: lifetime resolution, image improvement and wavelet analysis. *J. Fluoresc.*, **18**, 929–942.

91. Spring, B. Q., and Clegg, R. M. (2009). Image analysis for denoising full-field frequency-domain fluorescence lifetime images. *J. Microsc.*, **235**, 221–237.

92. Chen, Y.-C., Spring, B. Q., Buranachai, C., Malachowski, G., and Clegg, R. M. (2010). General concerns of FLIM data representation and analysis and frequency domain model-free analysis. In: Clegg, R. M. and Periasamy, A., ed. *FLIM Microscopy in Biology and Medicine*. Taylor & Francis, Chapman & Hall, CRC Press.
93. Eichorst, J., Huang, H., Clegg, R., and Wang, Y. (2011). Phase differential enhancement of FLIM to distinguish FRET Components of a biosensor for monitoring molecular activity of membrane type 1 matrix metalloproteinase in live cells. *J. Fluoresc.*, **21**, 1763–1777.
94. Chen, Y.-C., Spring, B. Q., Buranachai, C., Tong, B., Malachowski, G., and Clegg, R. M. (2010). *General Concerns of FLIM Data Representation and Analysis: Frequency-Domain Model-Free Analysis*. Chapman & Hall/CRC, New York.
95. Chen, Y. C., and Clegg, R. M. (2011). Spectral resolution in conjunction with polar plots improves the accuracy and reliability of FLIM measurements and estimates of FRET efficiencies. *J. Microsc.*, **244**, 21–37.
96. Lakowicz, J. R., Szmacinski, H., Nowaczyk, K., Lederer, W. J., Kirby, M. S., and Johnson, M. L. (1994). Fluorescence lifetime imaging of intracellular calcium in COS cells using Quin-2. *Cell Calcium*, **15**, 7–27.
97. Sanders, R., Draaijer, A., Gerritsen, H. C., Houpt, P. M., and Levine, Y. K. (1995). Quantitative pH imaging in cells using confocal fluorescence lifetime imaging microscopy. *Anal. Biochem.*, **227**, 302–308.
98. Gadella, B. M., Gadella, T. W., Jr., Colenbrander, B., van Golde, L. M., and Lopes-Cardozo, M. (1994). Visualization and quantification of glycolipid polarity dynamics in the plasma membrane of the mammalian spermatozoon. *J. Cell Sci.*, **107**, 2151–2163.
99. Gadella, T. W. J., Jovin, T. M., and Clegg, R. M. (1993). Fluorescence lifetime imaging microscopy (FLIM): spatial resolution of microstructures on the nanosecond time scale. *Biophys. Chem.*, **48**, 221–239.
100. Morgan, C. G., Mitchell, A. C., and Murray, J. G. (1990). Nanosecond time-resolved fluorescence microscopy: principles and practice. *Trans. R. Microsc. Soc.*, **1**, 463–466.
101. Morgan, C. G., Mitchell, A. C., and Murray, J. G. (1992). Prospects for confocal imaging on nanosecond fluorescence decay time. *J. Microsc.*, **165**, 49–60.
102. Gadella, T. W. J., Clegg, R. M., Arndt-Jovin, D. J., and Jovin, T. M. (1993). Epidermal growth-factor (EGF)-receptor clustering monitored by

fluorescence resonance energy transfer using donor photobleaching and lifetime-resolved fluorescence imaging microscopy. *Biophys. J.*, **64**, A130-A Part 2.

103. Oida, T., Sako, Y., and Kusumi, A. (1993). Fluorescence lifetime imaging microscopy (flimscopy). Methodology development and application to studies of endosome fusion in single cells. *Biophys. J.*, **64**, 676–685.
104. Chalfie, M., Tu, Y., Euskirchen, G., Ward, W. W., and Prasher, D. C. (1994). Green fluorescent protein as a marker for gene expression. *Science*, **263**, 802–805.
105. Perozzo, M. A., Ward, K. B., Thompson, R. B., and Ward, W. W. (1988). X-ray diffraction and time-resolved fluorescence analyses of aequorea green fluorescent protein crystals. *J. Biol. Chem.*, **263**, 7713–7716.
106. Chattoraj, M., King, B. A., Bublitz, G. U., and Boxer, S. G. (1996). Ultra-fast excited state dynamics in green fluorescent protein: Multiple states and proton transfer. *Proc. Natl. Acad. Sci. U.S.A.*, **93**, 8362–8367.
107. Patterson, G. H., Knobel, S. M., Sharif, W. D., Kain, S. R., and Piston, D. W. (1997). Use of the green fluorescent protein and its mutants in quantitative fluorescence microscopy. *Biophys. J.*, **73**, 2782–2790.
108. Kummer, A. D., Kompa, C., Lossau, H., et al. (1998). Dramatic reduction in fluorescence quantum yield in mutants of green fluoescent protein due to fast internal conversion. *Chem. Phys.*, **237**, 183–193.
109. Striker, G., Subramaniam, V., Seidel, C. A. M., and Volkmer, A. (1999). Photochromicity and fluorescence lifetimes of green fluorescent protein. *J. Phys. Chem. B*, **103**, 8612–8617.
110. Volkmer, A., Subramaniam, V., Birch, D. J. S., and Jovin, T. M. (2000). One- and two-photon excited fluorescence lifetimes and anisotropy decays of green fluorescent proteins. *Biophys. J.*, **78**, 1589–1598.
111. Subramaniam, V., Hanley, Q. S., Clayton, A. H. A., and Jovin, T. M. (2003). Photophysics of green and red fluorscent proteins: implications for quantitative microscopy. *Methods Enzymol.*, **360**, 178–201.
112. Rizzo, M. A., Springer, G. H., Granada, B., and Piston, D. W. (2004). An improved cyan fluorescent protein variant useful for FRET. *Nat. Biotechnol.*, **22**, 445–449.
113. Jung, G., Wiehler, J., and Zumbusch, A. (2005). The photophysics of green fluorescent protein: influence of the key amino acids at positions 65, 203, and 222. *Biophys. J.*, **88**, 1932–1947.
114. Kremers, G. J., Goedhart, J., van Munster, E. B., and Gadella, T. W. J. (2006). Cyan and yellow super fluorescent proteins with improved

brightness, protein folding, and fret förster radius. *Biochemistry*, **45**, 6570–6580.

115. Dickson, R. M., Cubitt, A. B., Tsien, R. Y., and Moerner, W. E. (1997). On/off blinking and switching behaviour of single molecules of green fluorscent protein. *Nature*, **388**, 355–358.
116. Lossau, H., Kummer, A., Heinecke, R., et al. (1996). Time-resolved spectroscopy of wild-type and mutant Green Fluorescent Proteins reveals excited state deprotonation consistent with fluorophore-protein interactions. *Chem. Phys.*, **213**, 1–16.
117. Patterson, G. H. (2004). A new harvest of fluorescent proteins. *Nat. Biotechnol.*, **22**, 1524–1525.
118. Tavaré, J. M., Fletcher, L. M., and Welsh, G. I. (2001). Using green fluorescent protein to study intracellular signalling. *J. Endocrinol.*, **170**, 297–306.
119. Wallrabe, H., and Periasamy, A. (2005). Imaging protein molecules using FRET and FLIM microscopy. *Curr. Opin. Biotechnol.*, **16**, 19–27.
120. Dixit, R., Cyr, R., and Gilroy, S. (2006). Using intrinsically fluorescent proteins for plant cell imaging. *Plant J.*, **45**, 599–615.
121. Piston, D. W., Masters, B. R., and Webb, W. W. (1995). Three-dimensionally resolved NAD(P)H cellular metabolic redox imaging of the in situ cornea with two-photon excitation laser scanning microscopy. *J. Microsc.*, **178**, 20–27.
122. Hanson, K. M., Behne, M. J., Barry, N. P., Mauro, T. M., Gratton, E., and Clegg, R. M. (2002). Two-photon fluorescence lifetime imaging of the skin stratum corneum pH gradient. *Biophys. J.*, **83**, 1682–1690.
123. Ameer-Beg, S. M., and Peter, M. (2004). Imaging molecular interactions by multiphoton FLIM. *Biol. Cell*, **96**, 231–236.
124. Yasuda, R., Harvey, C. D., Zhong, H., Sobczyk, A., van Aelst, L., and Svoboda, K. (2006). Supersensitive Ras activation in dendrites and spines revealed by two-photon fluorescence lifetime imaging. *Nat. Neurosci.*, **9**, 283–291.
125. Klucken, J., Outeiro, T. F., Nguyen, P., McLean, P. J., and Hyman, B. T. (2006). Detection of novel intracellular α-synuclein oligomeric species by fluorescence lifetime imaging. *FASEB J.*, **20**, 2050–2057.
126. Stubbs, C. D., Botchway, S. W., Slater, S. J., and Parker, A. W. (2005). The use of time-resolved fluorescence imaging in the study of protein kinase C localisation in cells. *BMC Cell Biol.*, **6**, 22.

127. Elangovan, M., Day, R. N., and Periasamy, A. (2002). Nanosecond fluorescence resonance energy transfer-fluorescence lifetime imaging microscopy to localize the protein interactions in a single living cell. *J. Microsc.*, **205**, 3–14.

128. Cremazy, F. G. E., Manders, E. M. M., Bastiaens, P. I. H., et al. (2005). Imaging in situ protein-DNA interactions in the cell nucleus using FRET-FLIM. *Exp. Cell Res.*, **309**, 390–396.

129. Bacskai, B. J., Skoch, J., Hickey, G. A., Allen, R., and Hyman, B. T. (2003). Fluorescence resonance energy transfer determinations using multiphoton fluorescence lifetime imaging microscopy to characterize amyloid-beta plaques. *J. Biomed. Opt.*, **8**, 368–375.

130. Calleja, V., Ameer-Beg, S. M., Vojnovic, B., Woscholski, R., Downward, J., and Larijani, B. (2003). Monitoring conformational changes of proteins in cells by fluorescence lifetime imaging microscopy. *Biochem. J.*, **372**, 33–40.

131. Biskup, C., Kelbauskas, L., Zimmer, T., et al. (2004). Interaction of PSD-95 with potassium channels visualized by fluorescence lifetime-based resonance energy transfer imaging. *J. Biomed. Opt.*, **9**, 753–759.

132. Suhling, K., Siegel, J., Phillips, D., et al. (2002). Imaging the environment of green fluorescent protein. *Biophys. J.*, **83**, 3589–3595.

133. Borst, J. W., Hink, M. A., van Hoek, A., and Visser, A. J. W. G. (2005). Effects of refractive index and viscosity on fluorescence and anisotropy decays of enhanced cyan and yellow fluorescent proteins. *J. Fluoresc.*, **15**, 153–160.

134. Diaspro, A., Krol, S., Campanini, B., Cannone, F., and Chirico, G. (2006). Enhanced green fluorescent protein (GFP) fluorescence after polyelectrolyte caging. *Opt. Express*, **14**, 9815–9824.

135. Liu, Y., Kim, H. R., and Heikal, A. A. (2006). Structural basis of fluorescence fluctuation dynamics of green fluorescent proteins in acidic environments. *J. Phys. Chem. B*, **110**(47), 24138–24146.

136. Treanor, B., Lanigan, P. M. P., Suhling, K., et al. (2005). Imaging fluorescence lifetime heterogeneity applied to GFP-tagged MHC protein at an immunological synapse. *J. Microsc.*, **217**, 36–43.

137. Szmacinski, H., and Lakowicz, J. R. (1995). Possibility of simultaneously measuring low and high calcium concentrations using Fura-2 and lifetime-based sensing. *Cell Calcium*, **18**, 64–75.

138. Thomas, D., Tovey, S. C., Collins, T. J., Bootman, M. D., Berridge, M. J., and Lipp, P. (2000). A comparison of fluorescent Ca^{2+} indicator properties

and their use in measuring elementary and global Ca^{2+} signals. *Cell Calcium*, **28**, 213–223.

139. Zhong, W., Urayama, P., and Mycek, M. A. (2003). Imaging fluorescence lifetime modulation of a ruthenium-based dye in living cells: the potential for oxygen sensing. *J. Phys. D: Appl. Phys.*, **36**, 1689–1695.
140. Jayaraman, S., Biwersi, J., and Verkman, A. S. (1999). Synthesis and characterization of dual-wavelength Cl^- sensitive fluorescent indicators for ratio imaging. *Am. J. Physiol.-Cell Physiol.*, **276**, C747–C757.
141. Kuner, T., and Augustine, G. J. (2000). A genetically encoded ratiometric indicator for chloride capturing chloride transients in cultured hippocampal neurons. *Neuron*, **27**, 447–459.
142. Miesenbock, G., De Angelis, D. A., and Rothman, J. E. (1998). Visualizing secretion and synaptic transmission with pH-sensitive green fluorescent proteins. *Nature*, **394**, 192–195.
143. Miyawaki, A., Llopis, J., Heim, R., et al. (1997). Fluorescent indicators for Ca^{2+} based on green fluorescent proteins and calmodulin. *Nature*, **388**, 882–887.
144. Miyawaki, A., and Tsien, R. Y. (2000). Monitoring protein conformations and interactions by fluorescene resonance energy transfer between mutants of green fluorescent protein. *Methods Enzymol.*, **327**, 472–500.
145. Sato, M. (2006). Imaging molecular events in single living cells. *Anal. Bioanal. Chem.*, **386**, 435–443.
146. Sato, M., Ueda, Y., and Umezawa, Y. (2006). Imaging diacylglycerol dynamics at organelle membranes. *Nat. Methods*, **3**, 797–799.
147. Miyawaki, A. (2005). Innovations in the imaging of brain functions using fluorescent proteins. *Neuron*, **48**, 189–199.
148. Knemeyer, J. P., Herten, D. P., and Sauer, M. (2003). Detection and identification of single molecules in living cells using spectrally resolved fluorescence lifetime imaging microscopy. *Anal. Chem.*, **75**, 2147–2153.
149. Eliceiri, K. W., Fan, C. H., Lyons, G. E., and White, J. G. (2003). Analysis of histology specimens using lifetime multiphoton microscopy. *J. Biomed. Opt.*, **8**, 376–380.
150. Owen, D. M., Lanigan, P. M. P., Dunsby, C., et al. (2006). Fluorescence lifetime imaging provides enhanced contrast when imaging the phase-sensitive dye di-4-ANEPPDHQ in model membranes and live cells. *Biophys. J.*, **90**, L80–L82.

151. Brismar, H., and Ulfhake, B. (1997). Fluorescence lifetime measurements in confocal microscopy of neurons labeled with multiple fluorophores. *Nat. Biotechnol.*, **15**, 373–377.

152. Carlsson, K., and Liljeborg, A. (1997). Confocal fluorescence microscopy using spectral and lifetime information to simultaneously record four fluorophores with high channel separation. *J. Microsc.*, **185**, 37–46.

153. Harpur, A. G., Wouters, F. S., and Bastiaens, P. I. H. (2000). Imaging FRET between spectrally similar GFP molecules in single cells. *Nat. Biotechnol.*, **19**, 167–169.

154. Breusegem, S. Y. (2002). *In vivo* investigation of protein interactions in *C. elegans* by foerster resonance energy transfer microscopy [PhD]. University of Illinois, Urbana-Champaign.

155. Williams, B. D., and Waterston, R. H. (1994). Genes critical for muscle development and function in *Caenorhabditis elegans* identified through lethal mutations. *J. Cell Biol.*, **124**, 475–490.

156. Baumgartner, R., Krammer, B., Pottier, R., and Stepp, H., eds. (2006). *Photodynamic Therapy with ALA a Clinical Handbook*. RSC Publishing, Cambridge.

157. Krieg, R. C., Fickweiler, S., Wolfbeis, O. S., Knuechel, R. (2000). Cell-type specific protoporphyrin IX metabolism in human bladder cancer in vitro. *Photochem. Photobiol.*, **72**, 226–233.

158. Scolaro, L. M., Castriciano, M., Romeo, A., Patane, S., Cefali, E., and Allegrin, M. (2002). Aggregation behavior of protoporphyrin IX in aqueous solutions: Clear evidence of vesicle formation. *J. Phys. Chem. B*, **106**, 2453–2459.

159. Seidlitz, H. K., Seidlitz, H., Stettmaier, K., Wessels, J., and Schneckenburger, H. (1992). Intracellular fluorescence polarization, picosecond kinetics, and light-induced reactions of photosensitizing porphyrins. *Opt. Eng.*, **31**, 1482–1486.

160. Russell, J. A., Diamond, K., Collins, T., et al. (2008). Characterization of fluorescence lifetime of Photofrin and delta-aminolevulinic acid induced protoporphyrin IX in living cells using single- and two-photon excitation. *IEEE J. Sel. Top. Quantum Electron.*, **14**, 158–166.

161. Sailer, R., Strauss, W., Wagner, M., Emmert, H., and Schneckenburger, H. (2007). Relation between intracellular location and photodynamic efficacy of 5-aminolevulinic acid-induced protoporphyrin IX in vitro. Comparison between human glioblastoma cells and other cancer cell lines. *Photochem. Photobiol. Sci.*, **6**, 145–151.

162. Graf, R., Rietdorf, J., and Zimmermann, T. (2005). Live cell spinning disk microscopy. *Adv. Biochem. Eng./Biotechnol.*, **95**, 57–75.

163. Wang, E., Babbey, C. M., and Dunn, K. W. (2005). Performance comparison between the high-speed Yokogawa spinning disc confocal system and single-point scanning confocal systems. *J. Microsc.*, **218**, 148–159.

Chapter 11

Advanced Fluorescence Techniques: FRAP, iFRAP, FLIP, FLAP, Photoconvertible, Photoactivatable, and Photoswitchable Proteins

Kelly Rogers[a] **and Sarah Ellis**[a,b]

[a] *The Walter and Eliza Hall Institute of Medical Research, 1G Royal Parade, Parkville Victoria 3052, Australia*

[b] *Cancer Research Division, Peter MacCallum Cancer Centre, 305 Grattan Street, Melbourne, Victoria 3000, Australia, and Sir Peter MacCallum Department of Oncology, University of Melbourne, Parkville, Australia*

sarah.ellis@petermac.org

11.1 Introduction

The aim of this chapter is to introduce biologists to the method known as fluorescence recovery after photobleaching (FRAP) and its derivations, fluorescence loss in photobleaching (FLIP) and fluorescence localization after photobleaching (FLAP). FRAP, FLIP, and FLAP are techniques for measuring the diffusion or movement of proteins within a cell. This information can be used in a variety of ways, including determining the connectivity between intracellular compartments and protein–protein binding dynamics. FRAP was

Fundamentals of Fluorescence Imaging
Edited by Guy Cox

ISBN 978-981-4774-85-7 (Hardcover), 978-1-351-12940-4 (eBook)
www.jennystanford.com

Table 11.1 Examples of FRAP applications, biological models, and methodologies

Fluorophore	Ex/Em (nm)	Application & Biological Model	Method Reported	Reference
GFP	488/509	On-target effect of bromodomain inhibitors U2OS (human osteosarcoma) cells	Pre-scans: x3 scans with 488 nm laser. Bleach: 488 nm laser Post-bleach: Pinhole set to 1.39 AU during acquisition and laser set to 1% of power used for bleaching to improve detection	Philpott et al. (2014)
GFP	488/509	Actin filament turnover beneath the apical membrane Reconstituted rabbit lacrimal acini	Pre-scan with 488 nm laser Bleach: 488 nm laser set at 60%, with 100% transmission Post-bleach: 0.1% of power used for bleaching	Jerdeva et al. (2005)
GFP	488/509	Actin filament turnover in hippocampal spines Rat hippocampal neuron cultures	Pre-scans: x3 scans with 488 nm laser Bleach: x5 scans (3.117 s) Photoactivation with PA-GFP also described	Koskinen and Hotulainen (2014)
GFP	488/509	Nuclear protein mobility Baby hampster kidney cells	Pre-scans: x 5 with 488 nm laser Bleach: Single bleach pulse of 200–500 ms Post-bleach: 1% of power used for bleaching FLIP also described	Phair and Misteli (2000)
Rhodamine-123 (Rh123)	507/529	Connexin36 gap junction coupling	GFP pre-scans: x3 scans with 488 nm laser Bleach: 150 cycles, 488 nm laser Bleaching time = 235.5 s	Farnsworth et al. (2014)

Venus	515/528	Murine pancreatic islets	Post-bleach: Laser set to <1% of power used for bleaching Venus: Multi-spectral detection Bleach: Infrared femtosecond laser at 720 nm for 120 cycles Post-bleach: Laser set to <0.004% of power used for bleaching	
Dendra2	490/507	Mitochondrial length and interconnectivity	Bleach: 405 nm laser until fluorescence was equal to ~10% of the pre-scanned value	Losón et al. (2013)
After photo-activation	553/573	Mouse embryonic fibroblasts	Photoconversion of Dendra2 also described	
FM1-43	510/626	Synaptic vesicle movement Frog motor nerve terminals	Pre-scans: 488 nm laser line Bleach: 1000–3000 line scans or spot scanning for 0.2–1.0 s	Henkel et al. (1996)
GFP	488/509	Diffusion within the cleavage furrow	GFP: Bleach – 25 scans with 488 nm laser at full power	Schmidt and Nichols (2004)
YFP	514/527	HeLa cells (human cervical cancer cells)	YFP: Bleach – 25 scans with 514 nm laser at full power Dil: Bleach – 40 scans with 514 nm and	
Dil	549/565	NRK cells (normal rat kidney epithelial cells)	546 nm lasers at full power Also describes photoactivation of PA- GFP	

Ex, excitation maxima; Em, emission maxima; GFP has a rate constant of 0.069 min^{-1} (Patterson et al. 1997).

first proposed by several research teams for studying the lateral transport of membrane proteins over 30 years ago (Peters et al. 1974, Poo and Cone 1974, Axelrod et al. 1976). During the ensuing years considerable advances in microscopy and the design and development of new fluorescent probes has resulted in an increase in the number of FRAP applications, which have been used to examine intracellular protein movement (Table 11.1).

The following pages will provide practical examples of how to measure the movement of fluorescent molecules by FRAP, FLIP, and FLAP in an easy-to-follow workflow with the aim of encouraging the reader to commence experiments and achieve meaningful data. The intention is to keep the dialogue simple with references provided for those wishing to pursue a higher level of understanding of the physics and algorithms involved in the measurements. The chapter will conclude with a discussion on the use of photoactivatable, photoconvertible, and photoswitchable proteins to measure protein movement within cells.

11.2 Fluorescence Recovery after Photobleaching (FRAP)

In a FRAP experiment, fluorescent molecules in a specific region of interest (ROI) are photobleached using a high-intensity laser beam. In this condition, the high-powered laser source causes irreversible modification to the molecular structure of a fluorophore, which renders it no longer capable of producing fluorescence. How fast a fluorophore bleaches, or how many excitation/emission cycles the fluorophore undergoes before becoming irreversibly bleached, depends on the properties of the fluorophore within a given environment and the dose of light delivered to that environment.

Most FRAP experiments are carried out with a confocal microscope equipped with an acoustic optical tunable filter (AOTF), which allows ROIs to be precisely defined and bleached. An AOTF is important because it enables the wavelength and intensity of illumination to be altered on a pixel-by-pixel basis without affecting the scan rate. It is also possible to perform FRAP experiments

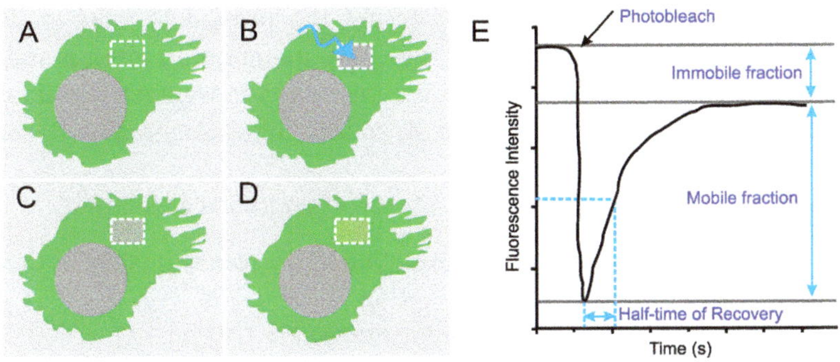

Figure 11.1 Schematic diagram illustrating each step of the FRAP procedure. (A) An initial sequence of frames are acquired to establish the intensity of the cell prior to bleaching and draw the ROI. (B) A bleaching cycle is executed to reduce the pool of fluorescent molecules within the ROI. This causes the intensity within the ROI to decrease by a defined intensity (e.g., to 50% of the pre-bleach value). (C and D) Image the recovery of fluorescence to the ROI until it equilibrates. (E) Plot the recovery of fluorescence within the ROI over time.

using wide-field microscopy or multiphoton microscopy. Wide-field microscopes have excellent temporal resolution and adaptors consisting of a laser, bleaching masks, and associated FRAP software are now available. Multiphoton microscopes that have a separate laser for FRAP can bleach a precisely defined ROI and allow continuous imaging throughout the bleaching and recovery process. For the purposes of this chapter, discussion will be limited to confocal microscopy applications.

FRAP operates on the basis that if a fluorescent molecule outside the ROI in the cytoplasm, organelle, or membrane is mobile, it will diffuse or be actively pumped into the bleached area. The rate of fluorescence recovery into the bleached area is then recorded in a time-lapse sequence using low laser intensity (down to 1% or below that of the laser power used for bleaching). A graph is then constructed of the temporal changes in fluorescence intensity within the bleached region (Fig. 11.1). FRAP depends on the principle that recovery of fluorescence within the bleached region is highly dependent on factors such as (i) the molecular weight of the fluorescent molecule or hybrid protein, (ii) its interaction with binding partners, (iii) whether or not the protein of interest is anchored to a less mobile structure (e.g., membrane protein), and (iv) whether or not the protein has restricted movement within a subcellular domain due to other structures, like membranes (Reits and Neefjes 2001).

A typical FRAP experimental protocol is outlined as follows:

(1) Take an initial series of baseline fluorescence images of the cell.
(2) Draw a ROI using the confocal software (Fig. 11.1A).
(3) Scan the ROI using high laser intensity until the fluorescence is markedly reduced (Fig. 11.1B). The bleaching process should be as short as possible, within a few seconds.
(4) Consecutively scan the cell to record non-bleached molecules diffusing into the ROI to replace the bleached molecules. In seconds to a few minutes, the fluorescence intensity of the ROI will equilibrate and eventually plateau (Fig. 11.1C,D).
(5) For the ROI, plot the recovery of fluorescence intensity over time (Fig. 11.1E). Most confocal software will perform this task.

Several parameters can be determined from the graph of fluorescence recovery (Fig. 11.1E). The fraction of the fluorescence that returns to the ROI is referred to as the mobile fraction (M_f). This is determined by calculating the ratio of the final fluorescence intensity of the ROI to the initial fluorescence intensity of the ROI. The fraction of fluorescence that does not move into the bleached ROI is referred to as the immobile fraction (IM_f) and is calculated by subtracting M_f from 1 (Ishikawa-Ankerhold et al. 2012). Additionally, the FRAP literature often refers to the half-life or half maximum to describe the speed of fluorescence recovery to half the maximum of the plateau intensity. This figure is calculated by drawing a line starting half way up the recovery curve from the final plateau and bisecting the *x* axis. Most confocal software contain a FRAP wizard and will automatically map the relative ROI fluorescence over time. If this option is not available, the ROI fluorescence intensity over time can be calculated using readily available third-party software such as Image J. Suitable controls must also be included in every FRAP experiment. For example, FRAP can be performed on the same cells that are fixed (no FRAP) or on an adjacent area when compartmental exchange of proteins is under investigation. FRAP experiments can also be performed at a low and high level (e.g., 10 iterations vs. 100 iterations) to verify that values for diffusion are not influenced by the conditions that have been set for photobleaching.

GFP is the most commonly reported fluorophore in the literature for FRAP and is relatively easy to bleach. Experimental protocols may require the use of other fluorophores, which are harder to bleach, particularly if the protein of interest is linked to a photoprotein or dye that has greater photostability than GFP. In such cases, higher laser powers can be utilized to ensure a suitable reduction in intensity at the rate required. Simultaneous illumination by a combination of the 405 nm laser together with any other relevant lasers (488 nm and 561 nm lasers) can sometimes facilitate bleaching in the time frame required. Applying a zoom during the bleaching step can also facilitate the bleaching process. Some FRAP wizards automatically apply a zoom to illuminate the ROI. All these methods increase the chance of photodamage and may compromise the viability of the specimen or the integrity of the target. For this reason, bleaching an ROI to approximately 50% of the initial intensity is usually adequate.

There are many excellent examples of the use of FRAP in the literature. In particular, FRAP has been utilized to examine the secretory pathways through the endoplasmic reticulum to the golgi (see Lippincott-Schwartz 2003 for an excellent overview). FRAP has also been used to study lateral diffusion or the mobility of different proteins and their mutants within the plasma membrane and the nucleus (Phair and Misteli 2000, Losón, Song et al. 2013, Philpott et al. 2014) and to study actin remodeling (Jerdeva et al. 2005). A small number of these publications have been outlined in Table 11.1 to provide the reader with a sense of how FRAP can be applied across different biological models. Where possible, publications that provide a clear description of the FRAP procedure or analysis have been selected.

Another variation of FRAP is inverse FRAP (iFRAP). As the name suggests, this is the reverse of FRAP. In an iFRAP experiment, the ROI

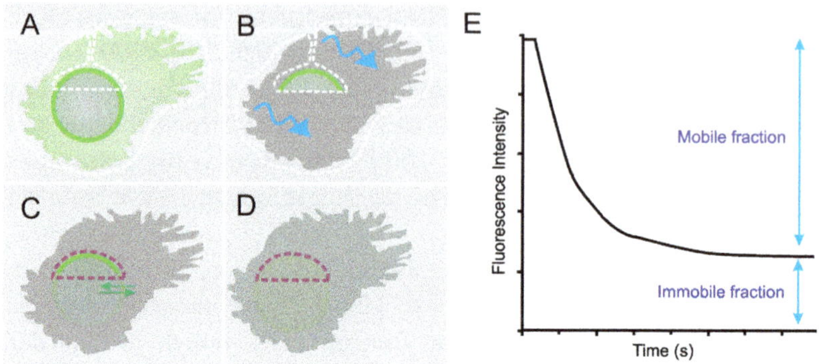

Figure 11.2 Schematic diagram showing an example of an inverse FRAP (iFRAP) experiment (adapted from Rabut et al. 2004). (A) An initial sequence of frames is acquired to establish the fluorescence intensity of the cell prior to bleaching. The area to be bleached is delineated (white dashed line). In the schematic, the entire cell will be bleached except the top half of the nucleus. (B) A bleaching cycle is executed to reduce the pool of fluorescent molecules within the cell. (C and D) Fluorescence in the ROI (purple region) decreases as the fluorescently tagged protein diffuses into surrounding areas. (E) The release kinetics of the GFP-tagged protein within the ROI can be measured by graphing the loss in fluorescence from this area.

is not bleached. Instead, an area surrounding the ROI is bleached. iFRAP has been used to study the time interval molecules reside within a particular organelle before moving out and to examine bind and release kinetics.

An iFRAP experiment is performed in the same manner as a FRAP experiment except the ROI records the loss in fluorescence over time (Fig. 11.2). No complicated analyses are required for iFRAP. However, a consideration with iFRAP is the duration required to bleach the areas surrounding the ROI. iFRAP was pivotal for studying the functional dynamics of nuclear pore complex (NPC) components (Rabut et al. 2004). In this study, the dissociation rate of 19 different GFP-tagged nucleoporins was examined with iFRAP. iFRAP experiments in conjunction with kinetic modeling highlighted three classes of nucleoporins: (i) those essential for correct NPC assembly, "scaffold" nucleoporins; (ii) those required for anchoring the NPC to the nuclear membrane, "adaptor" nucleoporins; and (iii) those whose main function was considered to be involved in transport or regulatory or adaptor functions, "dynamic" nucleoporins. Compared to FRAP, iFRAP is more suitable for studying the binding interactions of nucleoporins because the free pool of GFP-tagged protein is depleted in iFRAP leaving only the bound GFP-tagged protein. Depleting the free pool of GFP-tagged protein facilitates the measurement of changes in fluorescence of the bound molecules. The dissociation kinetics of RNA polymerase I components and pre-mRNAs have also been investigated using iFRAP (Dundr et al. 2002, Ishihama et al. 2008).

11.3 Fluorescence Loss in Photobleaching (FLIP)

FLIP is similar to FRAP as it involves bleaching a ROI. However, unlike FRAP the ROI is continually bleached over time and the loss of fluorescence in an area outside of the ROI is measured to determine the mobility of molecules. FLIP is a complementary method to FRAP and is therefore often used as a control for FRAP experiments. It measures what is happening outside the ROI whereas FRAP examines what is happening inside the ROI.

FLIP is also particularly useful for examining connectivity between compartments (Snapp et al. 2003). For example, compartments of the cell that are connected to the ROI being actively bleached will show a decrease in fluorescence as the fluorescent proteins move into the ROI. Compartments that are not connected to the ROI will retain their fluorescence. The technique has also been used to demonstrate the dissociation kinetics of proteins from particular compartments. Over a decade ago, Phair and Misteli utilized both FRAP and FLIP to elegantly demonstrate that proteins within nuclear compartments were highly dynamic. As part of this study, they created a GFP-fusion protein with the rRNA processing protein, fibrillarin, which is located in the nucleolus. FLIP experiments performed by repeatedly bleaching a ROI in the nucleoplasm or a ROI surrounding a single nucleolus whilst measuring the corresponding decrease in the pool of GFP in neighboring nucleoli, suggested there was a continuous and rapid exchange of fibrillarin between nucleoli and nucleoplasm (Phair and Misteli 2000). In a more recent study, both FLIP and FRAP were used

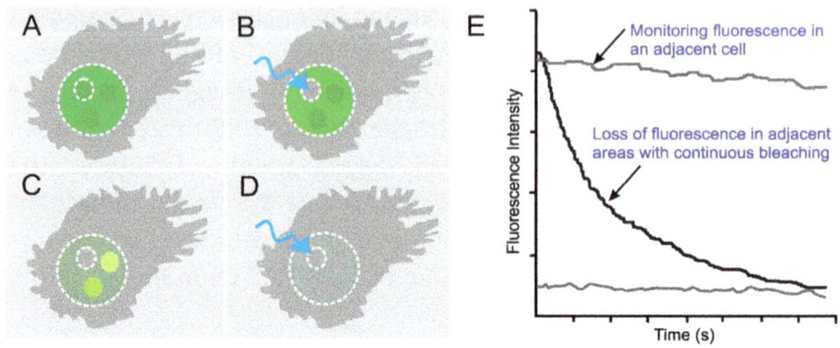

Figure 11.3 Schematic diagram showing an example of a FLIP experiment as described by Phair and Mistelli (2000). (A) An initial sequence of frames are acquired to establish the intensity of the cell prior to bleaching and to draw the ROIs. In the schematic, the ROI to be bleached is drawn around a nucleolus and another ROI is drawn around the nucleus. Yet another ROI should be drawn around the nucleolus in a neighboring cell to act as a control region (it should not fade). (B) A bleaching cycle is executed to reduce the pool of fluorescent molecules within the nucleolus ROI. (C and D) Continuous imaging and bleaching cycles follow. If the fluorescently tagged protein can rapidly exchange between neighboring nucleoli, there will be a resulting decrease in nucleolar fluorescence throughout the nucleus. (E) The loss of fluorescence with continuous bleaching can be monitored over time by acquiring a series of sequential imaging scans. Changes in fluorescence of control regions (e.g., nucleoli in neighboring cells, and background) should also be shown.

to study the translocation of the pro-apoptotic protein, Bax, between the cytosol and the mitochondrial membrane. Bax dissociation from the outer mitochondrial membrane in healthy cells was analyzed using FLIP; two small ROI were defined to bleach virtually all mobile cytosolic GFP-Bax. A third ROI was drawn around the mitochondria that lay outside the bleached regions to quantify the trafficking of mitochondrial GFP-Bax to the cytosol (Schellenberg et al. 2013). Both FLIP and FRAP were crucial in revealing that survival signals control the rate of Bax dissociation from mitochondria and not the targeting of Bax to mitochondria, as previously thought.

A typical FLIP experimental protocol is outlined as follows.

(1) Take an initial series of baseline fluorescence images of the cell including the ROI.

(2) Draw a ROI over the area of interest (the nucleus in the schematic below) and another ROI over the area to be bleached (a nucleolus in the schematic below) (Fig. 11.3A).
(3) Scan the ROI to be bleached using high laser intensity and alternate between bleaching and recording the image at low laser intensity (Fig. 11.3B). Generally, it is appropriate to capture 3–4 images at low laser intensity followed by a bleaching image. The low laser settings for image capture are optimized in step 1. If the confocal is equipped with a separate laser for bleaching, it will be possible to simultaneously bleach and collect the image.
(4) Continue until the fluorescence of the organelle/s (nucleus and nucleoli in the schematic below) being examined for connectivity fades, which will only occur if fluorophores continuously move into the area being bleached (Fig. 11.3C,D).
(5) Plot the fluorescence intensity of the area of interest (nucleus in the schematic) over time (Fig. 11.3E).

11.4 Fluorescence Localization after Photobleaching (FLAP)

In 2002, Dunn and colleagues introduced a new method for studying protein dynamics in living cells termed FLAP (Dunn et al. 2002). In contrast to FRAP and FLIP, the distribution of both the bleached and unbleached molecules can be tracked. The technique is relatively easy to implement. The molecule to be tracked is tagged with two different fluorophores, one that will be photobleached with the other serving as a reference. Following an experimental design similar to FRAP, the bleached molecules can be identified with simple image processing.

A typical FLIP experimental protocol is outlined as follows:

(1) Optimize the setup for the two fluorophores using offset and gain. Make sure there are no saturated pixels (Fig. 11.4B). For optimal FLAP, image the two fluorophores simultaneously. If the emission spectra of the two fluorophores overlap too

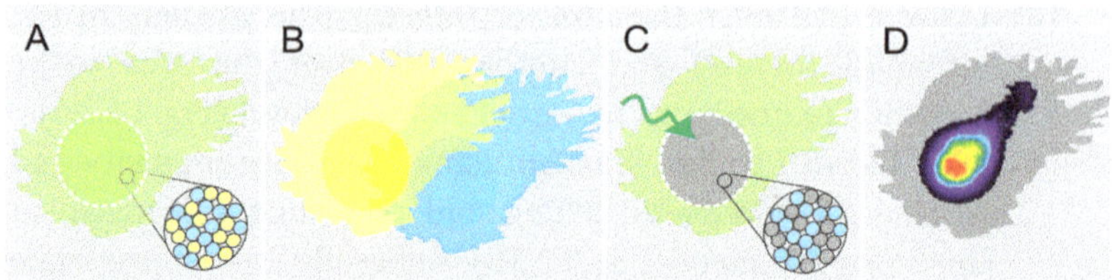

Figure 11.4 Schematic diagram illustrating an example of a FLAP experiment as described by Dunn et al. (2002). (A) Molecules to be tracked are labeled with two different fluorophores. CFP and YFP are used in this schematic. (B) Adjust the gain and offset for each fluorophore to match intensities with no saturation. (C) Draw a ROI and bleach YFP. (D) An "absolute" FLAP image is obtained by subtracting the YFP image from the CFP image and applying a pseudocolor lookup table to reveal the movement of the bleached molecules within the cell.

much to visualize separate emissions, capture the images sequentially using a line scan.

(2) Take an initial series of baseline fluorescence images of the cell including the ROI.
(3) Bleach one of the fluorophores within the ROI using a high intensity laser setting until the fluorescence is irreversibly bleached (Fig. 11.4C). The bleaching process should be as short as possible.
(4) Image the recovery of the fluorescence as per FRAP.
(5) The bleached molecules can be tracked by virtue of the difference in fluorescence intensity between the two fluorophores. In practice, the absolute FLAP image is acquired by subtracting the intensity of the target (bleached fluorophore) from the intensity of the unbleached "reference" fluorophore at each pixel. The FLAP image reveals the location of photobleached molecules. Prior to bleaching, the absolute FLAP signal will be close to 0. A pseudocolor look up table can be applied to the resulting time series after applying this basic image processing to examine the distribution pattern of the bleached molecules (Fig. 11.4D).

There are a number of publications, which detail FLAP image calculations in more detail (Dunn et al. 2002, Zicha et al. 2003, Dunn et al. 2004).

11.5 Potential Pitfalls in FRAP, FLIP, and FLAP Experiments

As with any imaging technique, there are a number of pitfalls to be aware of when utilizing bleaching techniques. Several of the main pitfalls are described below:

(1) The bleaching process may result in a decrease in the overall fluorescence intensity of the cellular compartment under investigation. For example, if the fluorescent molecules are "locked" within a cellular compartment and the ROI is within that compartment then the total fluorescence recovery will be affected and will influence both M_f and IM_f. Conversely, if a large ROI is selected (e.g., an entire nucleus), the full recovery of a freely mobile molecule will be less than the initial fluorescence of the ROI (nucleus). It is possible to correct for the loss in intensity produced by the FRAP experiment (see Phair and Misteli 2000).

(2) Another factor that can affect the overall fluorescence intensity and thus FRAP measurements is the photobleaching that occurs during image acquisition over time. This can be minimized by optimizing image capture parameters such as the laser intensity, pixel size, speed of acquisition, pinhole size, and choice of objective, and choosing robust fluorophores. To ascertain if photobleaching is occurring, the total fluorescence intensity of a cell distal from the one being examined but within the field of view can be measured over time. If the amount of photobleaching is very small, it can be tolerated. Confocal FRAP wizards usually offer an automatic correction for background photobleaching or the amount of background photobleaching occurring at each time point can be subtracted from the measurements using Excel or similar.

(3) There are several drawbacks associated with methods that bleach molecules: fluorescent proteins such as GFP are continually produced by living cells and can be slow to bleach. Cycloheximide can be added to the medium to inhibit the synthesis of new proteins (Rabut et al. 2004). Additionally, GFP is capable of reversible photobleaching,

which can result in an apparent recovery of GFP fluorescence despite no diffusion. The effect of the latter can be examined by repeating the FRAP experiment in fixed samples, which do not exhibit diffusion.

(4) Laser induced damage can occur in the area that has been bleached. Appropriate methods or controls should be applied in these experiments to ensure structural damage does not affect recovery rates (for examples, see Henkel et al. 1996). It is also important to minimize the production of reactive oxygen species and oxidative stress, which are caused by high laser intensities, such as those commonly used to bleach.

(5) Movement artifacts can be particularly problematic when working on non-adherent cells, like T cells or red blood cells. Possible workarounds include methods to stabilize the cells, such as using a small percentage of low melting agarose. Alternatively, an ultrafast bleaching protocol could be utilized together with a moving ROI to analyze recovery.

(6) Careful consideration also needs to be given to the three-dimensional imaging volume. Low numerical aperture objectives will bleach a column but high numerical apertures will produce a more complex bleaching profile.

For a more comprehensive discussion of the pitfalls associated with FRAP, see Henkel et al. (1996), Lippincott-Schwartz et al. (2003), Houtsmuller (2005), and Ishikawa-Ankerhold et al. (2012).

11.6 Photoactivable, Photoconvertible, and Photoswitchable Probes

FRAP and associated methods have proven to be highly valuable techniques for understanding the behavior of proteins in cells and cellular domains. However, these methods tend to be limited to studies on a particular protein population, and although they provide information about the mobility of proteins and how they diffuse in and out of a specific cellular domain, these techniques do not allow individual proteins or protein clusters to be monitored directly.

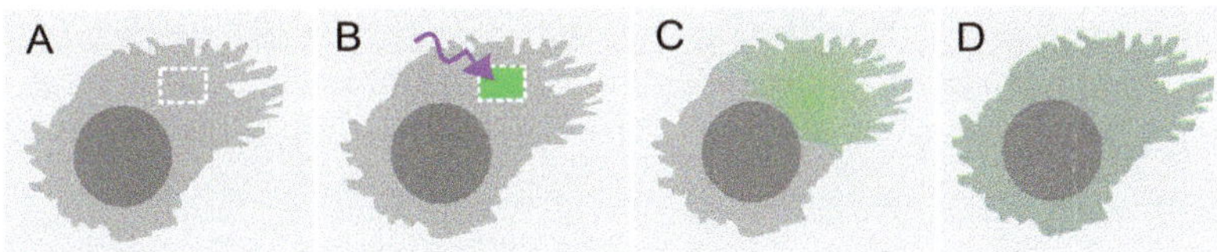

Figure 11.5 Schematic diagram illustrating the photoactivation of PA-GFP. (A) A ROI is drawn. (B) PA-GFP within the ROI is activated by violet light (405 nm) and is imaged using a 488 nm laser (Em. max. 517 nm). (C and D) The movement of the photoactivated molecules can be observed within the cell.

In 2002, Jennifer Lippincott-Schwartz and George Patterson introduced the first photoactivable GFP protein (PA-GFP) (Patterson and Lippincott-Schwartz 2002). PA-GFP is activated with light at 405 nm (Fig. 11.5). Once activated, PA-GFP can be excited with a 488 nm laser. This landmark study paved the way for a series of studies that tracked photoactivated molecules within the cell. Some years later, a photoactivatable version of mCherry (PA-mCherry) was described (Subach et al. 2009). Both PA-GFP and PA-mCherry fluoresce strongly after activation with light at 405 nm. Once activated, PA-GFP has an emission maximum of 517 nm and PA-mCherry has an emission maximum at 595 nm after excitation with a 561 nm laser.

Photoconvertible probes began to make their mark around the same time as PA-GFP. Photoconvertible proteins such as Kaede (Ando et al. 2002), Dendra2 (Gurskaya et al. 2006), mEosFP2 (McKinney et al. 2009), Kikume/KikGR (Tsutsui et al. 2005) and mKikGR (Habuchi et al. 2008) are irreversibly converted from one color (usually green) to red by activation with a particular wavelength of light. For example, the native version of Kaede can be imaged prior to conversion with a 488 nm laser and irreversibly converted from green to red after excitation with a 405 nm laser (Fig. 11.7). The resulting red version of the protein is usually very stable.

Another alternative to these approaches are photoswitchable proteins, whose fluorescence can be switched on or off. These include the blue-green or green emitting photoswitchable proteins Dronpa (Ando et al. 2004, Habuchi et al. 2005) and Padron (Andresen et al. 2008), and the red emitting photoswitchable

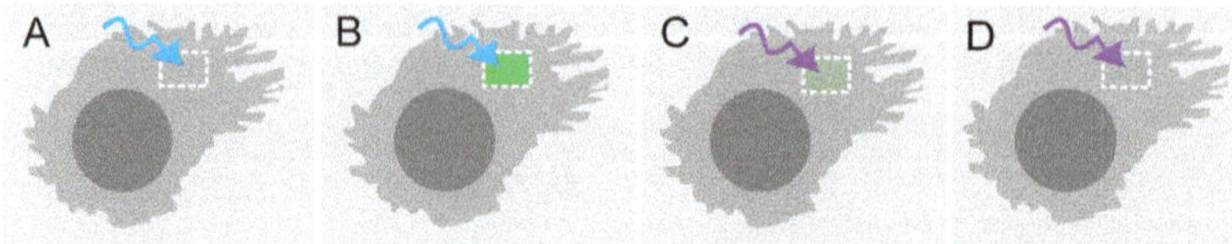

Figure 11.6 Schematic diagram illustrating the process involved in photoswitching Padron. (A) A ROI is drawn. (B) Blue light (488 nm) induces fluorescence and simultaneously turns Padron from the off state to the on state. (C and D) Violet light (405 nm) turns Padron off.

proteins Kindling (Chudakov et al. 2003), rsCherry, and rsCherryRev (Stiel et al. 2008). Both rsCherry and rsCherryRev emit at 610 nm and can be switched on or off with the 488 nm and 561 nm lasers. RsCherry is switched on with the yellow 561 nm laser and off with the blue 488 nm laser. The reverse is true for rsCherryRev (Stiel et al. 2008). The photoswitchable protein, Padron, is switched on by 488 nm light, emitting at 522 nm (Em. max.), and switched off by 405 nm light (Andresen et al. 2008) (Fig. 11.6).

The introduction of fluorescent proteins that can be photoactivated, photoconverted or photoswitched offers a powerful alternative to methods based on FRAP. A major advantage of these proteins is that they can be converted from one color to another or switched on or off to examine molecular diffusion using lower intensities of laser power compared to those required for FRAP.

In general, experimental methods for measuring molecular movement with these classes of fluorescent proteins are similar to FRAP or iFRAP depending on the type of fluorophore selected. For example, photoswitchable fluorescent proteins can be switched off following illumination with a particular wavelength and can be used in a manner similar to FRAP. On the other hand, photoactivated fluorescent proteins that are fluorescent only after illumination by a particular wavelength can be used in a manner similar to iFRAP. Perhaps the fluorescent proteins offering the greatest scope are the photoconvertible proteins, which enable fluorescent cells to be switched from green to red wavelengths and molecular trajectories to be continuously tracked during time. Overall, these tools have increased the biologists' capacity to monitor protein dynamics, dissociation kinetics of macromolecular complexes and studies on

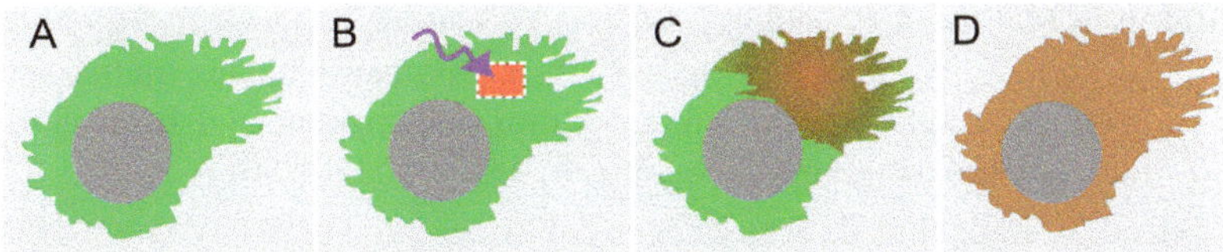

Figure 11.7 Schematic to illustrate the steps involved in photoconverting Kaede. (A) Capture a series of images prior to conversion using the 488 nm laser for excitation. (B) Draw a ROI and photoconvert Kaede from green to red using a 405 nm laser. (C and D) Image the movement of the photoconverted molecules.

cellular migration in live organisms, which has opened new avenues for studying the behavior and functioning of complex organisms.

A typical experimental design for a photoconvertible probe is outlined as follows:

(1) Prepare cells or tissues expressing a photoconvertible probe for imaging (e.g., Kaede). Minimize exposure of the samples to UV light to avoid any background conversion (Fig. 11.7A).
(2) Set up the red and green channels on the confocal. For the red channel, optimize the settings on a control sample that has been photoconverted.
(3) Draw a ROI over the region of interest.
(4) Photoconvert the fluorophores within the ROI using a 405 laser (Fig. 11.7B). Monitor both the green and red channels throughout the process until the probe is irreversibly converted from green to red.
(5) The photoconverted molecules or cells can then be tracked in a time-lapse experiment (Fig. 11.7C,D).

References

Ando, R., Hama, H., Yamamoto-Hino, M., Mizuno, H., and Miyawaki, A. (2002). An optical marker based on the UV-induced green-to-red photoconversion of a fluorescent protein. *Proc. Natl. Acad. Sci. U.S.A.*, **99**(20), 12651–12656.

Ando, R., Mizuno, H., and Miyawaki, A. (2004). Regulated fast nucleocytoplasmic shuttling observed by reversible protein highlighting. *Science*, **306**(5700), 1370–1373.

Andresen, M., Stiel, A. C., Folling, J., Wenzel, D., Schonle, A., Egner, A., Eggeling, C., Hell, S. W., and Jakobs, S. (2008). Photoswitchable fluorescent proteins enable monochromatic multilabel imaging and dual color fluorescence nanoscopy. *Nat. Biotech.*, **26**(9), 1035–1040.

Axelrod, D., Koppel, D. E., Schlessinger, J., Elson, E., and Webb, W. W. (1976). Mobility measurement by analysis of fluorescence photobleaching recovery kinetics. *Biophys. J.*, **16**(9), 1055–1069.

Chudakov, D. M., Belousov, V. V., Zaraisky, A. G., Novoselov, V. V., Staroverov, D. B., Zorov, D. B., Lukyanov, S., and Lukyanov, K. A. (2003). Kindling fluorescent proteins for precise in vivo photolabeling. *Nat. Biotech.*, **21**(2), 191–194.

Dundr, M., Hoffmann-Rohrer, U., Hu, Q., Grummt, I., Rothblum, L. I., Phair, R. D., and Misteli, T. (2002). A kinetic framework for a mammalian rna polymerase in vivo. *Science*, **298**(5598), 1623–1626.

Dunn, G. A., Dobbie, I. M., Monypenny, J., Holt, M. R., and Zicha, D. (2002). Fluorescence localization after photobleaching (FLAP): a new method for studying protein dynamics in living cells. *J. Microsc.*, **205**, 109–112.

Dunn, G. A., Holt, M. R., Soong, D. Y. H., Gray, C., and Zicha, D. (2004). Fluorescence localization after photobleaching (FLAP). *Curr. Protoc. Cell Biol.*, **24**(1), 21.2.1–21.2.16.

Farnsworth, N. L., Hemmati, A., Pozzoli, M., and Benninger, R. K. P. (2014). Fluorescence recovery after photobleaching reveals regulation and distribution of connexin36 gap junction coupling within mouse islets of Langerhans. *J. Physiol.*, **592**(Pt 20), 4431–4446.

Gurskaya, N. G., Verkhusha, V. V., Shcheglov, A. S., Staroverov, D. B., Chepurnykh, T. V., Fradkov, A. F., Lukyanov, S., and Lukyanov, K. A. (2006). Engineering of a monomeric green-to-red photoactivatable fluorescent protein induced by blue light. *Nat. Biotech.*, **24**(4), 461–465.

Habuchi, S., Ando, R., Dedecker, P., Verheijen, W., Mizuno, H., Miyawaki, A., and Hofkens, J. (2005). Reversible single-molecule photoswitching in the GFP-like fluorescent protein Dronpa. *Proc. Natl. Acad. Sci. U.S.A.*, **102**(27), 9511–9516.

Habuchi, S., Tsutsui, H., Kochaniak, A. B., Miyawaki, A., and van Oijen, A. M. (2008). mKikGR, a monomeric photoswitchable fluorescent protein. *PLoS One*, **3**(12), e3944.

Henkel, A. W., Simpson, L. L., Ridge, R. M. A. P., and Betz, W. J. (1996). Synaptic vesicle movements monitored by fluorescence recovery after photobleaching in nerve terminals stained with FM1-43. *J. Neurosci.*, **16**(12), 3960–3967.

Houtsmuller, A. B. (2005). Fluorescence recovery after photobleaching: application to nuclear proteins. *Adv. Biochem. Eng./Biotechnol.*, **95**, 177–199.

Ishihama, Y., Tadakuma, H., Tani, T., and Funatsu, T. (2008). The dynamics of pre-mRNAs and poly(A)+ RNA at speckles in living cells revealed by iFRAP studies. *Exp. Cell Res.*, **314**(4), 748–762.

Ishikawa-Ankerhold, H. C., Ankerhold, R., and Drummen, G. P. C. (2012). Advanced fluorescence microscopy techniques–FRAP, FLIP, FLAP, FRET and FLIM. *Molecules*, **17**(4), 4047–4132.

Jerdeva, G. V., Wu, K., Yarber, F. A., Rhodes, C. J., Kalman, D., Schechter, J. E., and Hamm-Alvarez, S. F. (2005). Actin and non-muscle myosin II facilitate apical exocytosis of tear proteins in rabbit lacrimal acinar epithelial cells. *J. Cell Sci.*, **118**(Pt 20), 4797–4812.

Koskinen, M., and Hotulainen, P. (2014). Measuring F-actin properties in dendritic spines. *Front. Neuroanat.*, **8**, 74.

Lippincott-Schwartz, J. (2003). Development and use of fluorescent protein markers in living cells. *Science*, **300**(5616), 87–91.

Lippincott-Schwartz, J., Altan-Bonnet, N., and Patterson, G. H. (2003). Review: Photobleaching and photoactivation: following protein dynamics in living cells. *Nat. Cell Biol.*, **5**, S7–S14.

Losón, O. C., Song, Z., Chen, H., and Chan, D. C. (2013). Fis1, Mff, MiD49, and MiD51 mediate Drp1 recruitment in mitochondrial fission. *Mol. Biol. Cell*, **24**(5), 659–667.

McKinney, S. A., Murphy, C. S., Hazelwood, K. L., Davidson, M. W., and Looger, L. L. (2009). A bright and photostable photoconvertible fluorescent protein for fusion tags. *Nat. Methods*, **6**(2), 131–133.

Patterson, G. H., Knobel, S. M., Sharif, W. D., Kain, S. R., and Piston, D. W. (1997). Use of the green fluorescent protein and its mutants in quantitative fluorescence microscopy. *Biophys. J.*, **73**(5), 2782–2790.

Patterson, G. H., and Lippincott-Schwartz, J. (2002). A photoactivatable GFP for selective photolabeling of proteins and cells. *Science*, **297**(5588), 1873–1877.

Peters, R., Peters, J., Tews, K. H., and Bähr, W. (1974). A microfluorimetric study of translational diffusion in erythrocyte membranes. *Biochim. Biophys. Acta, Biomembr.*, **367**(3), 282–294.

Phair, R. D., and Misteli, T. (2000). High mobility of proteins in the mammalian cell nucleus. *Nature*, **404**(6778), 604–609.

Philpott, M., Rogers, C. M., Yapp, C., Wells, C., Lambert, J.-P., Strain-Damerell, C., Burgess-Brown, N. A., Gingras, A.-C., Knapp, S., and Müller, S. (2014). Assessing cellular efficacy of bromodomain inhibitors using fluorescence recovery after photobleaching. *Epigenet. Chromatin*, **7**(1), 1–12.

Poo, M.-M., and Cone, R. A. (1974). Lateral diffusion of rhodopsin in the photoreceptor membrane. *Nature*, **247**(5441), 438–441.

Rabut, G., Doye, V., and Ellenberg, J. (2004). Mapping the dynamic organization of the nuclear pore complex inside single living cells. *Nat. Cell Biol.*, **6**(11), 1114–1121.

Reits, E. A. J., and Neefjes, J. J. (2001). From fixed to FRAP: measuring protein mobility and activity in living cells. *Nat. Cell Biol.*, **3**(6), E145–E147.

Schellenberg, B., Wang, P., Keeble, J. A., Rodriguez-Enriquez, R., Walker, S., Owens, T. W., Foster, F., Tanianis-Hughes, J., Brennan, K., Streuli, C. H., and Gilmore, A. P. (2013). Bax exists in a dynamic equilibrium between the cytosol and mitochondria to control apoptotic priming. *Mol. Cell*, **49**(5), 959–971.

Schmidt, K., and Nichols, B. J. (2004). A barrier to lateral diffusion in the cleavage furrow of dividing mammalian cells. *Curr. Biol.*, **14**(11), 1002–1006.

Snapp, E. L., Altan, N., and Lippincott-Schwartz, J. (2003). Measuring protein mobility by photobleaching GFP chimeras in living cells. *Curr. Protoc. Cell Biol.*, **19**(1), 21.1.1–21.1.24.

Stiel, A. C., Andresen, M., Bock, H., Hilbert, M., Schilde, J., Schönle, A., Eggeling, C., Egner, A., Hell, S. W., and Jakobs, S. (2008). Generation of monomeric reversibly switchable red fluorescent proteins for far-field fluorescence nanoscopy. *Biophys. J.*, **95**(6), 2989–2997.

Subach, F. V., Patterson, G. H., Manley, S., Gillette, J. M., Lippincott-Schwartz, J., and Verkhusha, V. V. (2009). Photoactivatable mCherry for high-resolution two-color fluorescence microscopy. *Nat. Methods*, **6**(2), 153–159.

Tsutsui, H., Karasawa, S., Shimizu, H., Nukina, N., and Miyawaki, A. (2005). Semi-rational engineering of a coral fluorescent protein into an efficient highlighter. *EMBO Rep.*, **6**(3), 233–238.

Zicha, D., Dobbie, I. M., Holt, M. R., Monypenny, J., Soong, D. Y. H., Gray, C., and Dunn, G. A. (2003). Rapid actin transport during cell protrusion. *Science*, **300**(5616), 142–145.

Chapter 12

Förster Resonance Energy Transfer Microscopy for Monitoring Molecular Dynamics in Living Cells

Vinod Jyothikumar, Yuansheng Sun, and Ammasi Periasamy
W.M. Keck Center for Cellular Imaging, University of Virginia, Biology, Gilmer Hall, 485 McCormick Road, Charlottesville, VA 22904, USA
ap3t@virginia.edu

Förster resonance energy transfer (FRET) is an effective and high-resolution method to monitor protein–protein interactions in live or fixed specimens. FRET microscopy is increasingly employed to access the molecular mechanisms governing diverse cellular processes such as vesicular transport, signal transduction, and the regulation of gene expression. FRET can be used to estimate the distance between interacting protein molecules in vivo or in vitro using light microscopy systems. In this chapter we provide the basics of FRET, FRET pairs, hardware involved in FRET imaging, issues involved in FRET imaging using light microscopy systems such as wide-field or confocal, spectral bleed-through corrections in FRET imaging, and the importance of fluorescence lifetime imaging microscopy (FLIM).

Fundamentals of Fluorescence Imaging
Edited by Guy Cox

ISBN 978-981-4774-85-7 (Hardcover), 978-1-351-12940-4 (eBook)
www.jennystanford.com

12.1 Introductory Physics of FRET

The concept of luminescence, which was introduced by the physicist Eilhardt Wiedemann in 1888, has opened a new area in science. Luminescence is an emission of ultraviolet, visible, or infrared photons from an electronically excited state. The types of luminescence can vary according to the mode of excitation. Fluorescence is a particular case of luminescence where the mode of excitation is the absorption of light. The absorbing species are excited into electronic excited states, and the emission of photons occurs within an average lifetime (typically 10^{-10}–10^{-7} s for most of the fluorophore molecules). In fact, several decay pathways exist after the absorption of light such as internal conversion, intersystem crossing, intramolecular charge transfer, and conformational changes (Chapter 1). The molecule in an excited state can interact with molecules in its close proximity. If the interaction takes place on a time scale comparable with the average excited state lifetime, the interactions can be in the form of electron transfer or energy transfer.

The historical precursors for the theory of FRET date back to the 19th and early 20th centuries with emerging understanding of electromagnetic and quantum mechanics. The first quantum mechanical theories of FRET were developed concurrently with the new theories of Heisenberg, Schrödinger, and Dirac. In the 1920s Jean-Baptiste Perrin and his son Francis Perrin explained the energy transfer process between two identical molecules in solution involving intermolecular dipole–dipole interaction. Perrin was the first to note that the energy transfer is distance-dependent and occurs over a specific range, which he calculated to be 15–25 nm, much larger than Förster's estimation. It was Förster in the mid-1940s who established the correct distance (1–10 nm) over which incoherent energy transfer (named FRET) would happen and he provided the quantitative means to measure molecular distances with his now well-known equations. More historical background about FRET can be obtained from the literature [1].

12.1.1 Basics of FRET Microscopy

FRET is the radiationless transfer of resonance energy from an excited donor fluorophore separated by several atomic diameters [2, 3] to a suitable acceptor fluorophore, representing a powerful technique to investigate protein-protein interactions, protein–DNA interactions, protein–membrane interactions, and three-dimensional structures of molecules in living cells [4, 5]. Fluorescence spectroscopy can provide structural and functional information and FRET spectroscopy is a valuable tool for measuring distances that can interpret conformational changes as small as 1 Å. There are several basic preconditions in order for FRET to occur. (i) A sufficiently large spectral overlap of the emission spectrum of a donor fluorophore (>30%) with the absorption spectrum of an acceptor [6]. If the spectrum is not overlapped, there cannot be any FRET. (ii) The donor molecule must be fluorescent with a high quantum yield. (iii) The distance between the donor and acceptor molecule should be within 100 Å. (iv) The emission dipole of the donor and the acceptor absorption dipole must be oriented to each other and must not be oriented perpendicular to each other. Dipole is an electromagnetic field that exists in a molecule that has one region with oppositely charged areas, a negative charge and another region with a positive charge. If the spectra are overlapped, the donor's oscillating emission dipole will look for a matching absorption dipole of an acceptor to oscillate in synchrony [7].

As mentioned above, this nonradiative energy transfer occurs via a long-range dipole-dipole coupling mechanism. The efficiency of energy transfer (E) from the donor to the acceptor is dependent upon the inverse of the sixth power of the distance (d) separating them, subject to the Förster distance (R_0) at which E is equal to 50% (Eq. 12.1). Defined by Eq. 12.2 in ångstrom, the R_0 of a FRET pair depends upon (i) the relative orientation between the dipoles of the donor emission and the acceptor absorption, κ^2 ranging from 1 to 4; (ii) the medium refractive index, n; (iii) the donor quantum yield, QY_D; (iv) the acceptor extinction coefficient, ε_A; and (v) the overlap between the donor emission and acceptor absorption spectra. For the majority of FRET pairs, the R_0 values are on the order of a few nanometers (3 $\sim$ 7 nm). Therefore, FRET is typically limited within 10 nm, providing a sensitive tool for investigating a variety of phenomena that produce changes in molecular proximity.

$$E = \frac{R_o^6}{R_o^6 + d^6} \tag{12.1}$$

$$R_o = 0.211 \cdot \left\{ \kappa^2 \cdot n^{-4} \cdot \mathrm{QY_D} \cdot \epsilon_A \cdot \frac{\int_0^\infty f_D(\lambda)\, f_A(\lambda)\, \lambda^4 d\lambda}{\int_0^\infty f_D(\lambda)\, d\lambda} \right\}^{\frac{1}{6}} \tag{12.2}$$

FRET imaging is a powerful tool in life-science research. FRET experiments have been carried out from the single-molecule level all the way to tissues and animals [8]. Combined with light microscopy techniques, FRET imaging can provide spatial and temporal information of protein-protein interactions in living cells under physiological conditions to address fundamental biological questions. Energy transfer is a dynamic process providing information on a molecule's close proximity by affecting the characteristics of fluorescence such as spectrum, quantum yield [9, 10]. FRET is a widely used technique to investigate the protein associations in a biological system on a nanometer scale. FRET occurs over distances similar to the Förster radius, which is 5 nm for common fluorophores [6]. FRET can function as a molecular ruler because of being very sensitive to distances. In 1948, Förster proved that FRET efficiency between two fluorophores decreases with the sixth power of distance [2].

The measurement energy transfer efficiency (E%)—an expression of distance—is at the center of any quantitative evaluation of FRET imaging. It is called "apparent" E%, as several factors influence the actual level, nevertheless providing valuable information about relative distances between molecules, control vs. intervention, and the like. At the energy transfer event, the donor fluorescence is "quenched" (fluorescence and lifetime is reduced) and the acceptor is "sensitized" (fluorescence is increased). E% is calculated based on donor fluorescence (Eq. 12.3) or donor lifetime (Eq. 12.4) in the presence (DA) or absence (D) of an acceptor:

$$E = 100 \times \left[1 - \frac{I_{DA}}{I_D}\right] \tag{12.3}$$

$$E = \left\{1 - \frac{\tau_{DA}}{\tau_D}\right\} 100$$

where τ_D is the lifetime and the unquenched donor (D) is a specimen labeled only with the donor fluorophore, i.e., in the absence of a potential acceptor. Bleaching the acceptor in a region of interest (ROI) and measuring the now unquenched donor in the absence of an acceptor fluorophore is another method [11–13].

12.2 Overview on Microscope Setup and Fluorescent Imaging

12.2.1 Microscopy Setup

FRET imaging can be implemented with either an upright or inverted microscope depending on the particular application. Magnification, numerical aperture (NA), aberration, and spectral properties are some of the factors that need to be taken into account when choosing objectives for FRET. This selection depends on the specimen under investigation and the FRET pair used to study the protein–protein interactions.

In cellular microscopy, the majority of work is performed using magnifying powers greater than 40× to allow for the detection of intracellular localization of donor–acceptor pairs. NA of the objective is important for two reasons: First, a high NA maximizes the resolution of the system, which is an especially important feature when donor–acceptor pairs may be localized within intracellular domains or organelles. Second, the NA dictates the light gathering power of the objective lens, a feature especially important with fluorescence measurements since selected multiple wavelengths of light (visible spectrum) are transmitted through the objective lens.

12.2.2 Fluorescence Illumination

In a standard epifluorescence microscope, the light emitted from a source is directed by associated optics into the back aperture of the microscope and then through an objective. As in all elements of the microscope, the optical components in this path must be capable of transmission of the relevant light wavelengths (i.e., excitation and emission wavelengths for both the donor and acceptor fluorophores). Excitation wavelength selectors, dichroic mirrors, and emission selectors are all additional components in the delivery of the excitation wavelengths and recovery of emission wavelengths in the FRET microscope.

Light Sources: The most commonly used sources of light for wide-field fluorescence microscopy are mercury (Hg), xenon (Xe), or metal halide arc lamps (Chapter 2). Although Hg lamps have some usable intensity maxima (e.g., as with fluorescein, rhodamine, and BFP excitation wavelengths), Xe arc lamps provide relatively even spectral radiance and, as with ratio imaging, are the light source of choice for FRET applications. When these lamps are combined with monochromators or appropriate interference filters, one can select the required excitation wavelengths to meet the filter requirement for a particular FRET pair. Since these lamps produce large amounts of emission in the infrared (IR), it is necessary to use a heat filter prior to the entry of light into the microscope or filter wheel to protect the costly excitation filters. Such filters are most commonly mounted in a filter wheel at the back entrance port of the epifluorescence optics. Alternatively, optical fibers or liquid light guides can be used to deliver the light to the back of the microscope or directly to the sample. Lasers offer more coherent light whose intensities are at acceptable levels only at discrete wavelengths, used in laser scanning confocal microscopy (Chapter 3). Argon, supercontinuum, and diode lasers at various wavelengths have been used in the confocal microscopy. Ti:sapphire laser is used in multiphoton microscopy systems (Chapter 4).

Wavelength Selectors: In wide-field microscopy one requires excitation and emission filter wheels and a dichroic to separate the excitation and emission of a selected FRET pair. The filter wheel combination and dichroic allow users to excite the specimen with donor excitation wavelengths and collect donor and acceptor emission signals. It is also possible to use a monochromator for selecting excitation and emission wavelengths. In confocal microscopy, acousto-optical tunable filters (AOTFs) are used for the selection of excitation wavelengths. For the emissions, interference filters or acousto-optical beam splitters (AOBS) can be used instead.

Detectors: High quantum efficiency (QE) detectors are desirable for FRET imaging. In a wide-field microscopy configuration a high QE CCD camera is used [14]. A high QE detector allows the users to use less excitation light intensity to reduce photobleaching. In confocal and two-photon microscopy photomultiplier tubes (PMTs) are used at and the conventional PMTs typically have a peak QE of about 20% at 546 nm. Currently most commercial confocal microscopy systems are equipped with high QE: about 40% at 546 nm with gallium arsenide phosphide (GaAsP)-PMTs or hybrid detectors. High QE detectors are advantageous in many ways to reduce photobleaching and improve the signal-to-noise ratio of the signal. In spinning disk confocals, a high QE CCD camera is used as a detector, such as the notable CMOS CCD and EMCCD cameras (Chapter 13).

12.3 Fluorescent Probes for FRET

The main spectral characteristics that should be considered in the design of FRET experiments are as follows: (i) R_0 of the selected pair should be as close as possible to the expected mean distance between the two labeled sites (including the linker of the probes); (ii) the orientation of the probes are expected to be averaged during the lifetime of the fluorescence of the excited state of the donor; (iii) photostability; (iv) minimal environmental sensitivity of the spectral characteristics, e.g. pH; (v) large Stokes shifts of the acceptor to achieve minimal overlap of the emission spectra of the donor and the acceptor.

Optimal fluorophore FRET partners are the key to the success of a FRET application, although acceptable trade-offs between conflicting and desirable features are often necessary. Many visible fluorescent proteins (FPs) (Chapter 6) have been employed in combination with FRET microscopy to visualize dynamic protein interactions under physiological conditions [15–17]. The natural diversity of FPs has provided scientists with a rich palette of variants with different biochemical and spectral characteristics, which represent a huge source of potentially powerful molecular tools for numerous applications in the study of complex biological systems.

Below we will consider, one by one, the key characteristics of FPs that might be important for their practical use. These considerations, however, cannot be comprehensive since each particular assay or experimental setup may require a specific choice or optimization of an FP.

Brightness: Quantum yield (QY) and molar extinction coefficient (EC): Sensitivity and signal-to-noise ratio of any fluorescence detection technique clearly depends greatly on the brightness of the fluorophore employed. For fluorescent labeling in living systems, brighter FPs offer additional advantages, requiring lower levels of excitation light and thus lower phototoxic effects occur. Importantly, bright FPs or their fusion constructs can be used in lower amounts, minimizing interference in physiological processes. Many of the naturally occurring FPs is characterized by high brightness determined both by an EC exceeding 100,000 $M^{-1}cm^{-1}$ and a QY approaching the theoretical limit of 1. Also, for the enhancement of artificially derived FP variants (such as monomerized FPs), brightness is the easiest parameter to improve, through mutagenesis followed by screening of expression libraries for brighter clones [18, 19]. However, enhancement of FP brightness during its optimization can happen at the expense of other characteristics, such as its monomeric nature and pH- and photostability. Meanwhile, FP of lower brightness but higher photostability can result in better signal-to-noise ratio, especially for experiments involving long time series. Therefore, brightness should not be the only parameter considered while choosing an FP for a particular application. It should also be kept in mind that fluorescent brightness measured for an FP sample in vitro could not be directly extrapolated to its actual brightness in vivo. Indeed, the resulting signal brightness generated by an FP expressed in living cells is determined not only by its intrinsic spectral characteristics, but also by such parameters as transcription and translation efficiency, mRNA and protein stability, and chromophore maturation rate. These parameters can differ for various FPs and their fusion constructs. Importantly, at the single molecule level, an FP can demonstrate much higher brightness than that measured for the ensemble of FP molecules in a sample [20, 21].

Maturation Rate: FP fluorescence appears after protein folding and chromophore maturation, which includes several consecutive covalent modifications that are especially extensive for orange and red FPs. Usually, chromophore maturation is the rate-limiting step for the FP to become fluorescent. Depending on the particular FP, oxygen concentration, and temperature, maturation can take from several minutes to hours or even days, as in the case of Timer FPs [22–25]. For practical use, the maturation rate should be reasonably fast to obtain a high and stable fluorescent signal. Most FPs in use have a maturation half-time from 40 min to 1–2 h, which is sufficient to label cells, organelles, and proteins of interest and to perform various quantitative experiments. However, for some applications, such as early detection of promoter activation, labeling proteins of interest with a short lifetime, or monitoring single translational events, FPs with very fast maturation are needed. For example, fast-maturing yellow FPs allowed the direct observation of the production of a single protein molecule in real time [26]. Finally, a rough comparison of FP maturation rates can be performed in living eukaryotic cells. However, this latter approach can give results quite different from in vitro data, due to differences in FP expression efficiency, turnover rates, etc. Mis-folding mutations of proteins into monomeric variants of ECFP have resulted in the production of new variants featuring enhanced brightness, folding efficient, solubility, and FRET performance [27].

pH Stability: The key parameter that is usually employed to compare pH stability of FPs is pK_a, which is the pH value at which the brightness of fluorescence for a given FP equals 50% of the maximal brightness measured at an optimal pH. Fluorescence of FPs typically increases at higher pH and reaches its maximum at pH 8–9. A further increase in pH commonly does not alter fluorescence until reaching alkaline values of 10–13, which lead to protein denaturation and/or degradation of the chromophore. For most FPs developed to date, fluorescence brightness depends on pH changes in the physiological range, with a typical pK_a from 5.0 to 7.0. Generally, pH dependence of FPs is most pronounced for yellow FPs (pK_a from 5.5 to 6.5) and to a lesser extent for the green FPs (pK_a from 5 to 6), while blue, red, and far-red variants are often much less sensitive to pH, with pK_a values as low as 2.7 for TagBFP [28–30] and 3.8 for TagRFP [31]. Such pH-stable FPs can be visualized in acidic organelles such as endosomes, lysosomes, and Golgi and provide more reliable readouts for quantitative assays. In general, since substantial pH changes are common during many physiological processes [29–31], the pH dependence of FP fluorescence can notably distort the results of quantitative experiments. Therefore, pH stability is an important parameter for the choice of an FP for quantitative measurements, especially for ratiometric dual or multicolor imaging and for FRET techniques. Organic fluorescent molecules, conjugated to proteins of interest, such as the Alexa- and Cy-families have a wide range of pH stability.

12.3.1 Engineering of GFP-Like Proteins

Some wild-type proteins such as EosFP proved to be useful as marker proteins in several imaging application without further optimization [32]. However, several features of natural GFP-like proteins hamper their applicability as marker proteins [33]. Common drawbacks are:

- unspecific aggregation that prevents the correct localization in cells and might have cytotoxic effects
- oligomerization that can interfere with the functionality of fusion proteins and prevent the correct cellular localization
- lack of functional expression in mammalian cells requiring culturing temperatures around 37°C
- fast photobleaching

The fluorescence of the protein chosen for further optimization can often be used as read out for the success of alterations of the amino acid sequence. Consequently, even major genetic engineering projects that might require the alteration of up to 30 residues in more or less unpredictable positions are feasible for skilled experimenters. Currently, three mutagenesis methods are routinely applied for the optimization of GFP-like proteins:

12.3.1.1 Random mutagenesis

This method exploits the increased error rates of polymerases under suboptimal reaction conditions. During the amplification of the FP-encoding cDNA in the PCR, errors are introduced randomly in the newly generated DNA fragments. The modified amplicons are subsequently cloned in expression vectors and expressed in bacterial hosts. The fluorescence of the bacterial colonies growing on agar plates can be used as indicator for the effects of certain mutations. Random mutagenesis proved to be especially helpful to create variants with improved folding and maturation properties [34, 35].

12.3.1.2 Site-directed mutagenesis

Today, a considerable number of crystal structures of GFP-like proteins are available, which can help to identify target residues for knowledge-based genetic engineering. The primer overlap extension PCR is a very robust technique to perform directed amino acid exchanges [33–35]. Oligonucleotide primers can be also designed to randomize a distinct residue so that any possible amino acid will be introduced in this position. The randomization of defined residues can be used, for instance, to verify quickly if a mutation picked up during screening of a random mutant library is indeed the optimal exchange in a certain position. Examples of successful applications of site-directed mutagenesis include the disruption of tetrameric associations or the generation of variants with strongly red-shifted emission [36–38].

12.3.1.3 Multisite-directed mutagenesis

Up to six oligonucleotide primers can be applied in a single mutagenesis reaction to a mixture of up to several different mutant templates using the QuikChange MultiSite-Directed Mutagenesis Kit (Stratagene, La Jolla, CA, USA) [38]. Not all of the newly generated DNA fragments will carry all exchanges; many will contain different combinations of the individual mutations. Mutant libraries generated by this method can be subsequently screened in bacteria to spot mutations that act synergistically to improve desired properties, for example, folding of monomeric variants [38].

12.3.2 Cloning of Fluorescent Vectors

After deciding to use a FRET approach to study a protein of interest, fluorescent donor-acceptor pairs need to be generated.

The most common approach is to obtain the coding regions for the chosen proteins and use commercially available vectors with multiple cloning sites (e.g., BD Biosciences Clontech, Stratagene, Qbiogene, and other companies) to generate in-frame fusions with the CFP, YFP, GFP, or mRFPl genes. GFP fusions can be generated at the N or C terminus of proteins of interest, or even internally, with GFP being interspaced between domains of a protein coding sequence. It is generally a good idea to simultaneously prepare both N- and C-terminal fusions, hoping that at least one of these chimeras retains functional activity. One common problem is that the linker between the protein of interest and GFP may be recognized as a nuclear localization signal (NLS) by the cellular machinery and the chimera may thus unexpectedly be detected in the nucleus instead of its proper cellular location. This chance is particularly high if the linker contains NLS-like motifs such as repeats of lysines (K) or arginines (R). Such common biological errors in linker designs should be avoided. Combinations of other amino acids, especially secondary structure breakers such as glycine (G) or proline (P) or small amino acids such as serine (S), alanine (A), or threonine (T) are less likely to create problems. However, to confirm the construction of functional GFP chimeras certain functional tests are critical. After DNA sequencing to verify the constructs, DNA for mammalian cell transfections needs to be prepared with great care. In general, commercial kits (e.g., Qiagen Maxiprep) work well. Many transfection reagents are routinely used to introduce plasmid DNA into living cells.

12.3.3 Functional Activity of Expressed Constructs

Cloned vectors will be introduced into the cells selected for initial test experiments. First, the expression levels and protein stability of the donor and acceptor need to be tested in a time course, and the spectral properties of the vector established. Often, a newly generated vector may be brightly fluorescent but has lost functional activity, displays the wrong localization, and/or lost the ability to interact with its normal partners in a protein complex. Therefore, it is most important to check the functional activity of vectors before attempting to draw any conclusions from experiments.

12.3.4 Expression and Overexpression

Optimal expression levels of vectors are important. Experiments based on time courses of expression need to select the time when the expression is sufficient to measure FRET, when the vectors are still localized in the same areas as in their endogenous wild-type counterparts, where their functional activity can be established. The over expression of exogenous proteins in experiments can also lead to homo-FRET, i.e., FRET between like chromophores. This process can only be monitored by fluorescence anisotropy [39] and is used to study protein structures, oligomerization and the organization of membrane proteins in sub-microdomains at the cell surface.

12.3.5 Guidelines for FRET Pairs

Suitable fluorophore FRET partners are the key to the success of a FRET application. Many visible fluorescent proteins (FPs) have been employed in combination with FRET microscopy to visualize dynamic protein interactions under physiological conditions [40–42]. Equally, the development of novel organic dyes that exhibit improved photo- and pH stability, as well as excellent spectral characteristics, provides additional choices for FRET imaging [43]; a wide range of labeled antibodies, protein conjugates (e.g., ligands), organelle markers (e.g., lyso-tracker) are available from various suppliers such as Invitrogen Molecular Probes (Alexa-dyes) and Amersham/GE Healtcare (Cy dyes). The selection of the right donor–acceptor pair should include the actual biological question to be addressed, the type of biological specimen to be imaged, and the instrument available to measure FRET. Here we listed a few FP-FRET pairs widely used in the literature (Table 12.1), organic fluorophore choices are available from manufacturers' websites.

There are optimized FPs derived from the *Aequorea* GFP that are most often used for FRET-based imaging applications (Table 12.1). Many early FRET studies relied on the *Aequorea* GFP derived blue FP (BFP), used in combination with GFP derivatives [30]. Recent mutagenesis strategies have yielded new BFP variants with

Table 12.1 Visible fluorescent proteins FRET probes

FRET probes	Ex/Em	Comments	Reference
BFPs (EBFP, EBFP2)	383/448	Donor to CFPs or GFPs	30
T-Sapphire mAmetrine	399/511 406/526	Long Stokes shift donor	51
CFPs (ECFP, CyPet, mCerulean, mCerulean3, mTurquoise)	433/475	Donors to YFPs; both mCerulean3 and mTurquoise have very high quantum yields (>0.8).	28, 29
mTFP1	462/492	Donor to YFPs; has a high quantum yield (0.85)	52
GFPs EGFP, Emerald	488/507	Donor to OFPs (mOrange2, mKO2, tdTomato, TagRFP-T, etc.) or RFPs (mRFP1, mRuby, mApple, mStrawberry, mCherry, mKate, etc.)	28–30
YFPs (EYFP, Citrine, Ypet mVenus, REAch)	515/528	Donor to RFPs; Acceptor for CFPs; Ypet is typically used together with CyPet; REAch is a strong absorber but weak emitter, and can be a good acceptor for a FLIM-FRET study	53

much higher quantum yields and improved photostabilities. The cyan color variant, mCerulean, or Teal CFP variant [44], used in combination with either the mVenus [44], or mCitrine YFP variants are among the most popular FRET pairings. The mCerulean was recently engineered further to improve its brightness and photostability, yielding variants named mTurquoise [44, 45] and mCerulean3.

The vast array of fluorescent proteins currently available offers tools for numerous applications from simply localizing proteins of interest or monitoring gene expression to observing specific interactions between molecules within defined regions of a living cell.

12.3.6 FRET-Based Sensors

One area for which the improvement of FPs for FRET is especially important is the evolution of highly sensitive biosensor probes [46]. The biosensor probes incorporate a biologically active linker peptide to separate the donor and acceptor fluorophores. The post-translational modification of the linker or the binding of a substrate can induce a change in the linker conformation and change the spatial relationship of the donor and acceptor FPs [47] (Chapter 8). The development of biosensors with new FPs is yielding probes that can be used in different spectral windows. Improved FRET-based biosensors have tremendous potential for the development of rapid large-scale screening assays for the discovery of novel pharmaceuticals and the development of therapeutic strategies.

FRET-based biosensors labeled with both donor and acceptor potentially change their conformation in response to signaling events or changes in analyte concentration. This change in conformation alters the donor/acceptor distances and orientations, resulting in changes in FRET efficiency. The first example of a genetically encoded biosensor made use of FRET between blue fluorescent protein (BFP) and GFP linked by a peptide from myosin light chain kinase (MLCK) [50]. This molecule was an indicator of Ca^{2+}-mediated signaling by calmodulin such that when calmodulin bound to the MLCK peptide it altered the distance and orientation between BFP and GFP, thereby changing the efficiency of FRET. A very similar sensor was also constructed that was composed of CFP and YFP separated by calmodulin and a peptide that calmodulin binds in the presence of high calcium concentration [48–50].

In the years that followed, more linked biosensors were developed to detect signals other than those of calcium. Development of these sensors generally made use of the experimenter's knowledge of signal transduction to combine an activatable component and a domain that binds the activated component to alter the conformation of the molecule. Recent examples include phosphorylation sensors, small G-protein-activation sensors for Ras and Rac, and sensors for protein kinase C activity [54–56].

Linked biosensors have advantages and disadvantages. The principal advantage of the linked-biosensor approach is that analysis is simple. Second, the molar ratio of donor to acceptor is always 1, which means that FRET can be measured by the ratio of donor emission to acceptor emission, exciting at the donor's excitation maximum [57]. In the nomenclature of this chapter, data from a linked biosensor can be calculated as a simple ratio:

$$\text{Simple ratio} = \frac{I_F}{I_D}$$

Unfortunately, this goes to infinity as E approaches 1 and results in slight skewing of data owing to the relationship between I_D and I_F. I_F is the intensity or image at the donor excitation and acceptor emission, and I_D intensity or image at the donor excitation and donor emission.

A better expression is the FRET ratio,

$$\text{FRET ratio} = \frac{I_F}{I_F + I_D}$$

which, although this is not the FRET efficiency, at least has the same functional form as E_D.

The principal disadvantage of linked biosensors is that they often lack structural components that are important for proper localization, and they are often signaling deficient molecules, which may disrupt certain pathways. Furthermore, these sensors are often difficult to construct, requiring many iterations to produce a functional sensor.

12.4 Practical Considerations for All FRET Measurements

Whether using FRET for quantitatively detecting interactions between complementary molecules in solution, cells, tissue or live animals, a number of measurements and controls must always be performed in order to minimize artifacts and maximize signal-to-noise ratios to obtain the highest quality images. It is important to note that in the absence of these kinds of controls, modulation of fluorescence intensities or lifetimes using a microscope system may be due to other factors besides FRET. FLIM-FRET may be an ideal technique to confirm the occurrence of intensity-based FRET [58]. A number of measurements and controls must be performed in order to minimize artifacts and maximize signal-to-noise ratios:

- Determine the background fluorescence of the sample. Measure the background fluorescence intensities of all samples in the absence of labeled molecules at the identical experimental and microscopy settings used for FRET measurements.
- Determine the fluorescence intensity range of a single-label donor. Ideally, this should be in a range where the fluorescence emission intensity varies linearly with fluorophores concentration. This reading will be the baseline value for determining the amount of quenching that occurs in the double-label specimen in the presence of a complementary molecule that contains the acceptor fluorophore.
- Determine the fluorescence intensity of the donor in the presence of the acceptor-labeled complement. The amount of quenching in the presence of acceptor is used along with the amount of quenching in the absence of acceptor labeled complement to determine the transfer efficiency. It is important to note that in the absence of these kinds of controls, modulation of fluorescence intensities or lifetimes using a microscope system may be due to other factors besides FRET.
- One of the important conditions for FRET to occur is the overlap of the emission spectrum of the donor (D) with the absorption spectrum of the acceptor (A). Because of spectral overlap, the FRET signal is contaminated by donor emission into the acceptor channel and by the excitation of acceptor molecules by the donor excitation wavelength. The combination of these signals is called the spectral bleedthrough (SBT) signal into the acceptor (or FRET) channel. In principle, the SET signal is same for one-photon or two-photon FRET microscopy [59–63]. In addition to SBT, other sources of noise contaminate the FRET signals, including spectral sensitivity variations in donor and acceptor channels, autofluorescence, and detector and optical noise.

In this chapter we have used PFRET software to remove both the donor and acceptor SBT signals and correct for variation in fluorophore expression level associated with FRET imaging, using single-label specimens as controls, imaged under identical conditions as the double-label samples. These corrections then allow determination of the energy transfer efficiency (E%) and provide an estimate of the distance between donor and acceptor molecules. Both the acceptor SBT and donor SBT are incorporated together in systematic mathematical based algorithm. Seven images are acquired with appropriate filters as shown in Table 12.2 for PFRET data analysis. Our approach works on the assumption that the double-labeled cells and single-labeled donor and acceptor cells imaged under the same conditions exhibit the same level of SBT [60, 61]. Thus, our algorithm follows fluorescence levels pixel by pixel and establishes the level of SBT in the single-labeled cells, and applies these values as a correction factor to the appropriate matching pixels of the double-labeled cell. PFRET then is

$$\text{PFRET} = \text{uFRET} - \text{ASBT} - \text{DSBT}$$

where uFRET is uncorrected FRET, ASBT is the acceptor spectral bleedthrough signal, and DSBT is the donor spectral bleed through signal. The algorithm also corrects for the expression level differences found between cells.

$$E = 1 - \left\{ \frac{q_{\text{D}}}{(c \times \text{PFRET} + q_{\text{D}})} \right\}, c = \left(\frac{\text{SS}_{\text{D}}}{\text{SS}_{\text{A}}} \right) \times \left(\frac{\text{QY}_{\text{D}}}{\text{QY}_{\text{A}}} \right)$$

E is determined using the PFRET signal and the quenched donor (q_{D}) signal. The PFRET signal is produced by the sensitized acceptor and measured in the "FRET" channel, while the q_{D} signal is produced by the donor and measured in the "donor" channel. Thus, a correction factor c is used to compensate for the difference between the donor and acceptor quantum yields (QY_{D}, QY_{A}) as well as different detector sensitivities for the donor and the acceptor (SS_{D}, SS_{A}). It should be noted that the PFRET method considers DSBT as part of q_{D} for a more accurate E estimation [64].

The PFRET data analysis allows us to characterize the 1p- and 2p-FRET microscopy system [40].

12.5 Application of FRET Microscopy

In the biological sciences, conformational changes in biomolecules pose a demanding challenge to understand the underlying bioprocess mechanisms. FRET has frequently been employed to measure conformations and related transitions of molecules at the molecular level [66–68]. It is noteworthy that the FRET efficiency depends not only on the distance of the two fluorophores, but also on their relative orientation, i.e., their absorption and emission dipole moments. Theoretically, the latter effect could be suppressed or reduced with rigidly connected donor and acceptor molecules of interest. Normally, this is not possible and flexible linkers are used for fluorophore labeling, thereby averaging over many possible orientations, and thus limiting the observable distance range to a few nanometers. Nevertheless, in spite of some limitations, FRET is a useful biophysical tool to quantify molecular dynamics, including protein–protein interactions, protein–DNA interactions, and protein conformational changes both in vitro and in vivo systems.

12.5.1 Wide-Field FRET Imaging

The simplest form of wide-field imaging is epi-illumination. Because the excitation volume is very large in the z direction in epi-illumination, the residual fluorescence from emitters out of the focal plane is also collected, leading to a low contrast in the resulting image. This hinders imaging of thick samples. However, if the sample is thin enough (e.g., Fig. 12.1), it is still possible to image single FP-labeled molecules inside the cell by localization enhancement with a reasonable signal-to-noise ratio [41].

12.5.2 Confocal FRET Imaging

Confocal imaging is a point-detection-based approach in which a tightly focused (diffraction limited) spot is scanned across the sample and the excited fluorescence is collected through the objective, spatially filtered with a pinhole to remove the out-of-focus signals, and detected by a PMT detector. The advantage of confocal imaging is that it can image deeper into a sample, the

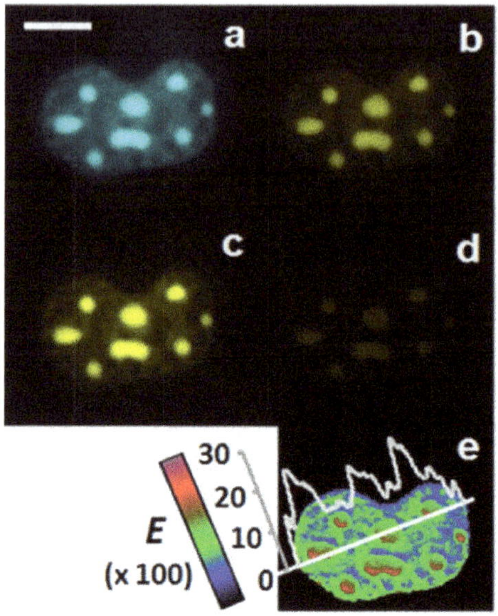

Figure 12.1 Demonstration of the homo-dimerization of CCAAT/enhancer-binding protein alpha (C/EBPα) in live mouse pituitary cell nucleus using wide-field FRET microscopy. Fluorescent signals from the nucleus of a cell co-expressing CFP-C/EBPα (donor (D)) and YFP-C/EBPα (acceptor (A)) were measured in the D (a), A (b), and FRET (c) channels in wide-field microscopy. The processed FRET (PFRET, d) and apparent FRET efficiency (E%, e) images were obtained after processing the data with the PFRET algorithm in combination of the images acquired from the single-label expressing cells. The interaction between CFP- and YFP-tagged C/EBPα is demonstrated by the E% images and indicates the homo-dimerization of C/EBPα in regions of centromeric heterochromatin of cell nucleus. Adapted with permission from Sun et al. [41].

depth depending on the objective and by multiple scanning the sample along the z direction, a three-dimensional (3D) image can be reconstructed. Because multiple regions of the sample cannot be probed simultaneously, the method is less practical for single-molecule imaging in cells. However, it is possible to track single-QD-labeled molecules (quantum dots) in living cells in three dimensions with a confocal-based setup involving multiple detectors and a feedback system. For example, we have shown confocal FRET using cerulean and Venus constructs in Fig. 12.2.

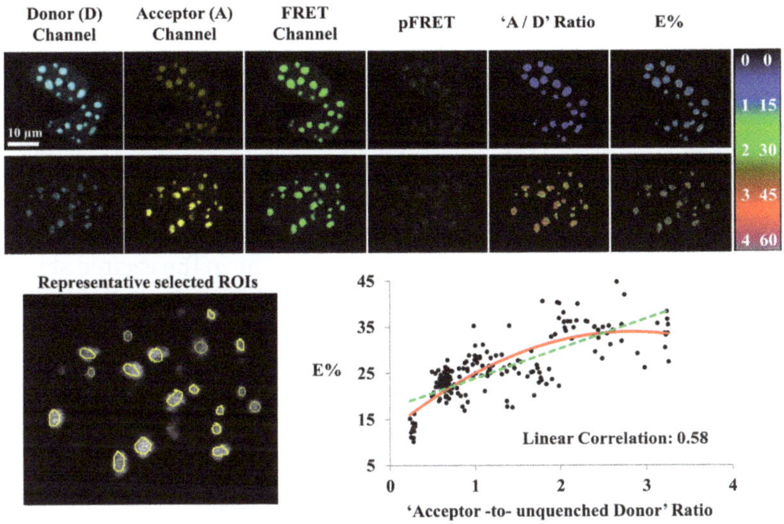

Figure 12.2 Confocal FRET microscopy using the PFRET method for analysis. Images of live cells co-expressing Cerulean- (donor) and Venus- (acceptor) tagged bZip were acquired in the donor, FRET, and acceptor imaging channels. The images of the double-expressing cells were processed using the PFRET algorithm to remove SBT contaminations and calculate the pFRET and the FRET efficiency (E%) images. Due to different acceptor: donor ("A/D") ratios, a range of E%s emerged. As an example, two sets of images of double expressing cells having different "A/D" ratios are shown for a comparison: for each set, the raw and pFRET images are contrasted in the same range (top: 0–1500, bottom: 0–3000, 12-bit); the "A/D" ratio image was obtained by calculating pixel-by-pixel ratios between the two images acquired in the acceptor and the donor imaging channels, respectively; the ratio (0–4) or E% (0–40) images of both sets are color contrasted in the same range, indicating larger "A/D" ratios yielding higher E%s. This is confirmed by plotting E% against the "acceptor-to-unquenched donor" ratios for 230 regions of interest (ROIs) selected from 15 cells. The unquenched donor is determined by adding the pFRET signal to the quenched donor signal. The graph shows an increasing trend of E% with an increased acceptor-to-unquenched donor ratio, by either a second-order polynomial (solid) or linear (dashed) curve fitting. (Zeiss 510 Meta confocal microscope and 63X/1.4NA oil immersion objective; the donor channel: the 458 nm laser line and the 470–500 nm bandpass filter; the FRET channel: the 458 nm laser line and the 535–590 nm bandpass filter; the acceptor channel: the 514 nm laser line and the 535–590 nm band-pass filter). Adapted with permission from Sun et al. [42].

12.5.3 FLIM-FRET Microscopy

Intensity-based FRET in wide-field and confocal microscopy requires spectral bleedthrough corrections. On the other hand, lifetime based FRET does not require any bleedthrough correction. The FLIM-FRET technique also provides a better estimate for the energy transfer efficiency [12]. FRET depopulates the excited state of the donor (D) and results in a decreased probability of photon emission from D and a shortening of D's lifetime in the excited state; meanwhile, the probability of photon emission from the acceptor (A) increases (is sensitized). Measurement of each event can provide direct proof of energy transfer. Combining these light microscopy techniques, FRET can provide two- or three-dimensional spatial and temporal information on protein–protein interactions in single living cells under physiological conditions (e.g, Fig. 12.3).

The development of these FRET microscopy methods has expanded the tool chest for biologists using FRET, which in turn

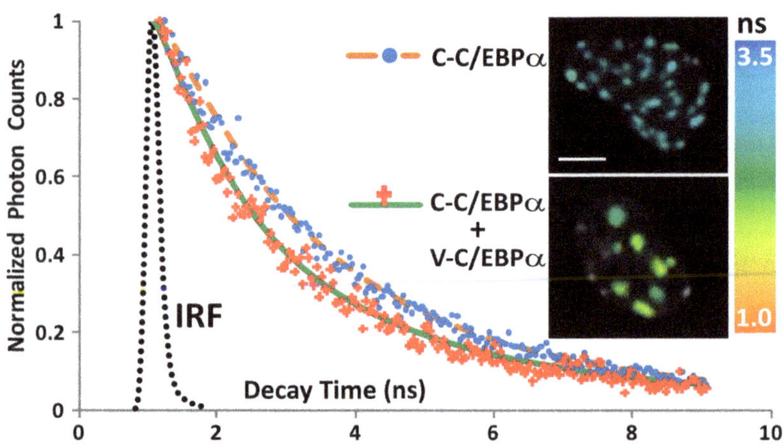

Figure 12.3 Localization of homo-dimerization of CCAAT/enhancer-binding protein alpha (C/EBPα) in living cell nucleus using 2-color TCSPC FLIM-FRET microscopy. C/EBPα was tagged with either Cerulean (C) or Venus (V). The fluorescent lifetime decay kinetics for the C-C/EBPα (FRET donor) in the absence and the presence of V-C/EBPα (FRET acceptor) were determined by fitting the measured decay data into a single or double exponential decay model, respectively, with the measured instrument function (IRF). Adapted with permission from Sun et al. [14].

has fostered the development of new and more sophisticated FRET tools. There is no single perfect method that is suitable for all FRET applications. Ideally, a biological question addressed by one FRET microscopy method should be verified by another FRET microscopy technique. For that reason international participants are trained on various FRET microscopy techniques in the annual, week-long, hands-on workshop on FRET microscopy during March of each year. (http://www.kcci.virginia.edu/workshop/index.php).

12.5.4 Three-Color FRET Microscopy

Employing a traditional two-color FRET method to determine the inter-relationships of three cellular components (A, B, C) requires sequential imaging and analysis of the FRET-pair combinations of AB, BC, and AC. A three-color FRET analysis system (Fig. 12.4) is of great benefit to determine protein complex assemblies in identical space and time during signaling events, trafficking dynamics, or cytokinesis, particularly in the context of treatments, which may alter the relationship of the three components of interest. Most publications on three-color FRET have dealt with in vitro studies in solution using fluorometry or spectroscopy to detect components labeled by organic dyes for medical diagnostics. Recently, a three-color FRET model based on the detection of the sensitized emissions from acceptors through steady-state imaging has been developed. This model has three symbolic fluorophores: F1 (donor), F2 (acceptor to F1 and donor to F3), and F3 (acceptor), and assumes that spectral overlap (donor emission–acceptor absorption) of each pair (F1–F2, F1–F3, F2–F3) is sufficient for FRET to occur [68].

12.5.5 Other FRET Microscopy Application

Biomolecular interactions: FRET and FLIM have been applied extensively to biomolecular interaction studies: in the analysis of protein–protein interactions with high spatial and temporal specificity (e.g. clustering), in the study of conformational changes, in the analysis of binding sequences, and in applications such as high-throughput screening. Recent advances in microscopy have allowed the development of new methods to analyze

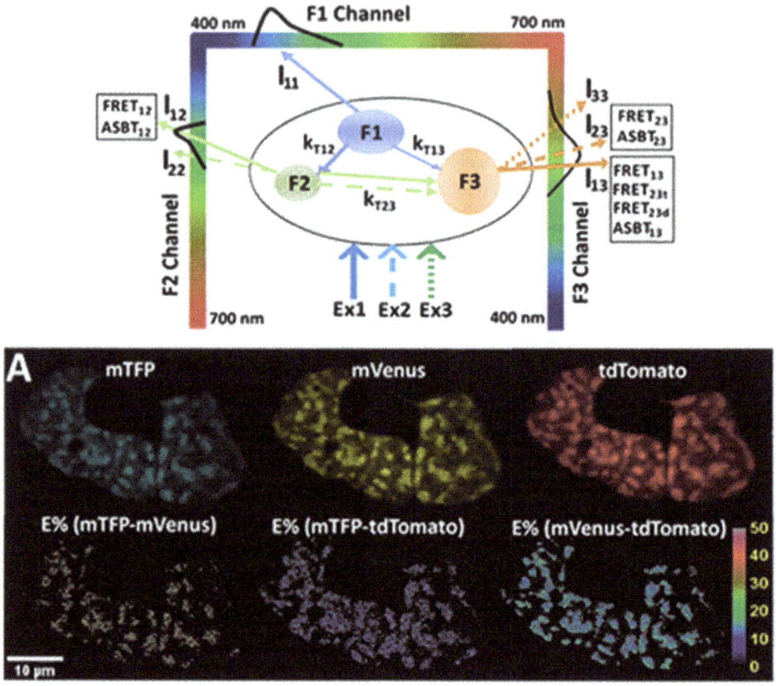

Figure 12.4 Demonstration of the homo-dimerization of C/EBPα and its interaction with H1Pα in live mouse pituitary GHFT1 cells by three-color spectral FRET (3sFRET) microscopy. The donor spectral bleedthrough (DSBT) is removed through linear unmixing; however, FRET signals still contain the acceptor spectral bleedthrough (ASBT). The λ-stack I_{1S} (excited by Ex1) is unmixed into the F1 (I_{11}), F2 (I_{12}), and F3 (I_{13}) channels: I_{11} is the quenched F1 signal; I_{12} has both FRET ($FRET_{12}$) and ASBT ($ASBT_{12}$) signals; I_{13} includes ASBT ($ASBT_{13}$) and several FRET signals: $FRET_{13}$, $FRET_{23t}$, $FRET_{23d}$. When excited by Ex2, the specimen becomes a two-color (F2–F3) FRET system. After unmixing the λ-stack I_{2S}, two images are obtained in the F2 (I_{22}) and F3 (I_{23}) channels: I_{22} is the quenched F2 signal; I_{23} is both FRET ($FRET_{23}$) and ASBT ($ASBT_{23}$). Ex3 does not produce FRET. The image I_{33} obtained through unmixing the λ-stack I_{3S} measures the F3 fluorescence intensities. Detailed explanation for each signal component is given in Ref. 68. (A) The unmixed images (I_{11}, I_{22}, I_{33}) shows C/EBPα (expressing mTFP or mVenus) and H1Pα (expressing tdTomato) are co-localized in regions of centromeric heterochromatin, where the dimerization of C/EBPα (mTFP-mVenus) and its interaction with H1Pα (mTFP-tdTomato and mVenus-tdTomato) are indicated by the $E\%$ (FRET efficiency in percentage) images obtained in 3sFRET microscopy. Adapted with permission from Milroy et al. [69].

protein–protein interactions at very high resolution in both fixed and live cells. Additionally, molecular interactions and conformational changes of various proteins involved in the regulation of cell adhesion and motility have been investigated.

HostPathology Interactions: Given the advantages and compatibility of FRET with other techniques, its use in understanding host–pathogen interplay has, to date, been surprisingly limited [69]. Fluorescence lifetime imaging is rapidly becoming a preferred method for making FRET measurements. This technique has been utilized to investigate plant cell invasion by an ascomycete fungus [70, 71]. FRET-based assays are valuable tools to resolve bacteria-induced protein–protein interactions in the context of the intact host cell.

Drug and Ligand Screening: TR-FRET (time-resolved FRET) brings together the low background benefits of TRF with the homogeneous assay format of FRET. This powerful combination provides significant benefits to drug discovery researchers including assay flexibility, reliability, increased assay sensitivity, higher throughput and fewer false-positive/false-negative results. Compared to other binding assays, the polarization [72–74] and HTRF (high-throughput TR-FRET) [75, 76] binding assays are nonradioactive, therefore safer to perform, yet very sensitive and homogeneous, therefore easier and faster to automate.

Lipid Membrane: Loura and coworkers have reviewed the application FRET in detecting and characterizing the phase separation in lipid bilayers (both in model systems and in cell membranes) [77, 78]. This review is also focused on models describing the rate and efficiency of FRET for both uniform probe distribution and phase separation, and recently reported methods for detection of membrane heterogeneity and determination of phase boundaries, probe partition coefficients, and domain size. FRET is also applied to characterize single-phase lipid systems, gel/fluid phase separation, liquid ordered/liquid disordered phase separation (lipid rafts), complex systems containing ceramide and cell membranes are presented to illustrate the wealth of information that can be inferred from carefully designed FRET studies of membrane domains. The intermembrane contact areas between single unilamellar vesicles and planar supported bilayers are quantified using FRET as basic tool [79, 80]. In the field of protein–lipid interactions, FRET has also been extensively used to characterize and quantify partition of membrane proteins to lipid membranes and particular lipid phases, and most notably to liquid-ordered (raft-like) phases.

12.6 Conclusions

The first FRET microscope was constructed over a quarter of a century ago. Recent advances in imaging technology and the development of wavelength selection devices have made this powerful technique popular among investigators interested in studying the interactions between molecules in solutions, cells, tissues, and live animals. With advances in optical instrumentation, FRET along with other existing optical techniques is used to reveal dynamics and interactions of fluorescently labeled and autofluorescent molecules with very high precision. Thus, FRET applications are exponentially growing, including single molecule fluorescence microscopy. Because of the non-invasiveness, simple, robust, and design of genetically encoded fluorophores, FRET is of tremendous use in life science research. FRET also enables the quantitative measurement of molecular interactions in living cells and organisms. Coupled with the advent of the development of new fluorescent probes such as the green fluorescent protein and other genetically engineered reporters, the future of this technology is sure to reveal molecular associations both within cells and on cell surfaces that were previously explored using only nonmicroscopic methods.

Major new FRET developments in the area of fluorescence lifetime determinations, either in the time or frequency domain will exploit techniques like FLIM, which can be combined with multispectral, polarization-sensitive, and optical-sectioning modalities, and as such offer the prospect of a "do-it-all" form of fluorescence microscopy [81]. Besides, combination of FRET with techniques like single-molecule and FCS exposes the dynamic processes in biomolecular processes [82].

Acknowledgments

We acknowledge funding from University of Virginia, National Center for Research Resources NCRR-NIH RR027409, and National Heart, Lung, and Blood Institute (NHLBI) PO1HL101871).

References

1. Clegg, R. M. (2006). In: *Reviews in Fluorescence*. Geddes, C. D., and Lakowicz, J. R., eds. Springer, New York, pp. 1–45.
2. Förster, T. (1965). In: *Modern Quantum Chemistry*. Sinanoglu, O., ed. Academic Press Inc., pp. 93–137.
3. Clegg, R. M. (1996). In: *Fluorescence Imaging Spectroscopy and Microscopy*. Wang, X. F., and Herman, B., eds. John Wiley & Sons Inc., New York, pp. 179–251.
4. Tavares, E., Macedo, J. A., Paulo, P. M., Tavares, C., Lopes, C., and Melo, E. P. (2014). Live-cell FRET imaging reveals clustering of the prion protein at the cell surface induced by infectious prions. *Biochim. Biophys. Acta, Mol. Basis Dis.*, **1842**(7), 981–991.
5. Engelborghs, Y., and Visser, A. J. (eds.) (2014). *Fluorescence Spectroscopy and Microscopy: Methods and Protocols*. Humana Press.
6. Lakowicz, J. R. (2006). *Principles of Fluorescence Spectroscopy*. Springer, New York.
7. Mondal, P. P., and Diaspro, A. (2014). General fluorescence imaging techniques. In: *Fundamentals of Fluorescence Microscopy*. Springer Netherlands, pp. 135–147.
8. Hohlbein, J., Craggs, T. D., and Cordes, T. (2014). Alternating-laser excitation: single-molecule FRET and beyond. *Chem. Soc. Rev.*, **43**, 1156–1171.
9. Stennett, E. M., Ciuba, M. A., and Levitus, M. (2014). Photophysical processes in single molecule organic fluorescent probes. *Chem. Soc. Rev.*, **43**(4), 1057–1075.
10. Kumagai, Y., Naoki, H., Nakasyo, E., Kamioka, Y., Kiyokawa, E., and Matsuda, M. (2014). Heterogeneity in ERK activity as visualized by in vivo FRET imaging of mammary tumor cells developed in MMTV-Neu mice. *Oncogene*, ***34***(8), 1051–1057.
11. Ebrecht, R., Paul, C. D., and Wouters, F. S. (2014). Fluorescence lifetime imaging microscopy in the medical sciences. *Protoplasma*, 1–13.

12. Periasamy, A., and Clegg, R. M. (eds.) (2009). *FLIM Microscopy in Biology and Medicine*. CRC Press, London.
13. Baird, L., Swift, S., Llères, D., and Dinkova-Kostova, A. T. (2014). Monitoring Keap1–Nrf2 interactions in single live cells. *Biotechnol. Adv.*, **32**(6), 1133–1144.
14. Sun, Y., Day, R. N., and Periasamy, A. (2011). Investigating protein-protein interactions in living cells using fluorescence lifetime imaging microscopy. *Nat. Protoc.*, **6**(9), 1324–1340.
15. King, C., Sarabipour, S., Byrne, P., Leahy, D. J., and Hristova, K. (2014). The FRET signatures of noninteracting proteins in membranes: simulations and experiments. *Biophys. J.*, **106**(6), 1309–1317.
16. Bindels, D. S., Goedhart, J., Hink, M. A., van Weeren, L., Joosen, L., and Gadella Jr, T. W. (2014). Optimization of fluorescent proteins. In: *Fluorescence Spectroscopy and Microscopy*. Humana Press, pp. 371–417.
17. Day, R. N., and Davidson, M. W. (eds.) (2014). *The Fluorescent Protein Revolution*. CRC Press.
18. Wang, L., et al. (2004). Evolution of new nonantibody proteins via iterative somatic hypermutation. *Proc. Natl. Acad. Sci.*, **101**(48), 16745–16749.
19. Padilla-Parra, S., Audugé, N., Coppey-Moisan, M., and Tramier, M. (2014). Quantitative study of protein–protein interactions in live cell by dual-color fluorescence correlation spectroscopy. In: *Fluorescence Spectroscopy and Microscopy*. Humana Press, pp. 683–698.
20. Fernández-Suárez, M., and Ting, A. Y. (2008). Fluorescent probes for super-resolution imaging in living cells. *Nat. Rev. Mol. Cell Biol.*, **9**(12), 929–943.
21. Achurra, P., Holden, S., Pengo, T., and Manley, S. (2014). Photoactivated localization microscopy for cellular imaging. In: *Super-Resolution Microscopy Techniques in the Neurosciences*. Humana Press, pp. 87–111.
22. Ma, L., Yang, F., and Zheng, J. (2014). Application of fluorescence resonance energy transfer in protein studies. *J. Mol. Struct.*, **1077**, 87–100.
23. Nienhaus, K., and Nienhaus, G. U. (2014). Fluorescent proteins for live-cell imaging with super-resolution. *Chem. Soc. Rev.*, **43**(4), 1088–1106.
24. Bindels, D. S., Goedhart, J., Hink, M. A., van Weeren, L., Joosen, L., and Gadella Jr, T. W. (2014). Optimization of fluorescent proteins. In: *Fluorescence Spectroscopy and Microscopy*. Humana Press, pp. 371–417.

25. Rice, W. L., and Kumar, A. T. (2014). Preclinical whole body time domain fluorescence lifetime multiplexing of fluorescent proteins. *J. Biomed. Opt.*, **19**(4), 046005–046005.

26. Prescott, M., Battad, J. M., Wilmann, P. G., Rossjohn, J., and Devenish, R. J. (2006). Recent advances in all-protein chromophore technology. *Biotechnol. Annu. Rev.*, **12**, 31–66.

27. Piston, D. W., and Kremers, G. J. (2007). Fluorescent protein FRET: the good, the bad and the ugly. *Trends Biochem. Sci.*, **32**(9), 407–414.

28. Newman, R. H., Fosbrink, M. D., and Zhang, J. (2011). Genetically encodable fluorescent biosensors for tracking signaling dynamics in living cells. *Chem. Rev.*, **111**(5), 3614–3666.

29. Dedecker, P., De Schryver, F. C., and Hofkens, J. (2013). Fluorescent proteins: shine on, you crazy diamond. *J. Am. Chem. Soc.*, **135**(7), 2387–2402.

30. Subach, O. M., Cranfill, P. J., Davidson, M. W., and Verkhusha, V. V. (2011). An enhanced monomeric blue fluorescent protein with the high chemical stability of the chromophore. *PLoS One*, **6**(12), e28674.

31. Merzlyak, E. M., Goedhart, J., Shcherbo, D., Bulina, M. E., Shcheglov, A. S., Fradkov, A. F., et al. (2007). Bright monomeric red fluorescent protein with an extended fluorescence lifetime. *Nat. Methods*, **4**(7), 555–557.

32. Wiedenmann, J., Ivanchenko, S., Oswald, F., Schmitt, F., Röcker, C., Salih, A., et al. (2004). EosFP, a fluorescent marker protein with UV-inducible green-to-red fluorescence conversion. *Proc. Natl. Acad. Sci. U.S.A.*, **101**(45), 15905–15910.

33. Wiedenmann, J., Oswald, F., and Nienhaus, G. U. (2009). Fluorescent proteins for live cell imaging: opportunities, limitations, and challenges. *IUBMB Life*, **61**(11), 1029–1042.

34. Shaner, N. C., Lin, M. Z., McKeown, M. R., Steinbach, P. A., Hazelwood, K. L., Davidson, M. W., and Tsien, R. Y. (2009). Evaluating and improving the photostability of fluorescent proteins. *Proc. SPIE*, **7191**, 719105.

35. Ai, H. W., Baird, M. A., Shen, Y., Davidson, M. W., and Campbell, R. E. (2014). Engineering and characterizing monomeric fluorescent proteins for live-cell imaging applications. *Nat. Protoc.*, **9**(4), 910–928.

36. Newman, R. H., and Zhang, J. (2014). The design and application of genetically encodable biosensors based on fluorescent proteins. In: *Fluorescent Protein-Based Biosensors*. Humana Press, pp. 1–16.

37. Campbell, R. E., Tour, O., Palmer, A. E., Steinbach, P. A., Baird, G. S., Zacharias, D. A., and Tsien, R. Y. (2002) A monomeric red fluorescent protein. *Proc. Natl. Acad. Sci. U.S.A.*, **99**, 7877–7882.

38. Kredel, S., Nienhaus, K., Oswald, F., Wolff, M., Ivanchenko, S., Cymer, F., Jeromin, A., Michel, F. J., Spindler, K. D., Heilker, R., Nienhaus, G. U., and Wiedenmann, J. (2008) Optimized and far-redemitting variants of fluorescent protein eqFP611. *Chem. Biol.*, **15**, 224–233.
39. Chan, F. T., Kaminski, C. F., and Kaminski Schierle, G. S. (2011). HomoFRET fluorescence anisotropy imaging as a tool to study molecular self-assembly in live cells. *ChemPhysChem*, **12**(3), 500–509.
40. Periasamy, A., and Day, R. (eds.) (2011). *Molecular Imaging: FRET Microscopy and Spectroscopy*. Elsevier.
41. Sun, Y., Wallrabe, H., Seo, S. A., and Periasamy, A. (2011). FRET microscopy in 2010: the legacy of Theodor Förster on the 100th anniversary of his birth. *ChemPhysChem*, **12**(3), 462–474.
42. Sun, Y., Rombola, C., Jyothikumar, V., and Periasamy, A. (2013). Förster resonance energy transfer microscopy and spectroscopy to localize protein-protein interactions in live cells. *Cytometry A*, **83A**(9), 780–793.
43. Wallrabe, H., and Periasamy, A. (2005). Imaging protein molecules using FRET and FLIM microscopy. *Curr. Opin. Biotechnol.*, **16**(1), 19–27.
44. Hazelwood, K. L., Ramko, E. B., Ozarowska, A. P., Olenych, S. G., Worthy, P. N., Guan, A., et al. (2008). Searching the fluorescent protein color palette for new FRET pairs. *Proc. SPIE*, **6868**, pp. 68680C.
45. Kremers, G. J., and Goedhart, J. (2009). Visible fluorescent proteins for FRET. *FRET FLIM Tech.*, **33**, 171–224.
46. Komatsu, N., Aoki, K., Yamada, M., Yukinaga, H., Fujita, Y., Kamioka, Y., and Matsuda, M. (2011). Development of an optimized backbone of FRET biosensors for kinases and GTPases. *Mol. Biol. Cell*, **22**(23), 4647–4656.
47. Conway, J. R., Carragher, N. O., and Timpson, P. (2014). Developments in preclinical cancer imaging: innovating the discovery of therapeutics. *Nat. Rev. Cancer*, **14**(5), 314–328.
48. Hertel, F., and Zhang, J. (2014). Monitoring of post-translational modification dynamics with genetically encoded fluorescent reporters. *Biopolymers*, **101**(2), 180–187.
49. Whitaker, M. (2010). Genetically-encoded probes for measurement of intracellular calcium. *Methods Biol.*, **99**, 153.
50. Palmer, A. E., and Tsien, R. Y. (2006). Measuring calcium signaling using genetically targetable fluorescent indicators. *Nat. Protoc.*, **1**(3), 1057–1065.

51. Shcherbakova, D. M., and Verkhusha, V. V. (2013). Near-infrared fluorescent proteins for multicolor in vivo imaging. *Nat. Methods*, **10**(8), 751–754.

52. Padilla-Parra, S., Audugé, N., Coppey-Moisan, M., and Tramier, M. (2014). Quantitative study of protein–protein interactions in live cell by dual-color fluorescence correlation spectroscopy. In: *Fluorescence Spectroscopy and Microscopy*. Humana Press, pp. 683–698.

53. Ganesan, S., Ameer-Beg, S. M., Ng, T. T., Vojnovic, B., and Wouters, F. S. (2006). A dark yellow fluorescent protein (YFP)-based Resonance Energy-Accepting Chromoprotein (REACh) for Förster resonance energy transfer with GFP. *Proc. Natl. Acad. Sci. U.S.A.*, **103**(11), 4089–4094.

54. Okumoto, S., Jones, A., and Frommer, W. B. (2012). Quantitative imaging with fluorescent biosensors. *Annu. Rev. Plant Biol.*, **63**, 663–706.

55. Balla, T. (2009). Green light to illuminate signal transduction events. *Trends Cell Biol.*, **19**(11), 575–586.

56. Hoppe, A. D. (2007). Quantitative FRET microscopy of live cells. In: *Imaging Cellular and Molecular Biological Functions*. Springer Berlin Heidelberg, pp. 157–181.

57. Miyawaki, A., Llopis, J., Heim, R., McCaffery, J. M., Adams, J. A., Ikura, M., and Tsien, R. Y. (1997). Fluorescent indicators for Ca^{2+} based on green fluorescent proteins and calmodulin. *Nature*, **388**(6645), 882–887.

58. Periasamy, A., Wallrabe, H., Chen, Y., and Barroso, M. (2008). Quantitation of protein–protein interactions: confocal FRET microscopy. *Methods Cell Biol.*, **89**, 569–598.

59. Elangovan, M., Wallrabe, H., Chen, Y., Day, R. N., Barroso, M., and Periasamy, A. (2003). Characterization of one-and two-photon excitation fluorescence resonance energy transfer microscopy. *Methods*, **29**(1), 58–73.

60. Elangovan, M., Day, R. N., and Periasamy, A. (2002). Nanosecond fluorescence resonance energy transfer-fluorescence lifetime imaging microscopy to localize the protein interactions in a single living cell. *J. Microsc.*, **205**(1), 3–14.

61. Chen, Y., Elangovan, M., and Periasamy, A. (2005). FRET data analysis: the algorithm. *Mol. Imag.: FRET Microsc. Spectrosc.*, **7**, 126–145.

62. Wallrabe, H., and Periasamy, A. (2005). Imaging protein molecules using FRET and FLIM microscopy. *Curr. Opin. Biotechnol.*, **16**(1), 19–27.

63. Cox, G. (2012). *Optical Imaging Techniques in Cell Biology*. CRC Press.

64. Sun, Y., and Periasamy, A. (2010). Additional correction for energy transfer efficiency calculation in filter-based Förster resonance energy transfer microscopy for more accurate results. *J. Biomed. Opt.*, **15**(2), 020513.

65. Brucale, M., Schuler, B., and Samorì, B. (2014). Single-molecule studies of intrinsically disordered proteins. *Chem. Rev.*, **114**(6), 3281–3317.

66. Lee, W., von Hippel, P. H., and Marcus, A. H. (2014). Internally labeled Cy3/Cy5 DNA constructs show greatly enhanced photo-stability in single-molecule FRET experiments. *Nucleic Acids Res.*, **42**(9), 5967–5977.

67. Kragh, S. L., and Birkedal, V. (2014). The power of single-molecule FRET microscopy applied to DNA nanotechnology. In: *Nucleic Acid Nanotechnology*. Springer Berlin Heidelberg, pp. 53–68.

68. Sun, Y., Wallrabe, H. Booker, C., Day, R. N., and Periasamy, A. (2010). Three-color spectral FRET microscopy localizes three interacting proteins in living cells. *Biophys. J.* **99**, 1274–1283.

69. Milroy, L. G., Grossmann, T. N., Hennig, S., Brunsveld, L., and Ottmann, C. (2014). Modulators of protein–protein interactions. *Chem. Rev.*, **114**(9), 4695–4748.

70. Hayward, R. D., Goguen, J. D., and Leong, J. M. (2010). No better time to FRET: shedding light on host pathogen interactions. *J. Biol.*, **9**(2), 1–4.

71. Czymmek, K. J., Bourett, T. M., and Howard, R. J. (2004). Fluorescent protein probes in fungi. *Methods Microbiol.*, **34**, 27–62.

72. Albizu, L., Cottet, M., Kralikova, M., Stoev, S., Seyer, R., Brabet, I., and Durroux, T. (2010). Time-resolved FRET between GPCR ligands reveals oligomers in native tissues. *Nat. Chem. Biol.*, **6**(8), 587–594.

73. Glickman, J. F., Wu, X., Mercuri, R., Illy, C., Bowen, B. R., He, Y., and Sills, M. (2002). A comparison of ALPHAScreen, TR-FRET, and TRF as assay methods for FXR nuclear receptors. *J. Biomol. Screening*, **7**(1), 3–10.

74. Inglese, J., Johnson, R. L., Simeonov, A., Xia, M., Zheng, W., Austin, C. P., and Auld, D. S. (2007). High-throughput screening assays for the identification of chemical probes. *Nat. Chem. Biol.*, **3**(8), 466–479.

75. Wyhs, N., Walker, D., Giovinazzo, H., Yegnasubramanian, S., and Nelson, W. G. (2014). Time-resolved fluorescence resonance energy transfer assay for discovery of small-molecule inhibitors of methyl-CpG binding domain protein 2. *J. Biomol. Screening*, **19**(7), 1060–1069.

76. Cywinski, P., Nono, K. N., Charbonnière, L., Hammann, T., and Löhmannsröben, H. G. (2014). Photophysical evaluation of a new

functional terbium complex in FRET-based time-resolved homogenous fluoroassays. *Phys. Chem. Chem. Phys.*, **16**, 6060–6067

77. Ruan, G., Agrawal, A., Marcus, A. I., and Nie, S. (2007). Imaging and tracking of tat peptide-conjugated quantum dots in living cells: new insights into nanoparticle uptake, intracellular transport, and vesicle shedding. *J. Am. Chem. Soc.*, **129**(47), 14759–14766.
78. Zhai, X., Boldyrev, I. A., Mizuno, N. K., Momsen, M. M., Molotkovsky, J. G., Brockman, H. L., and Brown, R. E. (2014). Nanoscale packing differences in sphingomyelin and phosphatidylcholine revealed by BODIPY fluorescence in monolayers: physiological implications. *Langmuir*, **30**(11), 3154–3164.
79. Groves, J. T., and Dustin, M. L. (2003). Supported planar bilayers in studies on immune cell adhesion and communication. *J. Immunol. Methods*, **278**(1), 19–32.
80. Kaizuka, Y., and Groves, J. T. (2004). Structure and dynamics of supported intermembrane junctions. *Biophys. J.*, **86**(2), 905–912.
81. Jares-Erijman, E. A., and Jovin, T. M. (2006). Imaging molecular interactions in living cells by FRET microscopy. *Curr. Opin. Chem. Biol.*, **10**(5), 409–16.
82. Sahoo, H. (2011). Förster resonance energy transfer–a spectroscopic nanoruler: principle and applications. *J. Photochem. Photobiol., C*, **12**(1), 20–30.

Chapter 13

Photodetectors for Fluorescence Microscopy

James Pawley
Zoology Department, University of Wisconsin,
Madison, WI, 53706, USA
jbpawley@wisc.edu

While the recent rapid growth in the use of fluorescence microscopy in biological research has been largely driven by the introduction of new microscopical techniques such as confocal microscopy, and the development of ever more clever and informative forms of fluorescent markers, it remains true that all of the information that can be gleaned through these techniques necessarily passes through the photodetector. The purpose of the photodetector is to produce an electronic output that can then be digitized and that is proportional to the photon signal emitted from a given area (or volume) of the specimen [1].

This detection process can be broken down into two steps. In the first step, the photon is absorbed to create an excited or unbound photoelectron (PE) in the sensitive volume of the photodetector. The fraction of incoming photons that actually produce a PE is called the raw quantum efficiency (QE_{raw}).

Fundamentals of Fluorescence Imaging
Edited by Guy Cox

ISBN 978-981-4774-85-7 (Hardcover), 978-1-351-12940-4 (eBook)
www.jennystanford.com

In the second step, this PE, and any others in the same pixel, are amplified sufficiently to be fed to a digital-to-analog converter (DAC). The DAC then senses the signal level and conveys a number proportional to this level to the image memory in the computer. Performance depends on the extent to which this process can be performed rapidly, accurately and without the addition of any spurious or extraneous values (noise) to the digitized end result [2, 3].

This chapter will describe the operating principles, and performance limitations of the types of photodetector commonly found in modern microscopes, while paying particular attention to their quantum efficiency and noise.

13.1 Historical Photodetectors

The first practical photodetector was photographic film. While it would be wrong not to acknowledge the tremendous contribution that images recorded on film have made to our understanding of biological systems, it remains true that, compared to modern electronic detectors, its QE and noise performance left much to be desired. Its major, and so far unique, capability is its stability: one can today still view images recorded photographically over 100 years ago. One wonders what type of digital storage medium might ever come close to this level of longevity.

Although film was briefly supplanted as the optimal medium for recording image data from microscopes using various types of vidicon video camera coupled to a video recorder, these have now been replaced by electronic sensors based on various semiconductor and photocathode materials. Photodetectors fall into two groups: single-channel devices, such as the avalanche photodiode (APD) and the various iterations of the photomultiplier tube (PMT, used to detect a time-varying signal in the low megahertz range as produced by the various single-beam laser-scanning) and silicon-based, image detectors such as the charge-coupled device (CCD), the electron-multiplier charge-coupled device (EM-CCD), and the scientific complementary metal–oxide silicon sensor (sCMOS) used to collect widefield image data.

13.1.1 The Photomultiplier Tube

The PMT not only has the high bandwidth required to track the time-varying output of a scanning microscope, it is also the first photodetector capable of producing a detectable signal from the absorption of a single input photon. While these facts seem to imply a high degree of perfection, closer inspection reveals considerable room for improvement [4].

In all PMTs, the initial step involves the absorption of a photon within an opaque or semitransparent photocathode (K) positioned in a vacuum in such a way that any photoelectron emitted from the surface will be attracted to a nearby positive electrode (the first dynode, d_1, Fig. 13.1a). The dynode material, shape, and orientation are chosen so that, on arrival at d_1, each PE will produce 3–20 secondary electrons (Fig. 13.1b). These in turn proceed to d_2, d_3, etc., and at each impact, additional charge amplification occurs. The total amplification depends on the voltage applied at each step but when a total of 1000–1500 V is split between 10–14 stages, gains of over one million are common before the negative charge pulse reaches the anode and is fed to a current-to-voltage converter for measurement. As a result, a single PE can make a pulse that is easily discriminated from the electronic noise of the converter. In addition, as long as the photocathode is kept cool, the output signal in the absence of light is almost zero (at least compared the data rate from a scanning light microscope: i.e., <0.1 count/pixel). However, as the main source of dark counts is PE emitted by the thermal interactions between the electrons in the PC, this count rate will increase if the PC is large or warm or made out of material that is particularly sensitive to the low-energy photons of red light. In summary, under normal circumstances, the PMT has a very low dark count and can "see" the signal from a single photon. What could be better?

The major problem occurs at the first step. Figure 13.2 shows the QE_r vs. wavelength (λ) of a number of PC materials. It shows that, even with even the best photocathodes, less than 40% of the incoming photons produce a PE and that this figure is a strong function of λ. In practice, the situation is actually worse than this, because the measuring technique used to make these QE curves usually measures all the PE leaving the PC, whether or not they actually multiply at d_1. Some miss d_1 and others fail to multiply. As

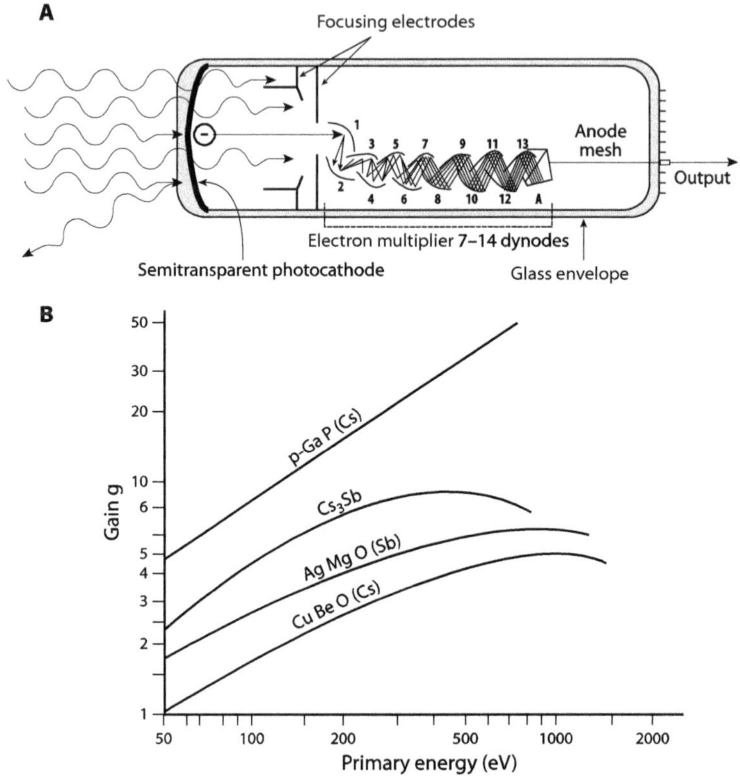

Figure 13.1 (a) Diagram of photomultiplier tube (PMT). A small fraction of the photons approaching the photocathode (PC) from the left produce photoelectrons (PE), which are then attracted by the positive voltage on the first dynode (d_1) where the resulting collision produces an average of 3–30 secondary electrons (SE). These then proceed to the second and subsequent dynodes, with the single-PE pulse becoming larger at each stage. (b) The number of SE produced at the first dynode varies with the voltage between it and the PC and the dynode surface composition. Higher first-stage gain reduces the effect of Poisson noise in the electron multiplier, reducing the excess noise factor.

result, the effective QE (QE_{ef}) is normally at least 30% lower than the raw QE curves suggest. It should also be remembered that making a photocathode is a delicate art that usually involves evaporation in a vacuum after the glass envelope has been sealed. Each PC will be slightly different and QE curves tend to show the performance of the

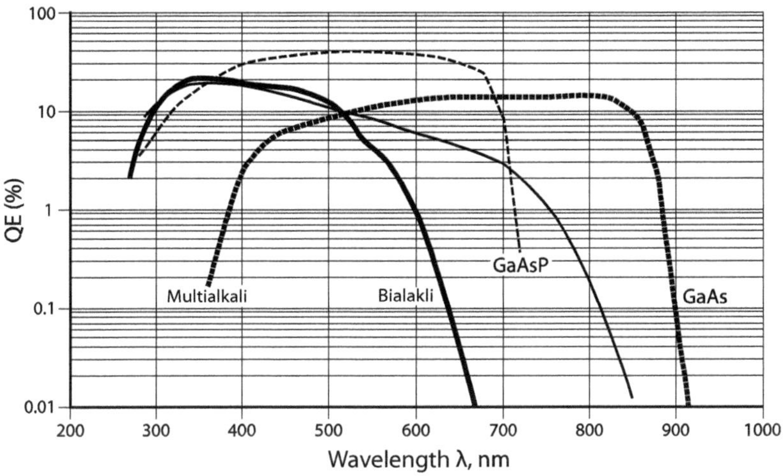

Figure 13.2 Quantum efficiency (QE_{raw}) of various photocathode materials.

best tubes of a given type. The tube in your microscope may have 10–30% lower QE_{raw} at a given λ.

Apart from low QE_{ef}, the PMT suffers from multiplicative noise (more recently called "excess noise"). Multiplicative noise is not the same as additive noise, which may come from the environment or the electronics, and comes about because statistical variations in the actual amount of charge multiplication associated with the amplification of each PE. Like any quantum-mechanical process, impact-mediated charge multiplication is governed by Poisson statistics that decrees that any process involving measuring an average of n quantum-mechanical events will be subject to an uncertainty of $\sqrt{n}$. If we assume that, on average first stage emits 9 secondary electrons for every arriving PE, the best that we can hope for is that, 67% of the time a particular PE will be multiplied somewhere in the range of 9 ± 3 and 33% of incoming PE will produce output pulses outside of this range. Normally, all the single-PE pulses are integrated during the pixel period. In this case, some photons will count more than others, adding another uncertainty in the final stored data in addition to that caused by Poisson statistics applied to the original number of PE. The excess noise factor (ENF) is the ratio of the Poisson uncertainty of these PE to the uncertainty of the pulse leaving the detector. The best PMTs operating with the K–d_1 voltage high enough (400–600 V) to produce a first-stage gain of at least 25, are characterized by an ENF of 1.2–1.3. However, commercial confocals are more likely to have a maximum first stage gain of 4 or 5 even when the at the maximum PMT voltage. This produces a correspondingly higher ENF.

Under these conditions, a measurement of 100PE will have a standard deviation of $\pm$ 12e or $\pm$ 14e not the expected $\sqrt{100}\text{e} = \pm\ 10\text{e}$. The effect of this is that one must collect 144PE in order to get 10% accuracy. In other words, the ENF has effectively reduced QE by about 30%. For PMTs optimized for other purposes (such as the linear arrays of mini-PMTs used in many spectral detectors), or when used with lower K–d_1 voltage, the reduction in QE_{eff} can be as much as a 70%.

13.1.1.1 Photon counting

The ENF can be eliminated by using photon-counting. In this process, each single-PE pulse is amplified, discriminated from background (and from other pulses) and then counted so the fact that some PEs produce twice the signal current ($9 + 3 = 12$) of others ($9 - 3 = 6$) is of no consequence. However, PMT output pulses are just too long to count more than ~10 of them arriving at random times during a 1 μs pixel dwell time without losing more than 20% to pulse pileup. As a result, pulse-counting PMTs are seldom used in scanning light microscopy.

As the reduction in QE_{eff} caused by ENF occurs in addition to those produced by the mechanisms mentioned previously, the final QE_{eff} of the PMTs used in most confocal microscopes will vary from 4% to 8% in the green and generally much lower in the red. Instruments employing GaAs and GaAsP photocathodes have higher QE_{eff}, especially in the red, but these must usually be cooled to keep the dark count down.

13.1.1.2 Hybrid PMTs

The only other way to reduce the effect of multiplicative noise is to greatly increase the gain of the first stage of charge amplification. Figure 13.3a shows how this can be achieved in what is called a

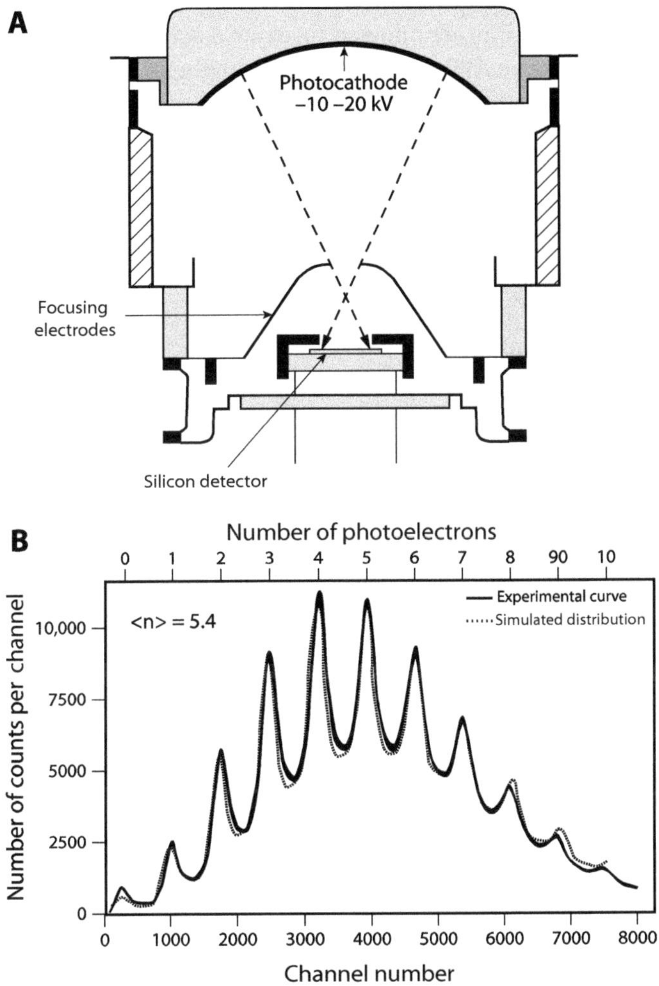

Figure 13.3 (a) Layout of hybrid PMT. Electrons ejected from the photocathode (PE) by impinging photons are focused onto an avalanche photodiode (APD). Because of the high potential difference, each PE will put 1000–2000 electrons into the conduction band of the APD where the signal is then further amplified. (b) Spectrum of output-pulse heights from a hybrid-PMT exposed to light pulses large enough to produce an average of 4.5 ± 2 PE/pulse. The high gain of the first stage reduces the excess noise factor sufficiently to permit pulses that actually produced 10 PE to be distinguished from those that produce 9 PE.

hybrid PMT [5–7]. Here the K–d_1 voltage is increased to 8–20 kV and the energetic PEs are allowed to strike a silicon PIN photodiode (or alternatively, an APD, operating in the low-gain, linear, mode; see below). After accounting for losses in the nonproductive outer layers of the silicon, a single, 10 keV electron arriving at the diode can knock $\sim$10,000 $\pm$ 100 electrons into the valence band of the diode. In other words the error introduced by Poisson noise is reduced to 1% from the 33% that characterized a first-stage gain of only 9. As a result, pulses representing 5PE can be discriminated from those representing 6PE with $>$50% probability (Fig. 13.3b).

This is a significant improvement but it comes at the cost of a much higher unit cost, increased circuit complexity and the necessity of using semi-transparent "end-window" photocathodes, which have somewhat lower QE_r than some of the solid-PC, side-window versions. In addition, accidentally exposing early hybrid PMTs to excessive light while the K–d_1 voltage was present caused severe damage. Over time the overload protection circuitry has markedly improved but problems can still occur when scanning specimens with massive variation in stain density because these protection circuits may turn off the detector whenever the beam scans over a very bright object such as a crystal of stain. Apart from this problem, cooled, hybrid PMTs with GaAs or GaAsP photocathodes are probably the most effective photodetectors currently employed in commercial, single-beam scanning light microscopes.

13.2 Solid-State Single-Channel Detectors

Every forward-biased P/N junction emits photons at a λ proportional to its bandgap. Therefore, we should not be surprised that the process can work in reverse. Any photon with an energy greater than that of the bandgap and impinging upon the depletion region of a reverse-biased P/N junction will be absorbed and promote one electron from the valence band to the conduction band with very high QE. In principle, this event can be detected as a current. The PIN photodiode is a device in which the physical size of the sensitive

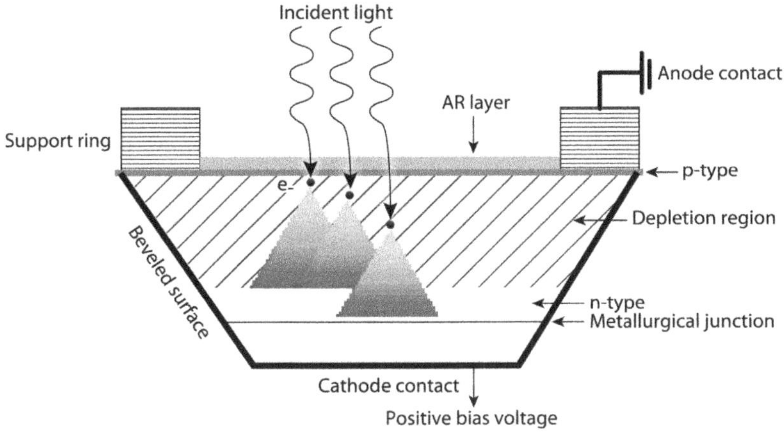

Figure 13.4 Schematic diagram of avalanche photodiode (APD). Photons approaching from the top are absorbed in the intrinsic silicon, producing conduction electrons. The high electric field then accelerates these electrons sufficiently to enable them to knock other valence electrons into the conduction band. Further collisions produce an average gain of up to 10^8, but with very high excess noise.

depletion region has been enlarged by inserting a layer of non-doped intrinsic (I) semiconductor between the P- and N-doped regions.

As it is very difficult to sense the current related to the motion of a single electron, especially in the presence of a much larger leakage current present in all reverse-biased PIN junctions, PIN photodiodes are seldom used to detect photon signals as small as those present in a scanning fluorescence light microscope.

However, much as the PMT developed from the earlier technology of the vacuum photodiode, efforts to add charge-multiplication to the PIN diode resulted in the avalanche photodiode (APD, Fig. 13.4) [8]. If the reverse bias is increased on any semiconductor diode, a point is reached (the peak-inverse voltage, or PIV) where the device breaks down and it starts to conduct. This occurs either because the higher reverse bias has so diminished the thickness of the depletion zone that electrons can "tunnel through" the potential barrier or because the voltage gradient in the intrinsic layer is so high that the any leakage electron can pick up sufficient energy from the bias field to ionize the first valence electron it collides with.

This collision mechanism initiates an avalanche cascade, effectively amplifying the single PE.

At low overvoltages, this amplification is 10–100× and is fairly linear, but unfortunately, the gain mechanism is subject to very high multiplicative noise (ENF > 2, producing a 4-fold reduction in QE_{eff}). More to the point the most probable gain is zero (i.e., many PEs are simply reabsorbed without multiplying, further reducing the QE_{eff}).

At higher overvoltage, the avalanche continues producing a high current until the applied voltage is reduced below the avalanche threshold. The device can be protected from overheating by adding a series resistor that reduces the voltage across the diode when the current increases, an arrangement called passive quenching.

Alternatively, one can employ active quenching in which transistor circuits are used to sense the start of the current pulse and then reduce the applied voltage. Active quenching permits slightly faster counting at the price of increased complexity and lower fill-factor.

Because all P/N diodes are effectively also capacitors, the time required (t_t) for the voltage on a passively quenched APD to return above the avalanche threshold is defined by, $t_t = RC$, where R is the resistance of the series resistor and C the capacitance of the diode.

An APD with passive quenching operated at a gain of 10^5–10^8 is said to be in the Geiger mode. Because each pulse grows until it is quenched, single-PE pulses are very similar in size and one can obtain an accurate estimate of the number of photoelectrons created by simply integrating these pulses.

The problem that limits the use of simple APDs in scanning light microscopes is that when used in the low voltage, linear mode, the high ENF reduces QE_{eff}, and when used in the Geiger node, it is insensitive to light for a short period following each avalanche. The dead time depends on the capacitance of the diode, which in turn varies with its area: an area of 0.01 mm^2, produces a deadtime of about 100 ns, or 1/10th of a 1 μs pixel. This means that, even at a signal level of 1PE/pixel, the sensor would be dead 10% of the time, and at 5PE/pixel, only half the potential PE signal could be counted. Clearly this is unacceptable: single APDs are just too slow.

13.2.1 Multipixel Photon Counters

One solution to this problem is to make an array of n, passively-quenched APDs, all connected in parallel to form a Multi-Pixel Photon Counter (MPPC, Fig. 13.5) [9–12]. As long as the light signal is evenly spread over the n sensors, the fact that any one sensor may be dead for 10% of the pixel-dwell time has no effect on the remaining $n - 1$ pixels. If $n = 100 - 1600$, pulse-pileup ceases to be a problem for most microscopical applications. Figure 13.6 shows a spectrum of output pulses produced by an MPPC device exposed to series of

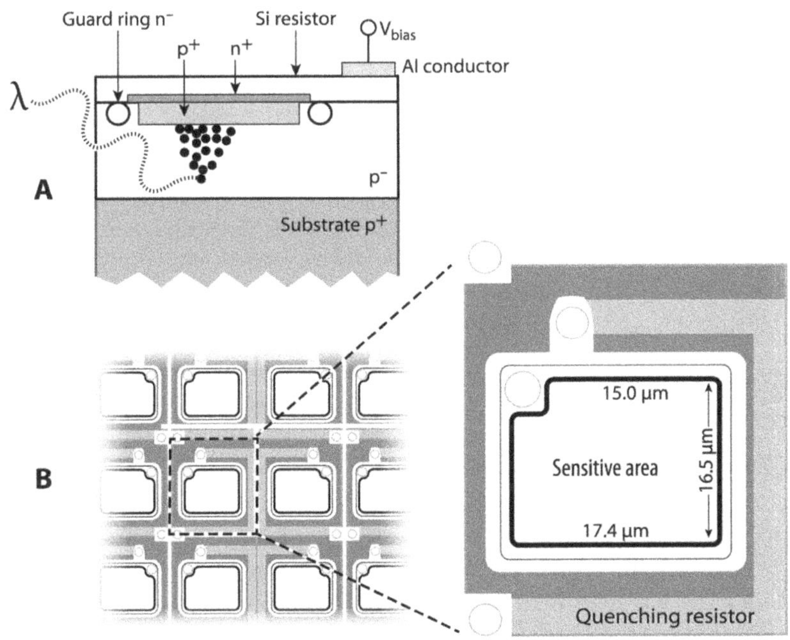

Figure 13.5 (a) APD subunit of a multipixel photon-counting silicon sensor (MPPC, also sometimes called a solid-state PMT). The subunit contains not only the APD but also the quenching resistor that allows the APD to operate in the Geiger mode, where all single-PE pulses are about the same size. (b) Sketch showing part of a 40 × 40 array of 25 × 25 μm APD subunits connected in parallel. Spreading the light signal over 400 subunits, reduces the chance of a PE being formed in an APD that is already in the process of making a Geiger pulse. This reduces pulse-pileup losses. Figures redrawn from Hamamatsu.

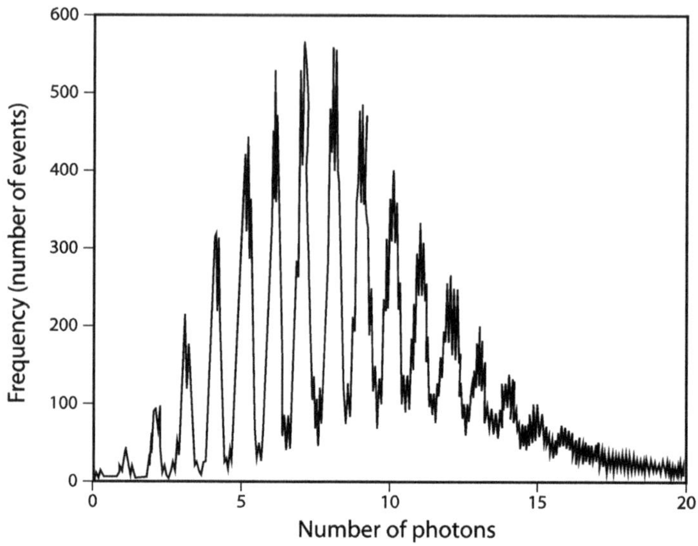

Figure 13.6 Spectrum of MPPC output-pulse heights exposed to light pulses large enough to produce an average of 9 ± 3 PE. Geiger-mode operation reduces the excess noise factor sufficiently to permit pulses that actually produced 15 PE from those that produce 16 PE. Figure redrawn from Hamamatsu.

short light pulses each with an average brightness of about 7 PE. The fact that pulses of 15 PE can be distinguished from those produced by 16 PE shows that the ENF is almost 1.0.

However, there are still important trade-offs. The sensors on present MPPC arrays are 25, 50, or 100 μm on a side. As part of the sensor area must be set aside for the quenching resistor as well as for electrical barriers to prevent adjacent sensors triggering each other, the light sensitive portion of a 25 μm sensor (the fill factor) will be much lower than that of the 100 μm sensor (Fig. 13.5b). Figure 13.7 plots the pulse-detection efficiency (PDE) against photon λ for 2 × 2 mm sensors with 25, 50, and 100 μm sensor spacings. The PDE is a very stringent measure of QE_{eff}, which includes the effects of many more types of signal loss than are discussed above. Although Fig. 13.7 seems to support the use of larger sensors, the capacitance of 100 × 100 μm sensors is more than 16× larger that of the 25 × 25 μm sensor. As a result,

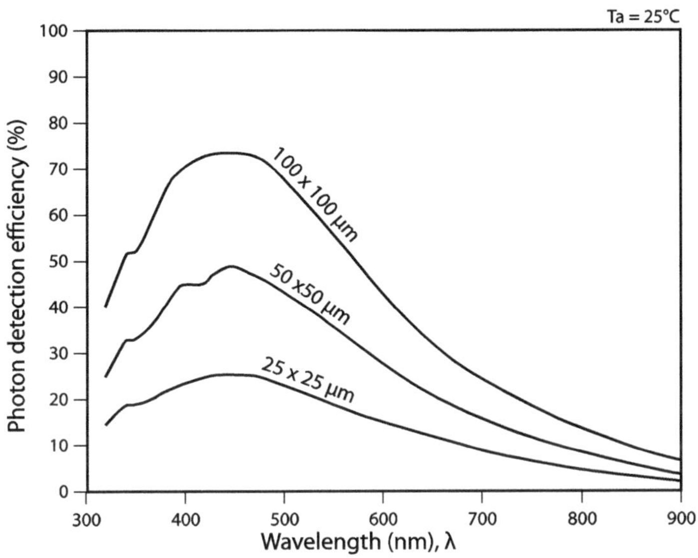

Figure 13.7 Dependence of pulse-detection efficiency (PDE) of MPPCs on cell size. MPPCs with larger cells (100 µm vs. 25 µm) devote proportionally less non-sensitive space to the guard zones and the quenching resistors and therefore have a higher fill factor and a higher PDE. On the other hand, having smaller cells increases both the pulse cycle time and the number of cells/area, reducing the risk of saturation due to pulse pileup. Figure redrawn from Hamamatsu.

the dead time of the smaller sensor is 16× shorter and there are also 16× more sensors in the array, so it can handle about 250× more photons/pixel before dead-time effects become important. On the other hand, dark counts are also proportional to diode area. Clearly, optimizing current MPPCs for confocal use will require the careful optimization of several parameters although matters would be simplified if it were found possible to add microlenses to improve the fill factor.

This optimization will probably involve changes in optical geometry. Present arrays are about 1 × 1 mm to 3 × 3 mm in size. As this is much larger than the pinholes in current commercial instruments, either optical elements must be introduced, or the optical axis extended, to allow the signal light to be spread evenly over the majority of the MPPC elements. To the extent that the MPPC is located in an image plane, optics that work well for one pinhole size will under- or overfill the detector at other pinhole sizes. If optics are added to place the sensor in an aperture plane, then the setup may have to change with the NA and magnification of the objective lens.

Clearly the MCCP is potentially a very attractive detector, particularly for FLIM [13] (Chapter 10), but there are a lot of details that still need to be worked out. Perhaps they can be configured in a circular array rather than a square, or better yet, as a bullseye, with separate readouts for each ring. This would produce a high-speed, high-QE, low-noise detector functioning at several pinhole sizes simultaneously.

It is also possible to read out arrays of single photon-counting APDs (SPAD arrays) so as to record images [14–17], Although current sensors have a low fill factor and the resulting images have a maximum bit depth of only 0–1 PE, arrays as large as 32 × 32 can be read out as fast as 100,000 frames/second permitting one to accumulate images with greater bit depth by summing a series of frames. Such sensors might be useful in a scanning version of structured illumination microcopy (SIM).

13.2.2 Single-Channel Photodetectors for Fluorescence-Lifetime Imaging (FLIM)

The discussion so far has stressed the importance of QE_{eff} and read noise in comparing different single-channel detectors. However, in instruments configured to separate different fluorescent probes on the basis of their fluorescence decay time, the time-domain performance of the photodetector becomes important. Because every amplification process takes some time, there will always be a delay between the production of a PE and the emergence of the single-PE pulse from the detector electronics. Ideally, this delay should be constant, however, because in the PMT each secondary electron follows a unique path when travelling from one dynode to the next, its transit time will vary and after 10–14 stages, there will be considerable spread among the electrons in the packet striking the anode. The variation in the delay of the rising edge of the current pulse is called the jitter and this parameter is particularly important when making fluorescence lifetime measurements using time-correlated single-photon counting (TCSPC) [18] (Chapter 10). The fact that the hybrid PMT has only a single bombardment stage greatly reduces this timing jitter.

Because TCSPC must be carried out at very low count rates, its accuracy is often limited by an additional factor called after-pulsing. After-pulsing occurs when an electron in the cascade collides with a residual gas molecule somewhere inside the PMT. When the resulting positive ion collides with a dynode, it can produce up to 20 secondary electrons that are then amplified by the remainder of the chain. Because an ion is much heavier than an electron, it takes longer to reach the nearest dynode and so the after-pulse will be delayed by a few microseconds. After-pulsing is more serious in large, old PMTs where the vacuum has degraded and the electron trajectories are longer. Although after-pulses seldom amount to more than a few percent of the single-PE pulses, these are still enough to distort the output of TCSPC/FLIM.

Because in the hybrid PMT only the PEs are accelerated, collisions are rare and after-pulsing is essentially absent. This provides a second argument for the use of this detector (or the MPPC) for TCSPC/FLIM.

13.3 Imaging Photodetectors

Although the history of the development of electronic image sensors is long and complex [19], in the past decade, sensors based on silicon semiconductors have become dominant. The image sensors found most useful for imaging sensitive biological specimens in the light microscope are the charge-coupled device (CCD), the electron-multiplier CCD (EM-CCD) and the scientific complementary metal–oxide-silicon (sCMOS) detector.

As mentioned above, any photon in the wavelength range of 350–700 nm striking silicon will be absorbed and produce an electron–hole pair within a few μm of the surface. Shorter wavelengths are absorbed so strongly that they tend to produce signal in the disordered layers near the surface. Longer wavelength photons have insufficient energy to force a valence electron into the conduction band.

13.3.1 The Charge-Coupled Device Detector

To use this effect in an image sensor [20, 21], it is necessary to not only detect the current that these photoelectrons represent but also to keep track of the location at which the photon was absorbed. For this to occur, the surface of the silicon must be divided into rectilinear array of subunits called pixels.

Figure 13.8 is a schematic diagram of the surface of a CCD. The PEs created in each pixel are attracted to the space under a series of overlying, positively charged, transparent, charge-transfer electrodes running horizontally (ϕ_1, ϕ_2, or ϕ_3) and are prevented from moving horizontally by repulsion from vertical bands of n-doped silicon called channel blocks.

Reading out the packets of PE trapped in each location requires many steps. First, a second set of charge-transfer electrodes, ϕ_2, lying just below each member of ϕ_1, goes positive allowing the charge packet to spread downward under ϕ_2. Then ϕ_1 goes to ground, trapping all the charge under ϕ_2. In similar manner, the packet is moved first under ϕ_3, and then under ϕ_1 of the pixel below. Nine different configurations of the voltages on the transfer electrodes will move the entire pattern of charge packets down by one line. The charge packets in the bottom line are transferred into a special area called the horizontal register. Here again the charge is constrained by channel blocks and voltages on sets of transfer electrodes (ϕ_3, ϕ_4, and ϕ_6). The difference is that, these voltages cycle at a faster rate because they need to transfer all the packets in a line of the image to the read node on the right-hand side before the packets from the next line are transferred down from the vertical register. At the read node, the charge in each packet is measured by sensing the voltage produced when the charge packet is coupled into a small capacitor. (The voltage on a capacitor is proportional to the charge it contains.) The capacitor voltage is then reset to zero, before the next packet arrives.

Because the noise performance of the CCD is so dependent on what happens at the read node, we will discuss it in more detail.

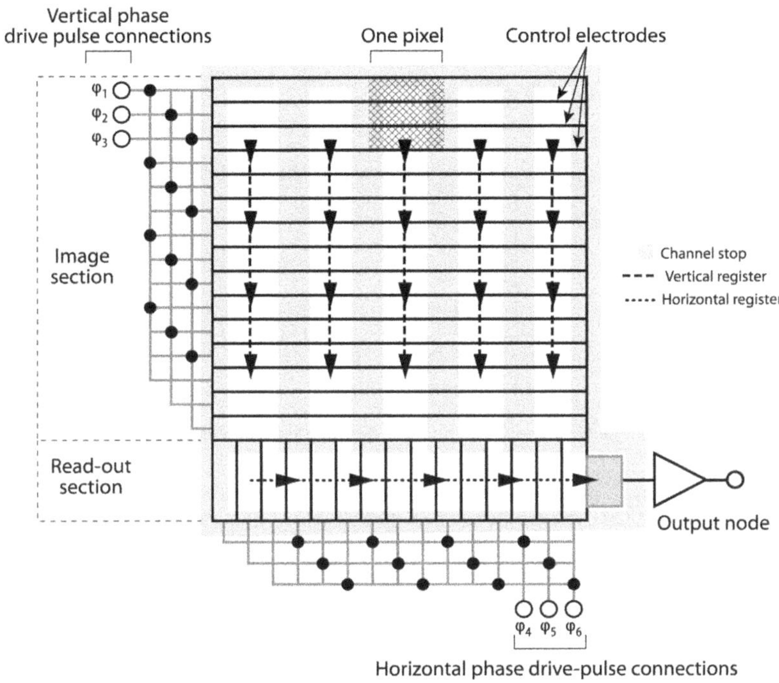

Figure 13.8 En-face view diagram of a CCD. Each pixel is bounded on the sides by n-doped channel stops, that prevent the PE in the charge packet from moving sideways, and sets of 3 overlying transparent polysilicon electrodes (ϕ_1, ϕ_2, and ϕ_3). By applying a positive bias to ϕ_1, a charge packet forms under this electrode. Sequential changes of the voltages on the three sets of electrodes move the packets in each line steadily lower until each is squeezed into the appropriate pixel of the horizontal register in the Readout Section. At the right hand side of this register, each charge packet is sequentially coupled into the read node, which is a small capacitor. The voltage on this capacitor is a measure of the size of each charge packet.

The important components can be seen in Fig. 13.9. A single electron placed into C_n will raise the voltage by a few microvolts. This change will be sensed by the READ field-effect transistor (FETs) causing it to conduct more current I_r and change the voltage across R_r. The problem is that I_r is very small and represents so few electrons/pixel that its accuracy is severely limited by Poisson statistics. This is also true of the current pulse that passes through the RESET FET. In other words, the charge packet in the read node can neither be measured

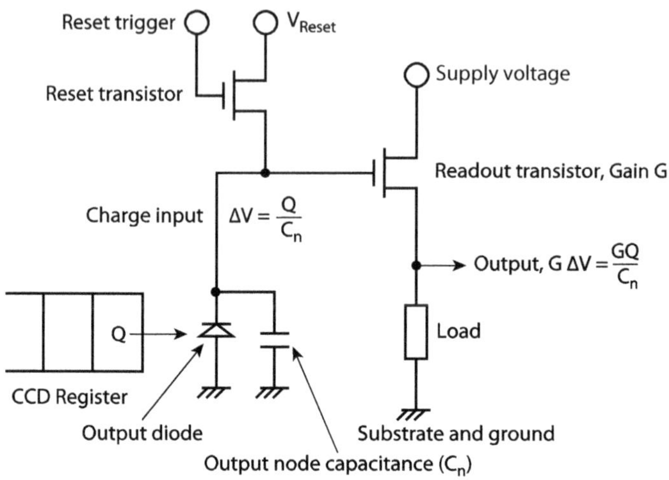

Figure 13.9 CCD Charge-detection circuit. Here the read node is represented by diode and capacitor in parallel. The readout transistor produces a signal across the load that is proportional to the size of the charge packet. The reset transistor then empties this charge, in preparation for the charge transfer circuitry to insert the next charge packet into the node. In correlated double-sampling, the readout circuit measures the voltage on the node both before and after each charge packet is inserted and records the difference.

nor zeroed with perfect accuracy. The accuracy can be improved by making C_n smaller or by taking more time to read each pixel. The former approach limits the maximum amount of charge (from bright areas of the image) that can be accurately measured and increasing the read time by a factor of 4 will reduce the read noise by $2\times$.

To further reduce the errors caused by low-frequency "flicker" noise in the reset and amplifier circuits, scientific CCDs employ correlated double-sampling. This simply means that the output of the READ FET is read both before and after the read node has been reset to zero. Employing CDS can reduce errors in the output as much as $10\times$, especially when recording high contrast images.

Although the QE of silicon is very high (90–95% depending on whether or not an antireflection coating is used) much of the signal light is scattered by the interfaces between the layers of polysilicon (RI: 4.0) and SiO_2 (RI: 1.55) that make up the charge transfer

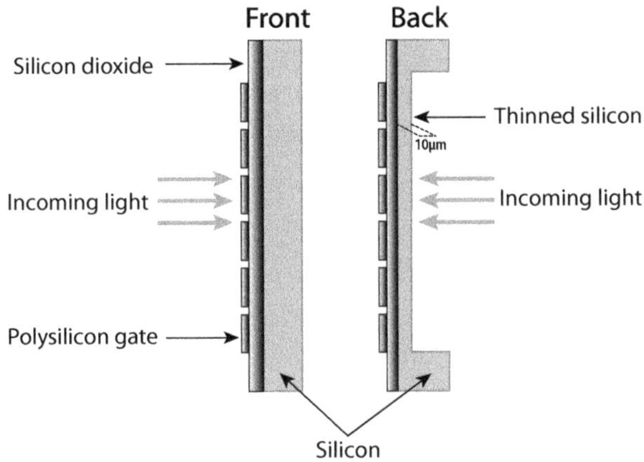

Figure 13.10 Edge view sketch of CCDs configured for either front-side (left) or back-side illumination. Illuminating from the back side increases the quantum efficiency (QE) by reducing the light lost to reflections that occur at interfaces between the various layers of SiO_2 and polysilicon that make up the charge-transfer electrodes. To insure that PE produced by the incoming light can be controlled by voltages on these electrodes, back-illuminated sensors must be thinned to about 10 μm and they also to use relatively large pixels.

electrodes covering the front surface of the CCD. As a result, only about 40% of the photons survive to be absorbed in the Si. This loss can be avoided by thinning the CCD substrate to 8–10 μm and turning it over so that the light approaches the sensitive area from the back side (Fig. 13.10). Back-illuminated CCDs have very high QE but, because PE created near the back surface are less affected by the voltages on coupling electrodes 10 μm away, they may diffuse to an adjacent pixel. Reducing the spatial resolution. To counter this, pixels tend to be larger (15 μm vs. 6.5 μm), and such cameras must be coupled to the microscope using a higher magnification photo tube to preserve Nyquist sampling.

In round figures, a good scientific CCD with 6 × 6 μm pixels will have a read noise of ±2 electrons at a read rate of 100k pixels/s. This performance requires not only a good chip but also very careful design in the associated electronics and the circuits needed to keep the chip cool. However, when such chips are read out at higher clock speeds of 5, 10, or even 20 MHz to provide a more real-time image, the read noise is likely to be in the range of $\pm$8–15e$^-$/pixel and in addition the shorter exposure time means that fewer photoelectrons will have been produced during each frame. In other words, high frame rates are incompatible with optimal, low noise performance.

13.3.1.1 CCD readout schemes

The readout scheme in Fig. 13.8 is called full-frame readout (Fig. 13.11a). Although often used in astronomical instruments where exposures are long and mechanical shutters are available, full-frame readout has the disadvantage that if the sensor isn't protected during readout, PE will be added to the pixels above any bright feature as the charge pattern is stepped down toward the horizontal register. This produces vertical streaks. As a result, full-frame CCDs usually employ a shutter (or a pulsed light source.)

Alternatively, the array can be divided into two parts to create a frame-transfer CCD. The upper "sensor" half is used to collect the photon signal while the lower "readout" half is shielded from light by a thick layer of aluminum (Fig. 13.11b). In operation, the image PEs are first transferred down vertically to the readout array. Because the horizontal register need not be emptied between each line transfer, this can be done quite rapidly. Although there will still be some streaking, it is smaller by a factor related to the number of columns in the array.

The third approach is to use interline transfer (Fig. 13.11c). In this system, each pixel is divided into a sensor half and a shrouded readout half. At the end of the exposure period, the signal charge packet is transferred to the readout node, and then read out from there. As the initial transfer takes only microseconds, there is no streaking and no shutter is needed. As a result, almost all CCDs now used for microscopy utilize interline transfer. Although initially such sensors had relatively poor QE_{eff} because many photons landed on the shrouded area of the readout pixel, this problem was solved when SONY introduced an overlying layer shaped in the form of an array of microlenses, with one lens located over each sensor. Converging all the light from the pixel onto the sensitive area improved the fill factor considerably (70% vs. 40% of the light

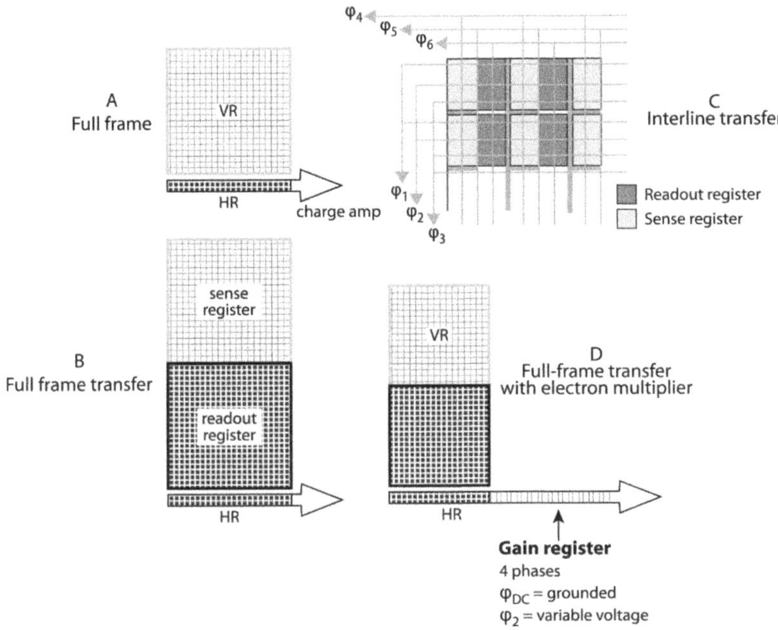

Figure 13.11 CCD readout schemes. Full-frame readout, (as diagrammed in Fig. 13.9) is shown top-left. While simple and having the advantage that almost all of the silicon is used for capturing the light signal, full-frame readout will produce vertical streaking of bright objects unless the signal it blanked during readout. This problem is reduced in full-frame transfer (bottom left) where the charge pattern is first moved to a readout register that is protected from light. As there is no need to employ the horizontal register during this process, the vertical transfer can be carried out at a much higher speed, reducing the amount of streaking. Interline-transfer chips (top right) have both a sensor and a storage area in each pixel. While this reduces the fill factor of the sensor, the ability to read out during acquisition and to use the transfer to storage as an electronic shutter has made this the preferred system, especially when it can be paired with a microlens array to increase the effective fill factor.

striking the chip was channeled to a sensitive area (Fig. 13.12)). This technique has become especially important as pixels shrank below 2 μm to fit into cell phones, etc.

Microlenses may solve the fill-factor problem but they do not increase the area of the sensitive element and this is what determines how much charge can be stored in a pixel before

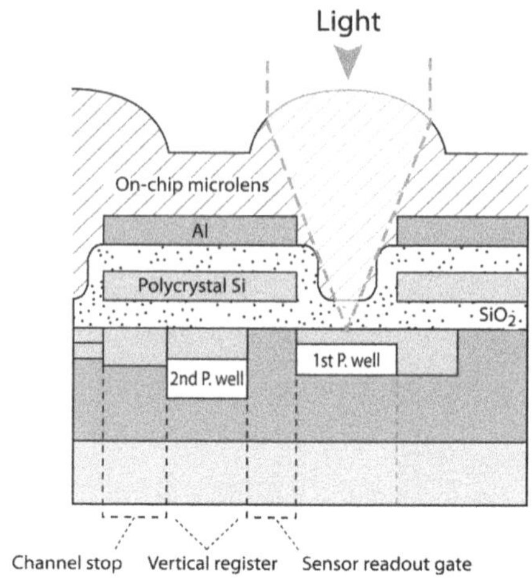

Figure 13.12 Side view of microlens array used to increase fill factor of interline-transfer CCDs. Because interline-transfer CCDs must devote much of the space in each pixel to channel blocks and storage areas, many photons strike parts of each pixel that fail to produce useful PEs. As long as the incoming light is fairly well collimated, small lenses located over each pixel can focus most of it onto the sensitive area. Figure redrawn from SONY.

mutual repulsion between the electrons forces them into adjoining pixels (or into the specially designed overflow drains used in consumer CCDs to prevent specular reflections from dominating the image of shiny objects.). In general, overflow occurs when > 600 electrons/μm^2 are stored in a CCD pixel. Assuming no read noise, Poisson noise limits the number of statistically distinct levels to the square root of n, or about 25 statistically distinct levels in the 1.2×1.2 μm pixels found in cell phones. Consequently, such images are so noisy that they are usually only digitized to 6 bits.

Although interline transfer and microlenses may seem to solve all CCD readout problems, one must remember that, when using such a device to record from a spinning disk microscope, it is important to synchronize the exposure period to exactly coincide with the amount of time needed for each beamlet to scan across the field of view an integral number of times. Failing to do so will artificially brighten or darken small areas of the image [22].

13.3.1.2 Dark signal

As mentioned above, P/N diodes have a leakage current and this is also true of the array of P/N diodes that make up a CCD.

To limit this, scientific CCDs are cooled and multi-pinned phase (MPP) charge-transfer voltages are used. This reduces leakage to $\sim$1 electron/pixel/s @ -40°C. Leakage signals of this size are only a problem in the EM-CCD, as discussed below.

13.3.2 Electron-Multiplier CCDs

In the previous discussion of the APD, it was noted that free photoelectrons that are subjected to a high electric field inside the silicon can pick up enough energy during their brief lifetime to create additional free electrons. In low gain APDs this process can produce $\sim$100 e^-/PE; in Geiger mode perhaps 10^8 e^-. Although the fields imposed by the charge transfer electrodes in a CCD are much lower than those used in an APD, by choosing a suitable electrode shape and transfer voltage, it is possible to produce a very slight amplification of the charge packet, say about 1.01$\times$/transfer (Fig. 13.13). Although this minute increase seems trivial in itself, if the process is repeated a few hundred times, significant gain is possible, and this is the operating principle of the electron multiplier CCD (EM-CCD).

In practice, an additional set of 300–400 charge-transfer electrodes, called the Gain Register, is introduced between the right-hand pixel of the horizontal register and the read node (Fig. 13.9d). This register often has 4 phases, with the second of these held permanently at ground potential. By varying the transfer voltage applied to the following phase, the electric field generated between it and the grounded electrode can be high enough to cause about one PE in a hundred to knock a valence electron into the conduction band. Because the probability of such an event is exponentially sensitive to the exact field strength and the device temperature, both these variables must be controlled very carefully. In addition, because the charge packet grows in size as it passes down the gain register, its "pixels" must become steadily larger in area with the result that the final read node has a large capacitance and

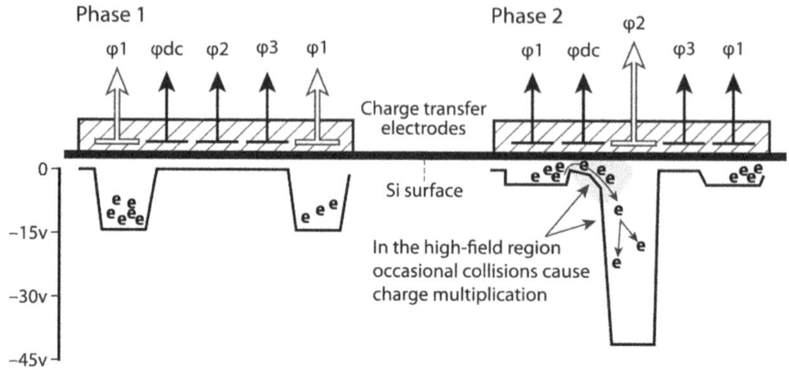

Figure 13.13 Operation of the gain register of an electron-multiplier CCD (EM-CCD). EM-CCDs employ a gain register between the end of the horizontal register and the read node. This register usually has 4 sets of charge transfer electrodes. The extra phase $\phi_{\chi\delta}$, is located between ϕ_1 and ϕ_2 and is held at earth potential. To produce gain, the voltage applied to ϕ_2 is increased above that normally used for charge transfer so that electrons passing from ϕ_1 to ϕ_2 must pass through a high-field region. In this area, the local field is high enough to cause about 1% of the electrons in the charge packet to knock a second electron out of the valence band. As a result, each transfer in the gain-register has a gain of about 1.01× and a register of 400 stages can have a gain of about 1000×. This is sufficient to raise the signal from a single PE above the noise floor of the read amplifier. Figure redrawn from Andor.

consequently a relatively high read noise (commonly $\pm$ 50 e^-/pixel @ 10 MHz vs. $\pm$ 6–10 e^-/pixel in a good quality normal CCD). However, as a gain of ~500× will easily separate the signal of a single PE from this read noise, the device itself can be almost noise free.

The "almost" has to again to do with multiplicative noise. Imagine a PMT with 400 stages and a gain of 1.01×/stage. The result is an "excess noise" of about 1.4: in other words, the device works as though it were noise free but that it had counted only half as many PE are were actually produced by the light hitting the pixel. A simple way to think of this is that when an EM-CCD is used with high electron-multiplier gain, it has no read noise but its QE_{eff} is only 50% of what would be the case were the same chip operated as a normal CCD. Indeed, the EM-CCD sensors made by E2V Inc.

incorporate a normal CCD amplifier at one end of the horizontal register and a gain-register amp at the other. The former is preferred when the lowest signal-pixel signal is greater than the square of the noise figure of that amplifier and the latter when it is not. It should also be noted that some manufacturers of EM-CCD camera provide a photon-counting option where single PE flashes are counted and stored. While this removes the excess noise factor, there can still be problems when the flashes corresponding to more than one photon overlap during the signal integration period. In addition, as the back-illuminated EM-CCD chips all read out by frame transfer, some streaking may occur.

Because read noise in the EM-CCD is effectively zero, otherwise neglected noise factors now become more important. A dark charge of even one electron/pixel becomes a major problem because, of course, Poisson Noise on this charge (1, ± 1) becomes an error source 10–100$\times$ larger than the read noise. In addition, there is coupling-induced charge (CIC) noise. CIC is actually just the EM gain produced by the application of normal charge-transfer voltages and geometries and it can be somewhat reduced by carefully shaping these voltage pulses. Both these factors seem to be less severe at lower temperatures (-80–90°C) and slower pixel clock rates. The crucial performance test is to record a short exposure with the sensor completely (!) protected from light. After adjusting to display a line-profile so that the one can see the noise level of the read amplifier, a single line of the recorded image from a 512 $\times$ 512 sensor should have between 2–6 prominent single-PE pulses, each one only about one pixel wide. In other words, less than one pixel in 100 should have even a single dark count.

13.3.3 sCMOS Imagers

The second common type of silicon image sensor is the CMOS sensor. As with the CCD, the surface of a CMOS sensor is divided into a raster of rows and columns. However, in this case each pixel contains not only sensor and storage areas but also a read amplifier. In the simplest case, power and ground voltages are conveyed to each line by a metal conductors running horizontally, while the output of each row of amplifiers is conveyed to the top or bottom edge of the chip by a vertical wire. Additional sets of vertical wires may be present to carry trigger signals to initiate actions such as resetting the pixel. In operation, power is supplied to only one row at a time. The output voltages of these amplifiers are conveyed to the edge of the chip where they are collected and digitized. Power is then supplied to the next line, etc.

The main advantages of this approach are that the time available for each amplifier to settle is longer than if a single amplifier reads all the pixels in a line sequentially and also that, not needing the buried channel and other features of the CCD, CMOS chips can be made using the same facilities used to fabricate computer chips. As a result, CMOS imagers are less expensive to make and can be read out at higher frame rates and added microlenses can largely overcome the reduction in QE_{eff} due to the loss of active sensor area.

The main disadvantages are that some of the pixel area must be devoted to the amplifier, resulting in a lower fill-factor and smaller full-well limits. More to the point, small variations in the gain and offset of each amplifier become a major source of "fixed-pattern noise." Particularly as a result of the latter, the performance of early CMOS sensors was markedly inferior to that of contemporaneous CCDs. More recently however, the demand for ever faster readout has led to the development of technical solutions to these limitations. Gain and offset variation is now corrected by on-the-fly comparison with stored "black" and "white" digital reference images (a process called flat-fielding). In addition, two entire, on-chip analog-to-digital converters (ADCs) are now attached to the output wire from every column of pixels. By operating the members of each pair of 11-bit ADCs at a different analog gain, it is possible to combine the digital output to produce 16-bit data even with very short pixel read times (Fig. 13.14). The result is the scientific CMOS (sCMOS), a sensor that now surpasses the read-noise performance of even the best scientific CCDs and which preserves this edge even when read out at hundreds of megapixels/s. The extremely low read noise ($\pm 1 - 2\ e^{-}$/pixel) more than compensates for fact that the full-well capacity of the sensor is reduced by having to use some of the pixel area for the read amplifier and associated circuitry. Current sCMOS sensors tend to have high pixel counts (current chips have 4.0 and 5.5 M pixels), a feature that permits imaging larger fields

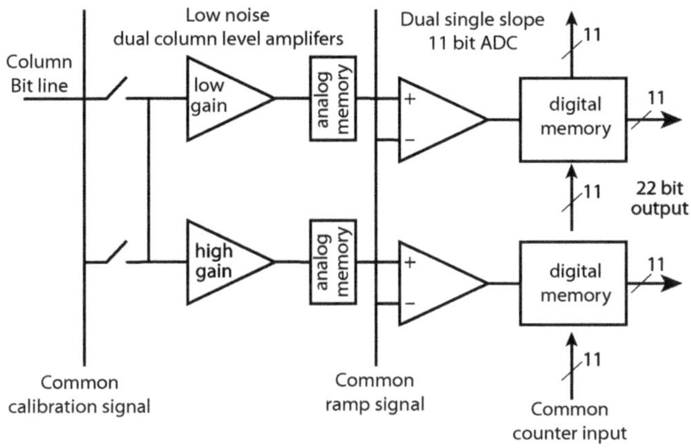

Figure 13.14 Readout schema of the scientific complementary metal–oxide silicon (sCMOS) sensor. The signal from each pixel of a particular line of the sCMOS sensor is simultaneously presented to both a high-gain and a low-gain 11-bit analog-to-digital converter (ADCs) located at the top of every column of pixels. With proper calibration, this arrangement permits accurate 16-bit digitization. Having a separate set of ADCs for each column increases the settling time available for the readout amplifier that is present in each sCMOS pixel, decreasing the read noise to ±1–2 electrons, even at read rates of up to 400 MHz. Figure redrawn from Fairchild.

of view. However, it is important to note that, probably because much of the noise is associated with inaccurate correction of the interpixel differences in gain and offset, the "read" noise spectrum is not Gaussian and there is a tail of relatively large errors.

Even better performance may be possible in the future. The image sensors in some recent cell-phone cameras use the SONY EXMOR R^{TM} chip that is constructed so that the sensor is located in silicon that has been deposited on top of the amplifier circuits. This not only permits the accumulation of a larger full-well signal with virtually no signal lost to scattering, it also makes it theoretically possible to use larger, more complex pixel amplifiers. It has been suggested that such circuits might monitor each pixel for overload, allowing it to be emptied and refilled several times while accounting circuits keep track of the total charge. Alternatively, it may become possible for such a chip to more closely mimic the performance of the eye by comparing the output of each pixel with that of its neighbors, a procedure that would facilitate edge detection etc. Such a sensor would in effect become a massively parallel image processor.

Fascinating though these advanced feature may seem in terms of increasing cell-phone utility, their capabilities are mentioned here only to warn against them. It is hard enough trying to gain an understanding of subcellular processes from the intensity patterns of a 3D fluorescent image without having to worry about how the sensor is distorting the intensity information because it thinks it knows what you need to see!

A final point about sCMOS cameras relates to their readout scheme. Because pixels can be addressed directly, it is possible to read out virtually any set of pixels in any order. In practice, the main options available are global and rolling readouts. In global readout, all pixels are zeroed then light is collected and the at the end of the exposure period, charge is transferred to the read register, where intensities are then probed, one line at a time while a second exposure is being accumulated in the sensors. In the rolling readout mode, exposure is continuous with each line zeroed just after it has been read out. This has the advantage of not requiring either a read register or a synchronously pulsed signal source. It also permits faster readout and simplifies the on-pixel circuitry, which increases the full-well capacity. On the other hand, it means that each line of the recorded image corresponds to a slightly different exposure period so a vertical line moving sideways will appear to be sloping. This can complicate subsequent image analysis.

13.3.4 Color Silicon Imagers

From the fact that the QE spectra for silicon sensors are very flat, it follows that they are effectively "colorblind." Color sensors can be created by placing a pattern of very small red, green or blue (RGB) colored filters above the sensitive areas of each pixel. By sorting the signals by color, it is possible to obtain independent R, G, and B images. A common pattern uses two green for each red and blue. A minor disadvantage of such 1-chip color sensors is that, because adjacent pixels represent different images, they may differ markedly in intensity. Consequently, the read amplifier must have a higher bandwidth and therefore somewhat higher read noise.

Though convenient for demonstrations, "one-chip" color CCDs or sCMOS devices are seldom used in critical studies for the simple reason that they "waste light": green or blue light that falls on a pixels with a red filter will be absorbed and lost. This problem can be reduced by employing a "3-chip" camera in which the incoming light is separated into R, G, and B streams using a system of prisms some of whose surfaces have been coated with dichroic mirrors. Although this solves the wasted light problem, such sensors are so much more complex and expensive that they are mostly used in consumer-level cameras where the prism axis can be very carefully aligned with the optical axis and they need not be cooled.

In general, except when viewing standard, stained histological sections using bright field imaging, it is more common to separate the images of different colors using the barrier filters in the microscope. And record each color sequentially using a monochrome detector.

13.4 What Imager to Choose?

For over a decade, the workhorses of biological fluorescence microscopy have employed the SONY ICX285 AL chip. It is a 1392 × 1024 microlens-coupled interline-transfer CCD chip that was originally mass-produced for the Japanese high-definition video standard. These chips are relatively inexpensive, and have been highly optimized because of the relatively vast market for such sensors (compared to the astronomy and microscopy market for which the CCD was first developed). Mounted in a well-built camera, this chip can provide $\pm$ 2.5 e^- read noise @ 1MHz and $\pm$ 6.5e^- @ 20MHz/–55°C. It has peak QE of about 60% and a full-well of 18k e^-/pixel. Somewhat higher QE_r can be obtained with a somewhat larger Kodak chip that uses transparent indium/tin oxide charge transfer electrodes. Those planning to use widefield imaging followed by 3D deconvolution over large fields of view (relatively low magnification and high NA) or looking for very faint signals may

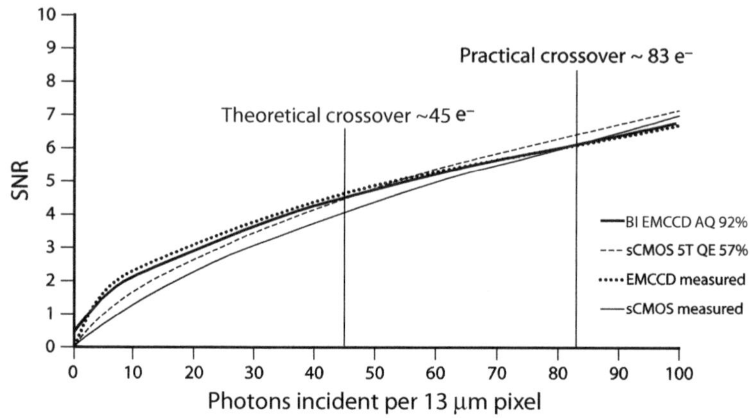

Figure 13.15 Relative S/N of EM-CCD and sCMOS cameras at low signal levels. As electron multiplication effectively halves the QE_{eff} of the EM-CCD, the sCMOS works better at higher signal levels where read noise is less of an issue than Poisson Noise. However, the lower read noise of the EM-CCD prevails at very low signal levels where read noise is important. Given the measured QE_{raw}, the crossover signal should be at about 45 photons/pixel. However actual measurements performed by Colin Coates (Andor, Belfast, UK) show the crossover to be closer to 80 photons/pixel, probably because the procedures needed to correct for differences in the gain and offset of the read amplifiers found in each pixel of an sCMOS sensor produce non-Gaussian errors (printed with permission of Andor, Belfast, UK).

find the sCMOS well suited to their needs because the QE is about the same while the read noise is about 3× lower.

However, those wanting the best performance in terms of QE_{eff}, read noise and high speed pixel clock, for use with a disk-scanning confocal microscope or PALM/STORM imaging (Chapter 14), should choose the a highly-cooled, back-illuminated EM-CCD. This is because in these techniques, about 99% of the pixels receive zero photons and the EM-CCD reads the zero-signal more accurately than the sCMOS, even one with a measured read noise of $\pm 1 - 2$ e^-/pixel. On the other hand, as soon as the LOWEST signal level in the field of view becomes greater than say 40 photons, the sCMOS gets the nod because its QE_{eff} is higher than that of a back-illuminated CCD with a multiplier. This relationship can easily be seen in Fig. 13.15, kindly provided by Colin Coates of Andor

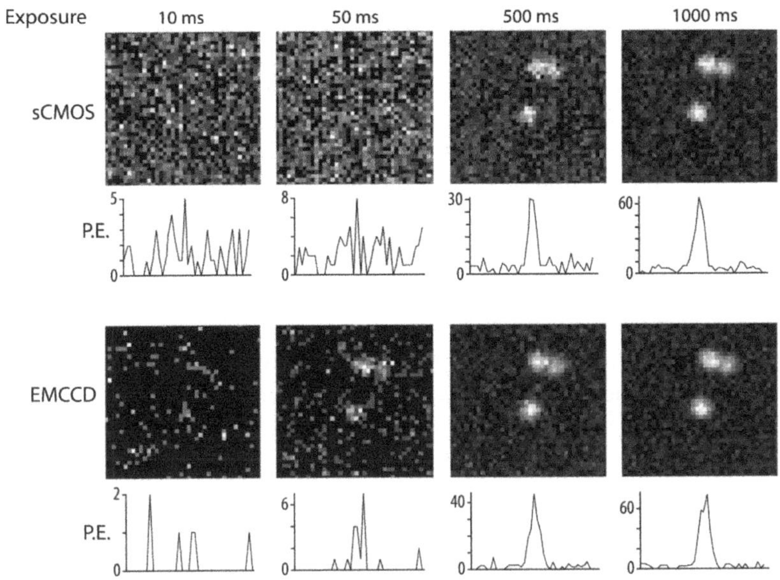

Figure 13.16 Comparison of low-light sCMOS and EM-CCD performance. Images obtained simultaneously from the two cameras set up on a beam-splitter specially modified to make pixels on both sensors have the same effective dimensions when referred to the specimen plane. Viewing images of fluorescent beads obtained at different exposure times, one can see that the EM-CCD signal emerges from the noise background first. This figure was kindly provided by Dr. John Oreopoulos (Spectral Applied Research, Richmond Hills, CA) and made using procedures described by Oreopoulos [23]. The line traces below each image show single-PE pulses.

Technologies (Belfast, Northern Ireland), which plots S/N vs. signal level in photon/pixel. At signal levels up to about 80PE, the BI EM-CCD is superior because of lower read noise. Above 80PE, the sCMOS wins because of higher effective QE. This is shown with even greater clarity in Fig. 13.16, kindly provided by John Oreopoulos (Spectral Applied Research, Richmond Hills, Canada) [23]. Here we see small segments of images obtained simultaneously from EM-CCD and sCMOS cameras arranged so that each had the same pixel size when referred to the fluorescent specimen. With the EM-CCD, the image begins to emerge from the background at an exposure of about 50 ms, while the sCMOS shows very little until the exposure is at least 10 ms.

One other difference relates to pixel binning. On the CCD, the signal from several adjacent pixels can be binned before the packet is readout. This reduces the spatial resolution but increases the effective pixel size and hence the signal level. On-chip binning is not possible with the CMOS where readout occurs at the pixel level and data can only be binned after it has been digitized.

The sCMOS is currently the optimal way of obtaining megapixel images at high frame rates and if the darkest pixel still registers at least 20PE, then its QE_{eff}, will be almost twice that of the best EM-CCD. This makes it suitable for rapidly screening large arrays of cells (high content screening), as long as one can solve the significant problems associated with handling and processing 16-bit data at hundreds of megahertz.

Acknowledgments

The discussion of the relative merits of BSI EM-CCD and sCMOS detectors when used to record very faint signal from fluorescent microscope specimens owes much to discussion between the author, Colin Coates (Andor Instruments, Belfast, Northern Ireland) and John Oreopoulos (Spectral Applied Research, Richmond Hills, Canada). The author would like to thank them both for their time and for providing the data presented in the figures related to this topic.

References

1. Art, J. (2005). Photon detectors for confocal microscopy, in *Handbook of Biological Confocal Microscopy*, 3rd edn. (Pawley, J. B., ed.), Springer, NY, pp. 251–264.
2. Pawley, J. B. (1994). The sources of noise in three-dimensional microscopical data sets, in *Three Dimensional Confocal Microscopy: Volume Investigation of Biological Specimens* (Stevens, J., ed.), Academic Press, New York, pp. 47–94.
3. Sabharwal, Y. (2012). Digital camera technologies for scientific bio-imaging. Part 4: Signal-to-noise ratio and image comparison of cameras. *Microsc. Anal.*, **26**(1), S4–S8.

4. Donati, S. (2006). *Photomultipliers*. Wiley Encyclopedia of Biomedical Engineering John Wiley & Sons, pp. 1–17.
5. Suyama, M., and Lares, M. (2008). Photomultipliers: hybrid detector combines PMT and semiconductor-diode technologies. *Laser Focus World*, March 1.
6. Borlinghaus, R., and Kappel, C. (2011). Detectors for sensitive detection. *Proc. Focus on Microscopy*, Konstanz, Germany.
7. Schroder, J., and Kappel, C. (2011). Leica HyD for confocal. *Proc. Focus on Microscopy*, Konstanz, Germany.
8. Daum, R. (2012). Descanned point detection of single fluorophore fluorescence. *Focus on Microscopy*, Singapore.
9. Yamamoto, K., Yamamura, K., Sato, K., Ota, T., Suzuki, H., and Ohsuka, S. (2006). Development of multi-pixel photon counter (MPPC). *Proc. IEEE Nuclear Science Symposium Conference Record*, N30-102, pp. 1094–1097.
10. Shushakov, D. A., et al. (2008). New approach to solid state photomultipliers. *SORMA West 08*, pp. 1–5.
11. Borlinghaus, R. (2013). Gated detection: improved super-resolution and zero excitation background in confocal imaging. *Proc. Focus on Microscopy*, Singapore.
12. Vinogradov, S., et al. (2009). Efficiency of solid state Photomultipliers in photon number resolution. *IEEE Nuclear Science Symposium Conference Record* (NSS/MIC).
13. Popleteeva, M., Esposito, A., and Venkitaraman, A. (2013). Innovating solid state detection technologies for biomedical imaging. *Proc. Focus on Microscopy*, Maastricht, NE.
14. Guerrieri, F., Tisa, S., and Zappa, F. (2009). Fast single-photon imager acquires 1024 pixels at 100 kframe/s. *Proc. IS&T/SPIE Electronic*, San Jose.
15. Buchholz, J., et al. (2012). FPGA implementation of a 32×32 autocorrelator array for analysis of fast image series. *Opt. Express*, **20**, 17767–17782.
16. Krieger, J., Buchholz, J., Pernus, A., Carrara, L., Charbon, E., and Langowski, J. (2013). Macromolecular in live cells imaged at 10-μs resolution single plane illumination and an avalanche photodiode array detector. *Proc. Focus on Microscopy*, Maastricht, NE.
17. Krishnaswami, V., Burri, S., Regazzoni, F., Bruschini, C., van Noorden, C. J. F., Charbon, E., and Hoebe, R. (2013). SPAD array camera

for localization based super resolution microscopy. *Proc. Focus on Microscopy*, Maastricht, NE.

18. Becker, W., Su, B., and Berman, A. (2010). Better FLIM and FCS data by GaAsP hybrid detectors. *Proc. SPIE* **7569**, *Multiphoton Microscopy in the Biomedical Sciences* (Periasamy, A., So, P.T.C., and König, K., eds.), San Francisco, California.
19. Inoue, S., and Spring, K. (1997). *Video Microscopy*, 2nd edn., Plenum, New York, pp. 1–741, particularly Chapters 5–9.
20. Janesick, J. R. (2001). *Scientific Charge-Coupled Devices.* SPIE Press, Bellingham.
21. Pawley, J. B. (2005). More than you ever really wanted to know about CCDs, in *Handbook of Biological Confocal Microscopy*, 3rd edn., Springer, NY, pp. 918–931.
22. Toomre, D., and Pawley, J. B. (2005). Disk-scanning confocal microscopy, in *Handbook of Biological Confocal Microscopy*, 3rd edn., Springer, NY, pp. 221–238.
23. Oreopoulos, J. (2011). A comparison of sCMOS and EMCCD digital camera technologies for spinning disk confocal microscopy. Application note, available from Spectral Technologies, Richmond Hill ON, Canada. http://www.spectral.ca/_files/file.php?fileid= fileWFBaXTDIaW&filename = file_4_CSU_Camera_Comparison__ sCMOSvsEMCCD.pdf

Chapter 14

Practical Aspects of Localization Microscopy

Mark B. Cannell,[a] Christian Soeller,[b] and David Baddeley[c]

[a]*Department Physiology & Pharmacology, Biomedical Sciences Building, University Walk, University of Bristol, Bristol BS8 1TD, UK*
[b]*Living Systems Institute, University of Exeter, Stocker Road, Exeter EX4 4QD, UK*
[c]*Department of Cell Biology/Nanobiology Institute, Yale University West Campus, 300 Heffernan Dr., West Haven, CT 06515, USA*
c.soeller@exeter.ac.uk

14.1 Introduction

The wavelength of light was long thought to impose a fundamental physical limit on the resolution that could be achieved with a light microscope (at least in the optical far field) [1, 2]. However, it was also recognized that for a self-luminous object, resolution was not limited by the wavelength of light per se—provided other information about the object is available [3]. Recent advances have seen a number of methods invented which allow us to circumvent the Abbe limit. These methods are almost exclusively built on fluorescent microscopy and rely on manipulating the sample such that only a part of the specimen is fluorescent at any given time. If the properties (dimensions) of that part of the specimen are known

Fundamentals of Fluorescence Imaging
Edited by Guy Cox

ISBN 978-981-4774-85-7 (Hardcover), 978-1-351-12940-4 (eBook)
www.jennystanford.com

the Abbe limit need not apply [4]. The first such method, STED (or stimulated emission depletion) microscopy, which is discussed in Chapter 16, takes an ensemble approach and uses a saturable depletion effect to selectively darken all molecules except those within the center of a laser spot that is scanned through the sample and is closely related to confocal microscopy [5] (Chapter 3). The effect of the depletion ring is akin to that of a small field stop at the specimen [4].

Localization microscopy[a] is a widefield technique which also relies on darkening all but a handful of fluorophores at any given time [6–8]. This allows the fluorophores that remain "on" to be observed as isolated, diffraction limited spots. The key insight in the method is that, although the image of a single fluorophore is still only visible with diffraction limited resolution, its position can be determined much more accurately, an idea that had been used for quite some time for the tracking of biomolecules in cells with nanometer resolution [9]. If one then changes which fluorophores are visible, and record each of these sets separately in a large number of images (camera frames), it is possible to independently observe all (or a large fraction of) the fluorophores within a sample, and build a list of their measured positions. In contrast to STED, in which the switching occurs in a targeted way, localization microscopies rely on a much simpler stochastic approach to switching. By adjusting the relative rates at which molecules are switched on and off such that the off-rate is much larger than the on-rate, an equilibrium can be obtained in which only a few molecules are visible at a given time.

There are a number of subtly different flavors of localization microscopy, sharing the same underlying principle but exploiting different mechanisms to switch the fluorescent molecules between fluorescent and dark states. The major approaches to switching and their respective acronyms are:

- PALM/fPALM [6, 8] (**p**hoto**a**ctivated **l**ocalization **m**icroscopy), which uses fluorescent probes that are initially dark (in

[a]We will use the term *localization microscopy* to mean the generic approach which includes methods known as PALM, STORM, dSTORM, etc., that all rely on the serial localization of individual (or small groups of) emitters and build up a complete sample image over time.

the channel of interest) such as PA-GFP, mEos, or caged fluorescein. Exposure to weak UV light (typically at 405 nm) stochastically turns some molecules on which are then rapidly bleached by stronger readout illumination.

- STORM [7] (stochastic optical reconstruction microscopy), which uses a pairing of two small molecule dyes, one of which is typically a Cy5 or Cy7 derivative and fluoresces at the far red end of the spectrum, the other is typically a shorter wavelength excited dye used for activation. Exposing the cyanine dye to intense illumination in the presence of thiols in the solution will cause a transition to a dark state, from which it can be effectively recovered by excitation of the shorter wavelength dye.
- dSTORM/RPM/GSDIM [10–12], which expand on the STORM concept, extending it both to a wider range of dyes and dispensing with the specific activator dye, instead relying on either thermal relaxation or a weak blue/UV absorption characteristic of the dark state for reactivation.

In the remainder of this chapter, we will draw on our experience of dSTORM-type imaging, with the recognition that the underlying principles are essentially the same for the other switching modalities described above. There are a number of good reviews highlighting the principles of localization microscopy (e.g., [13, 14]) and here we concentrate on the practical aspects which will be of most interest to those who want to get these methods working in their laboratories.

14.1.1 Anatomy of a Localization Microscope

The hardware needed to perform single-molecule-based super-resolution imaging is straightforward, consisting of little more than a conventional widefield fluorescence microscope equipped with a high numerical aperture objective, a very sensitive camera, and at least one laser source to provide intense illumination. This "simplicity" has contributed to the relative popularity of localization microscopies among optical super-resolution methods. Localization microscopy, however, demands significant improvements over the

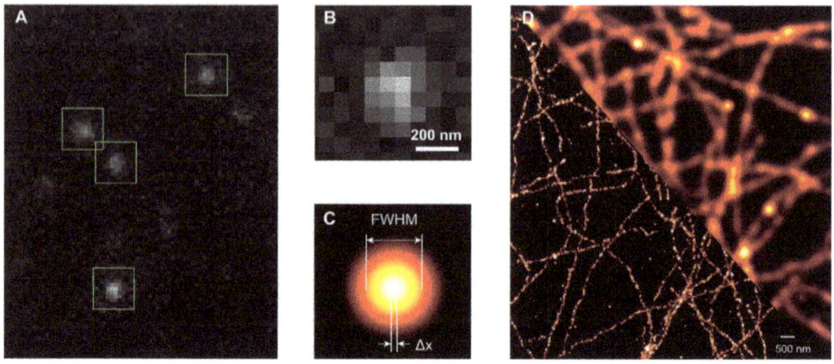

Figure 14.1 Principles of localization microscopy. (A) Detected single-molecule events from a single frame of a dSTORM-type acquisition. A detected event (B) is localized in this example by fitting a 2D Gaussian approximation (C) to the PSF. While the diameter of the image of the single emitter (as characterized by its FWHM diameter) is diffraction limited, the centre of the image can be estimated with a much better accuracy Δx that is typically $\sim$10 times smaller than the FWHM. (D) The resulting super-resolution image is generated from all marker positions detected in the whole series of frames. In this example a resolution of $\sim$30 nm was achieved, much better than that in the diffraction-limited image of the sample (upper right), here illustrated with microtubules labelled with an antibody against alpha-tubulin.

specifications of a standard fluorescence microscope in terms of detector sensitivity, noise, and mechanical stability.

14.1.1.1 Sensitivity

The need for sensitivity is intuitively obvious: the signal from a single molecule is relatively weak when compared to the fluorescence from a conventional sample in which all fluorescent molecules (possibly hundreds or thousands) within the point-spread function simultaneously contribute to the signal), and the accuracy of the position determination depends on how much of this weak signal we can collect [15]. As the optical train of a fluorescence microscope should be fairly efficient, the principal optimizations reside in the selection of objective lens and camera. As light collection efficiency scales with the square of the numerical aperture (NA), the highest NA objective available is preferred. Many

Table 14.1 Camera characteristics. Values given are typical of currently available models from a wide range of manufacturers

Technology	Readout noise	Effective QE	Speed (512 × 512 pixel ROI)
Conventional CCD	5 e-	65%	20 FPS
EMCCD	<1 e-	~45%*	56 FPS
sCMOS	1.5 e-	70%	400 FPS

*After correction for electron-multiplication noise.

localization microscopes therefore use TIRF objectives with NAs of 1.45 or higher—even when TIRF illumination is not used. The trade-off here is that these objectives are particularly sensitive to refractive index mismatch leading to aberrations when imaging deeper into a sample and a loss of localization accuracy may occur [16]. Therefore, in some applications, especially when using (e.g.) aqueous mounting media and imaging deep into a specimen, there may be a benefit in using a lower NA but better matched objective (e.g., water immersion lenses).

The other major determinant of sensitivity is the light detector, which should have both good quantum efficiency (QE) and low read noise (as one may only detect a few tens of photons per pixel). To allow rapid imaging, the detector should also offer fast readout without seriously compromising the signal to noise ratio. Three principal detector technologies have been used with localization microscopy: conventional CCD cameras, electron-multiplying CCDs (EMCCDs), and, more recently, scientific CMOS (sCMOS) (see Chapter 13). The indicative characteristics of the different camera technologies are summarized in Table 14.1. EMCCDs have been the preferred choice due to their extremely low readout noise and comparatively fast speed. However, the introduction of sCMOS cameras, which offer higher readout speed and improved QE with small readout noise, may change this preference. In connection with this point, it has been shown that provided the sCMOS camera sensor is calibrated/corrected for field non-uniformities (pixel-to-pixel variations in gain and readout noise) such cameras can perform at the theoretical limits of localization microscopy [17, 18].

14.1.1.2 Stability

In order to generate a quality, high-precision, super-resolution image, a large number of molecules must be independently localized. This requires the collection of a large number of image frames (typically on the order of 20–50,000). If collecting data at a typical rate of 20 camera frames per second (FPS), this corresponds to 15–40 minutes of continuous imaging. Obviously, any instrumental drift over this time period will result in degradation in the accuracy of the resulting image data. As conventional microscope hardware can experience a drift of several microns over this time period, this is a critical issue for localization microscopy. There are two principal approaches to minimizing or correcting for drift: (i) adapting the microscope hardware to minimize drift, and (ii) measuring and correcting the drift after data acquisition. Most localization microscopes combine aspects of both approaches. Postprocessing-based drift correction can either use fiduciary markers such as gold beads embedded in the sample to report drift (e.g. [6]), or attempt to infer drift from the structure being imaged itself [19]. Focus drift is especially problematic and difficult to correct in postprocessing, so some form of active focus control is often employed during imaging [20, 21].

With a basic microscope setup, as depicted in Fig. 14.2, single color 2D super-resolved images of well-labelled biological samples can be readily obtained with typical in-plane resolution between 20 and 30 nm, as illustrated in Fig. 14.3. The in-plane resolution improvement as compared to a comparable diffraction-limited image is obvious and typically approaches a factor of 10.

14.1.2 Multi-color and 3D Localization Microscopy

14.1.2.1 Multi-color imaging

Whilst super-resolved single color images are useful for some applications, many biological questions rely on the knowledge of the relative positioning of multiple different proteins. As with conventional widefield microscopy it is possible to image multiple color channels sequentially [22], with the caveat that, in order to minimize cross-bleaching, one should start with the longest

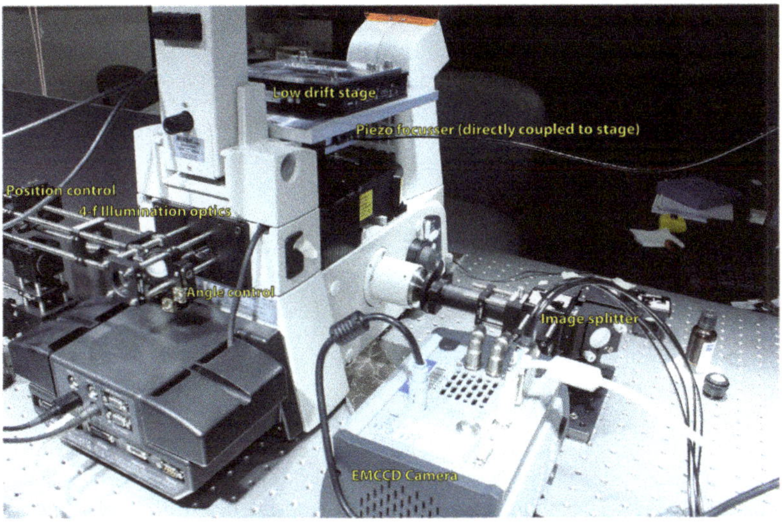

Figure 14.2 Our localization microscope is based on a conventional widefield microscope stand, with the addition of an EMCCD camera, a low-drift stage, and illumination optics to enable laser illumination of a small region ($\sim$17 $\times$ 17 μm) in the sample. On the illumination side, we use a 4-f configuration with mirrors in conjugate image and Fourier planes to allow independent control of illumination position and beam angle. An arc lamp illuminating a wider field of view is also used to locating regions of interest within the sample (in our case coupled in to the illumination path using a polarizing beam splitter).

wavelength label and subsequently image the shorter labels later. There are however, a number of factors which make this approach difficult to use in practice. These include:

Photochemistry differences: In order to have acceptably low crosstalk and cross-bleaching sequentially imaged dyes need to have good spectral separation. For dSTORM-type imaging, this typically implies completely different dye chemistries, e.g., a Fluorescein or Rhodamine derivative such as Alexa 488 and a Cyanine derivative such as Alexa 647. The different dye chemistries have different switching behaviors and may require different switching buffer compositions to optimize switching.

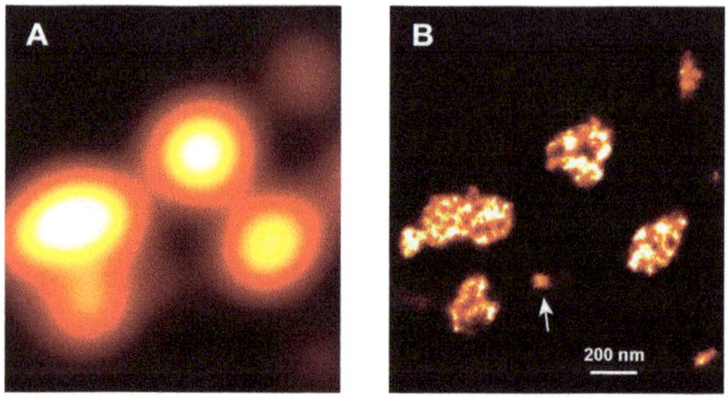

Figure 14.3 Illustration of resolution improvement in 2D single-color localization microscopy. The figure shows images recorded in a small region of a cardiac ventricular myocyte [12] in which antibodies against an intracellular calcium channel, the cardiac ryanodine receptor (RyR), were used to detect clusters of this receptor close to the cell surface membrane. (A) The panel on the right is a simulated diffraction-limited image obtained by superimposing the diffraction-limited images of all single marker molecules that were detected in the frame series. (B) The super-resolution image that was rendered from the estimated marker localization has about 10-fold higher resolution. Note the detailed outlines of the RyR clusters and also the presence of very small clusters (e.g. arrow) that were not detected in the diffraction-limited image.

Acquisition time, instrumental drift, and chromatic shifts: Sequential acquisition multiplies total acquisition time, magnifying the effect of any instrumental drift and making accurate registration of the color channels more difficult. The large wavelength difference between species coupled with nonreproducible shifts caused by changing the emission filters further complicate this. In addition, chromatic aberration in the objective requires correction [23].

Fortunately, an elegant solution is provided by imaging multiple color channels at the same time using spectrally and photochemically similar emitters. Two main approaches to simultaneous multicolor imaging have been demonstrated. The first of these uses classical STORM probes (see "dye choices" below) combining a single reporter molecule with one of a number of spectrally distinct activation complexes, with activation pulses interleaved allowing a fine-scale temporal multiplexing of the color channels [24]. By using the same emitter dye for all the channels this has the advantage

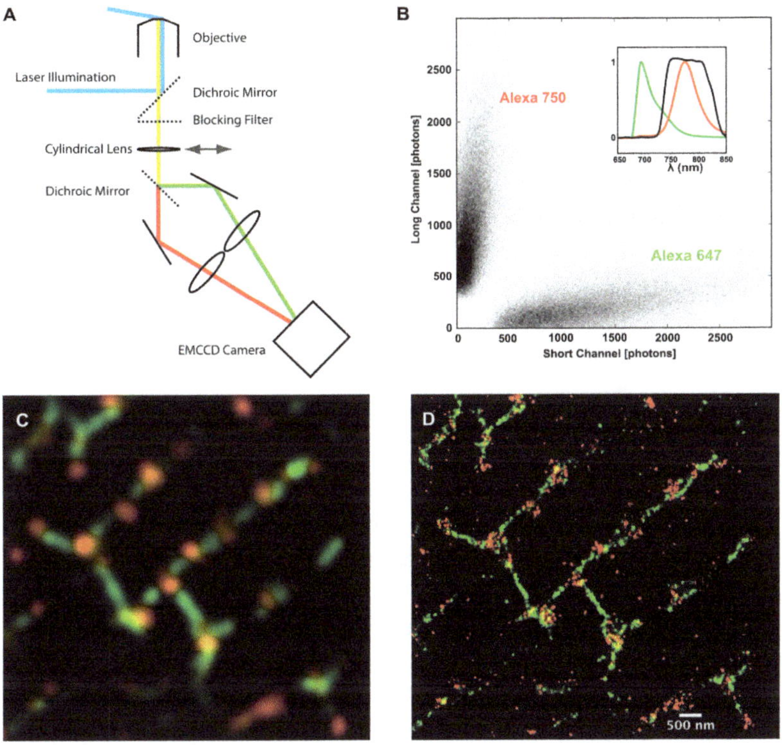

Figure 14.4 Multicolor imaging can be effectively performed by exciting two dyes with the same laser and imaging the detected fluorescence on two halves of an EMCCD camera chip (A). Each single molecule appears in both channels, with the ratio of intensities being characteristic of the dye species (B). As each species has a clearly defined ratio, one can make a clear assignment to one color channel, even in the case of significant spectral overlap. Super-resolution imaging using this configuration to image RyR (red) and Caveolin (green) in mouse cardiac myocytes is demonstrated in C (widefield) and D (dSTORM). Panels A and B modified from Baddeley et al. [23].

of having zero chromatic shift. The second approach is to use two distinct dyes whose emission characteristics are spectrally similar with similar photochemical properties. In this case, fluorescence is detected in two spectral channels simultaneously using an image splitting device that separates the emitted fluorescence into two channels [23, 25]. Each single fluorophore will spread its emission between the two channels in a known ratio and at any one point in time and space only one fluorophore is emitting. Using this additional knowledge that we are looking at a single molecule, it is possible to use dyes which would be spectrally too similar to reliably separate in a conventional widefield image and still separate them with high precision (crosstalk $<1\%$) [23]. Furthermore, the fixed ratios of emission at the two wavelengths even allows more than two species to be separated using only two color channels, which is impossible for spectral unmixing approaches in conventional microscopy.

14.1.2.2 3D localization

Localization microscopy, as described so far, improves lateral resolution, but leaves the axial resolution essentially unchanged at approximately 600 nm. Interestingly, unlike the widefield approach underlying localization microscopy,[b] 2D localization microscopy does give confocal-like optical sectioning due to the rejection of single-molecule events which arise outside the focal plane. These events are rejected either following detection on the basis of increased diameter, which indicates defocus, or may be simply undetected as, with increasing defocus, their peak amplitude becomes indistinguishable from the background. To further improve axial resolution, i.e., to achieve full 3D super-resolution, we must use a way to accurately localize molecules in the third, axial, direction. Because molecules blink on and off stochastically we cannot assume that a molecule will remain on (and emit with constant intensity) for a number of consecutive frames, rendering the z-stepping approach used to obtain depth information problematic. Instead we can derive the 3D information about a molecule in one instant, i.e., within a single camera frame. There are two subtly different classes of approach to achieve this: (i) Simultaneously acquiring several frames at different focal positions and infer axial position from the relative brightness and width of the molecule image.

---

[b]Note that we here refer to localization microscopy with non-TIRF illumination schemes to investigate biological samples away from the immediate cover-slip interface. This is further discussed in the section on Practical Considerations for Illumination.

(ii) Alternatively, engineer the PSF to encode molecule depth in its lateral appearance. In the latter case, it is preferable that the microscope PSF should change from a circularly symmetric Airy disk in a way that depends on defocus (and which is known from prior measurement).

The canonical example of a multifocal method is biplane imaging [26], which acquires two different images at different focal planes side by side on the same camera. Of the PSF engineering methods, the use of a weak astigmatic lens [27] to introduce a distortion in the lateral extent of the PSF as focus changes is currently the most common. However, there are other methods of altering the PSF to encode depth, two of which are the double helix method which uses a combination of Laguerre–Gaussian modes to produce a double-lobed PSF, which rotates in a helical spiral as the focus varies [28], and the phase-ramp method, which produces a linear shift in PSF lobes with a much simpler optical configuration [29].

Currently, the most accurate 3D localization approach may be the more complicated interferometric PALM (iPALM) method which utilizes interference between fluorescence collected from two objectives on opposing sides of the sample. This offers a large axial resolution improvement analogous to that obtained with 4Pi confocal microscopy [30, 31].

14.1.3 Sample Preparation

Sample preparation for PALM/STORM imaging shares many similarities to conventional fluorescence microscopy, with strategies such as immunofluorescence (generally using organic dyes as markers) and transgenic expression (employing fluorescent proteins [FPs] or fusion proteins) being used. The two principal requirements are that the dye/FP should be switchable, and that the achieved labelling density needs to be ~10–100 times higher than would be needed for diffraction limited imaging. Practically, this means that samples that look quite reasonable when imaged on a confocal microscope may give quite poor results with super-resolution imaging. This results from the need to sample structures appropriately according to the Nyquist criterion which demands at least two markers over a resolvable distance [32, 33]. With the potential resolution

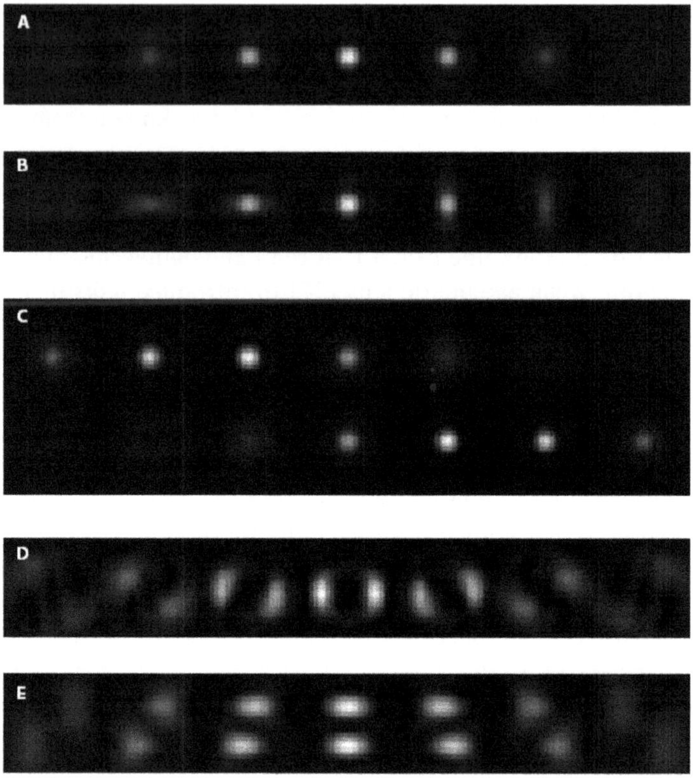

Figure 14.5 Various different 3D PSFs used in localization microscopy. (A) Widefield PSF. This provides some depth information in its width, but is slowly varying near the focus. Images above and below focus also are near identical, making it a poor choice for 3D localization. (B) An astigmatic PSF has the *x* and *y* foci axially shifted with respect to each other and allows depth to be determined from PSF aspect ratio. (C) The biplane method observes the sample at two different degrees of defocus resolving the symmetry and information problems of the widefield PSF. Multi-lobed PSFs such as the double-helix (D) and phase-ramp (E) methods change faster with changing defocus, offering better *z* resolution, albeit at the expense of a small loss in lateral resolution.

of localization microscopy increasing by a factor of ~10 in each lateral dimension we would therefore need ~100 times as many markers in the sample to achieve the performance limit compared to conventional imaging. Wide-field image brightness is the simplest indicator of labelling density and hence whether a dSTORM sample will work well in super-resolution imaging. This is why we also include a wide field illuminator in our microscopes and visual sample inspection to quickly select samples and regions of interest. One should also consider effects of the size of the label itself, particularly when using antibody labelling as the size of a primary/secondary complex can approach the localization accuracy. In practice, however, we have found that the labelling density tends to be the limiting factor, well before the size of the label becomes a concern.

14.1.3.1 Switchable dyes/buffers

As mentioned earlier, a key requirement for localization-based super-resolution is to be able to switch the emitters, to be able to ensure that the vast majority of emitters in the sample are off at any given point in time. A growing family of fluorescent proteins [34] are photo-switchable in standard physiological buffer conditions and a popular choice is mEos 3.2 [35]. If one can deviate from physiological conditions (such as when imaging fixed specimens) and use a low-oxygen reducing buffer, many more dyes become usable. Many (if not all) fluorescent dyes will show switching behavior under the right buffer and illumination conditions [10, 12]. So far, the most promising class of dyes for buffer-induced switching are the near-infrared emitting cyanine dyes such as Cy5 and Alexa 647 [7, 11, 36]. These exhibit good photon yields and sufficiently long-lived dark states when imaged in a buffer containing ~10–50 mM of a primary thiol (acting as a reducing agent) (see also [37]). In addition to the thiol reductant, it may be advantageous to reduce the level of dissolved oxygen in the buffer through the use of an oxygen scavenging system (such as glucose oxidase and catalyse [38]). A putative explanation for switching behavior in thiol/reducting conditions is shown in Fig. 14.6.

An extensive survey of the blinking properties of various dyes in thiol buffers is given in Dempsey et al. [39]. However, the list of potentially suitable dyes and fluorescent proteins is growing and the reader should consult both the recent literature and experiment with commercial markers as necessary; particularly if they have a

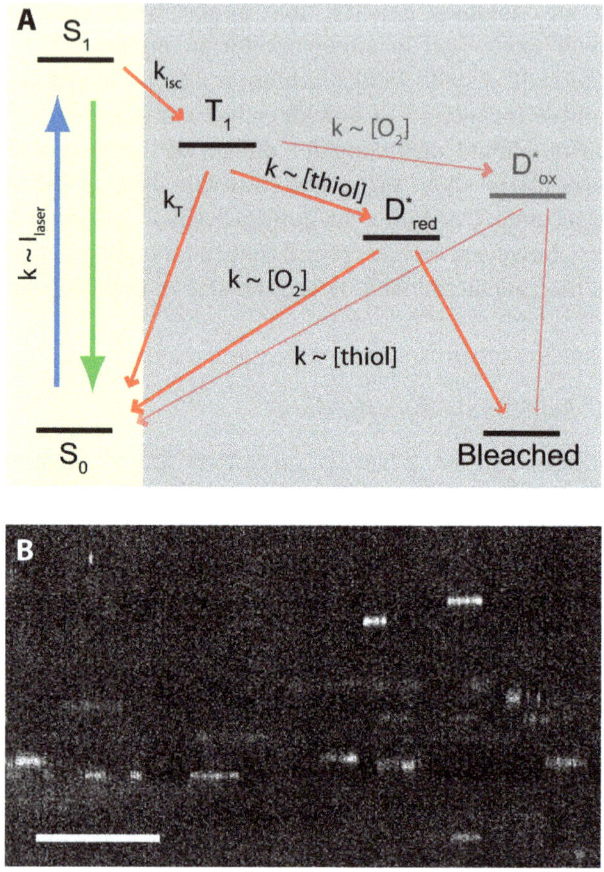

Figure 14.6 (A) The putative mechanism behind dSTORM switching. A long-lived dark state is created by chemical reduction of the triplet excited state. (B) Kymograph showing typical single molecule switching events. The scale bar is 4 seconds.

clear chemical relation to dyes/FPs that have already demonstrated promise for localization microscopy.

In our own work, we have frequently used Alexa 680 and Alexa 750 as a dye pair which have broadly compatible switching behavior in buffers containing MEA (mercaptoethylamine) as a reducing agent. This pair makes a suitable combination for dual-color super-resolution images with very low crosstalk [23]. As most of our work involves imaging thicker (2–15 μm) samples we also require better refractive index matching than offered by aqueous switching buffers and use a glycerol-based switching buffer. Our glycerol-based switching buffer is essentially just 10 mM MEA, in a 9:1 mix of glycerol and 10× PBS. The lack of an explicit oxygen scavenger may be partially compensated by the increased viscosity when compared to aqueous mounts which effectively lowers oxygen availability (enzymatic scavengers have no effect in glycerol). A stock 1 M solution of MEA in 10× PBS (pH 7.5) is stable for 1–2 weeks in the fridge. This stock is diluted to 100 mM with 10× PBS before use and then carefully mixed with 9 parts glycerol. Fresh buffer is made before each imaging session.

14.2 Practical Considerations

14.2.1 Illumination

For a localization microscope, a good illumination design should maximize the signal obtained from in-focus molecules, while minimizing the out-of-focus background, both of which improve detectability and localization accuracy of in-focus markers. TIRF illumination [40], which restricts fluorescent excitation to the immediate vicinity of the coverslip, is an obvious candidate to achieve this, and was used in the initial PALM and STORM demonstrations [6]. A major limitation of TIRF is that one can only image structures within ~200 nm of the coverslip. Perhaps surprisingly, TIRF may not always improve contrast for localization microscopy—even for samples at the coverslip and may even be significantly worse in many situations. The reason for this observation is that the contribution from background at a certain depth in the sample depends not only on the strength of the excitation at that depth but also on the number of molecules at that depth that are active (i.e., not in a dark state). STORM, dSTORM, GSDIM, etc., all require that the vast majority of labels are in the dark state before single molecules can be observed. Even in PALM, in which the molecules are initially

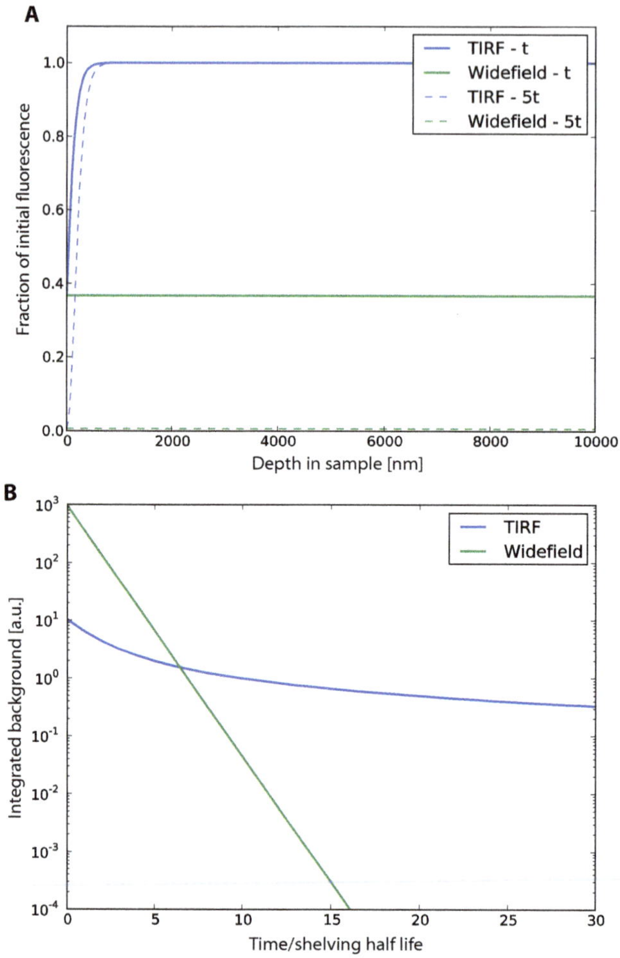

Figure 14.7 Comparison of TIRF and wide-field illumination and the effect on out-of-focus background. (A) Profile showing how many fluorophores remain in a fluorescent (rather than shelved state) as a function of depth at two different timepoints (t and $5t$) where 36% and 0.6% of the in-focus fluorescence remains. TIRF illumination is only effective at shelving the fluorescence in the immediate proximity of the coverslip. (B) Integrated background fluorescence (product of unshelved dye and illumination profile) as a function of time for TIRF and wide-field excitation. While TIRF initially offers better contrast, wide-field illumination is much more effective at shelving the out-of-focus fluorescence and results in better contrast in the long term.

in the dark state, the readout beam must effectively bleach/shelve the photo-activated molecules in order to maintain good contrast. Because TIRF illumination is only bright near the coverslip, it cannot effectively shelve molecules deeper in the sample. There is, however, sufficient energy in the exponential tail of the evanescent field to excite the distant molecules and cause a significant background when integrated over the entire out-of-focus signal (see Fig. 14.7). For dSTORM, this out-of-focus fluorescence will typically be larger than when using wide-field illumination, which effectively switches off the out-of-focus labels.

Figure 14.7 shows that widefield illumination can be clearly superior to TIRF for PALM/STORM-type imaging. The simple model used to generate these curves, however, does not account for any additional background from other sources (such as glass slide fluorescence). In the presence of a non-bleaching background component, the optimal readout illumination is somewhere between the extremes of widefield and TIRF, with both highly inclined (HILO, [41]) and light sheet illumination [42] being good options. Of the two, we tend to use HILO as it is easy to implement and is compatible with standard slide mounting techniques.

In cases where a separate laser is used to switch molecules on, e.g., PALM, STORM, and some forms of dSTORM/GSDIM the optimal method is to activate with a spatially confined illumination (e.g., TIRF if observing close to the coverslip), but read out in a way which still effectively supresses out-of-focus fluorescence.

14.2.2 Choice of 3D Method

Given the multitude of 3D localization methods available, it can be very difficult to decide which method to use. To help make this choice, several questions should be considered:

14.2.2.1 How much experimental complexity can I live with?

The different 3D methods require different levels of complexity in the optical setup. The astigmatism or phase ramp methods simply require an astigmatic lens or phase ramp to be inserted at a suitable location within the optical path. Biplane detection is somewhat more complicated, requiring two separate focal planes be imaged side by side on the CCD (or onto a second CCD camera). A double helix PSF generated using a spatial light modulator is still more complicated, although recent advances using a lithographically manufactured phase masks promise to simplify the technique. Finally, iPALM tops the list for complexity, but provides the highest achievable axial resolution.

In addition to the complexity of a monochromatic setup, it is worth considering how multicolor detection will interact with the 3D detection scheme. Ratiometric (or other simultaneous color detection modalities) will typically split the two color channels onto different halves of the same CCD. If combined with biplane 3D, this will require the CCD to be divided into four quadrants, and a considerably more complicated optical setup. Higher pixel counts will then be needed to maintain field of view at a given pixel resolution. However, the optical elements for astigmatism or phase ramp methods can be inserted independently of the color separation optics, resulting in no additional increase in complexity.

14.2.2.2 How many molecules will be on simultaneously in a given frame?

Each depth encoding 3D PSF has an associated 2D "footprint," with a given lateral extent. This lateral extent limits how close single molecules can be in any given frame whilst still being able to be independently isolated in space, and hence indirectly the speed with which data can be acquired. In terms of footprint, iPALM has the smallest molecular 2D footprint, both biplane and astigmatism have moderate 2D footprints, and the multi-lobed PSFs such as those of the double helix and phase ramp methods have relatively large 2D footprints. More recent data analysis approaches allow the requirement to avoid any overlap between marker images to be somewhat relaxed. Using so called "multi-emitter fitting" a small number of events may overlap and the fitting strategy is adjusted so that it simultaneously estimates the localization of several emitters whose images may overlap [43–45]. Nevertheless, while this method works quite well for 2D methods it remains more problematic with 3D approaches.

14.2.2.3 How important is depth vs. lateral information?

All 3D methods (with the exception of iPALM, which avoids this deterioration at the expense of significantly increased optical complexity) trade a certain amount of lateral resolution in exchange for achieving depth information. Depending on the biological problem at hand, you may be better off simply performing 2D imaging. Figure 14.8 shows two contrasting ways of quantifying the 3D resolution of a range of different 3D localization strategies. The 3D Cramér–Rao bound (CRB) is a measure that allows estimation of the best accuracy with which molecules can be localized in 3D for a given PSF shape [46]. The 3D CRB is the square root of the sum of the squares of the localization uncertainties over the individual axes, and is heavily weighted by the worst axis, as shown in Fig. 14.8A. This quantity allows an appreciation of how isotropic the localization resolution is and is probably the measure you want to use if you want to be able to resolve close features in 3D (regardless of the orientation of the underlying object). We propose that an alternative measure, shown in Fig. 14.8B and which is calculated from the product of the 3 axis localization uncertainties, should also be considered. This can be interpreted as the volume of the "localization ellipse" and more clearly illustrates the trade-offs in lateral and hence volumetric resolution inherent in obtaining 3D information. As such, it may be a more useful measure when the biological question requires a quantification of volume overlap (i.e., colocalization). Another feature made clear by these plots is that both the double helix and phase ramp methods exhibit an extended axial range over which good localization accuracy can be obtained (although this also degrades quite rapidly at high background levels; see below).

14.2.2.4 How much background will I have?

Background affects localization-based methods in two ways: it reduces both the localization accuracy and the detectability of events prior to localization. The different 3D methods are, unsurprisingly, affected differently, as illustrated in Fig. 14.9, where we again compare the measures of 3D resolution limits but now in the presence of significant background fluorescence levels. In

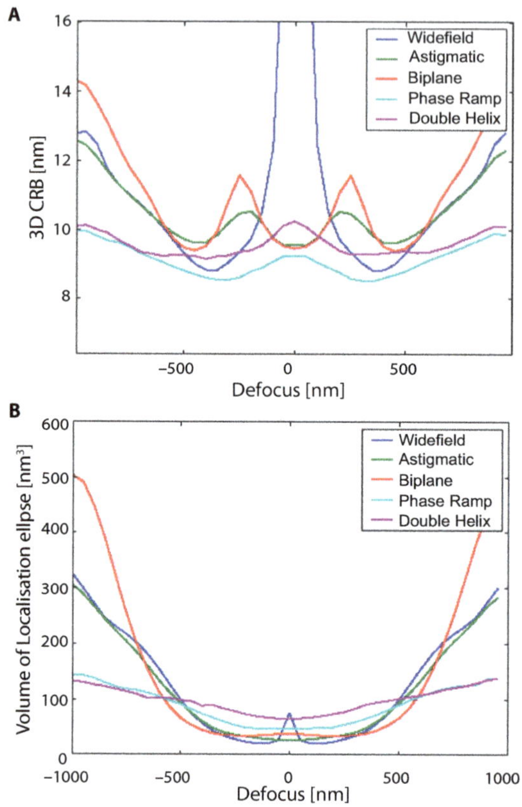

Figure 14.8 Two metrics which allow comparison of the 3D resolving power of different 3D localization microscopy techniques. The 3D CRB (A) is the geometric mean of the expected individual localization precisions. An alternative metric (B) is the product of individual axis precisions, or the volume of a "localization ellipse."

general, methods which maintain a compact PSF support (e.g., astigmatism and biplane) are more resistant to high backgrounds than those which separate the PSF into multiple lobes (i.e., double helix and phase ramp methods). This is perhaps not surprising as these extended support PSFs spread the same available total photon emission over a larger area and this makes associated events harder to detect and localize against the background. Note that

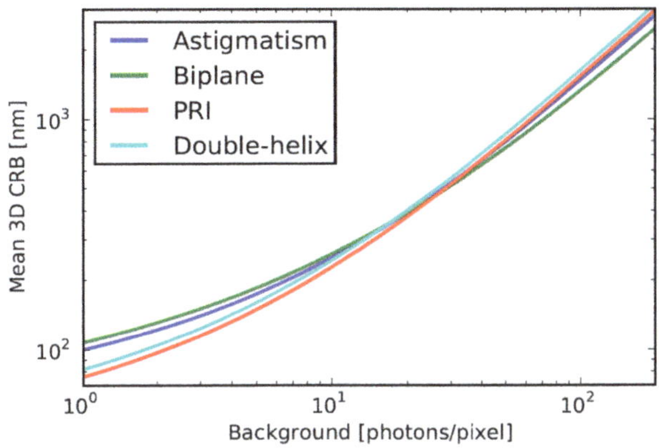

Figure 14.9 3D CRB in the presence of background. Calculations obtained using simulated PSFs show that the average 3D CRB over a 1.2 μm axial range for a fluorophore photon count of 2000. Note that the exact crossover point depends heavily on the choice of PSF parameters and axial range, but the general trend is for the two-lobed PSFs such as PRI and double helix to show better performance at very low background levels, but to not be as resilient to increasing background.

all of the methods can be tuned to optimize their response for a certain fixed background level and fluorophore brightness by tuning their parameters, e.g., the strength of the astigmatism or the axial separation of the two images in biplane.

Due to the tunability of the techniques it is difficult to determine a truly objective recommendation of when one should switch from lobed to compact PSFs. In our experience, there is always a background level above which compact methods such as astigmatism outperform the other methods, which has prompted us to use astigmatism when imaging thick tissue sections or other high background samples ($>\sim$50 counts/pixel background).

14.2.2.5 How well behaved is the sample refractive index?

Refractive index mismatch and other optical aberrations are generally going to negatively affect the accuracy of all 3D methods. Of the contemporary methods, iPALM is likely to be the worst behaved with its interferometric nature requiring very stringent refractive

Table 14.2 Qualitative comparison of 3D localization methods. More stars indicate better performance

	Simplicity	Footprint	Axial resolution	Lateral resolution	Background sensitivity	Aberration tolerance
No 3D	*****	*****	*	****	*****	*****
Astigmatism	****	****	***	***	***	**
Biplane	***	***	***	***	***	***
Double-helix	**	**	****	**	**	****
Phase ramp	****	**	****	**	**	****
iPALM	*	*****	*****	*****	*****	*

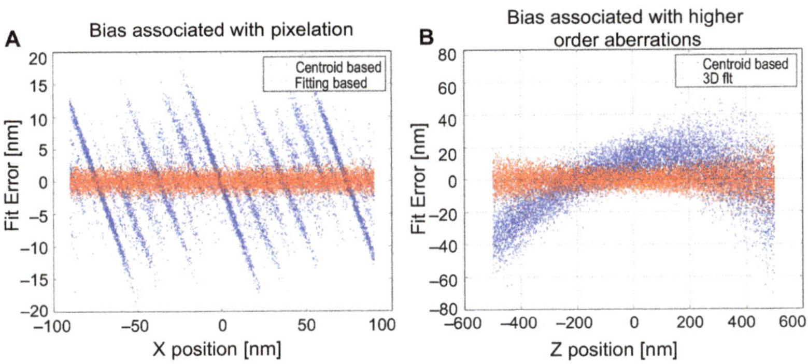

Figure 14.10 Bias in centroid based approaches. (A) The measured bias after fitting simulated single molecule events at various sub-pixel shifts using a centroid-based and a fitting-based approach. (B) The *x* position bias when performing 3D-fitting with a non-ideal PSF. In both cases events were simulated using an experimentally measured astigmatic PSF and assuming 2000 photons/event. The measured PSF had a small amount of coma in addition to the astigmatism, although this was only detectable on close examination. Note that 2.5D methods which perform a 2D fit and then use the relative widths of fitted spot to determine axial position will be subject to the bias illustrated in B (unless corrected using, e.g., a lookup table), but not that in A.

index matching and/or very thin samples. The traditional approach to fitting an astigmatic PSF using major and minor axis [32] ratios is very sensitive to spherical aberration and coma, with small amounts of spherical aberration leading to significant changes in the relationship between the difference of axial widths and depth (our unpublished data), and with coma leading to a *z*-position correlated lateral bias (see Fig. 14.10). Methods such as the double-helix and phase ramp PSFs which rely on the relative position of two lobes come out best, although they are by no means unaffected.

Once you have answered these questions for your super-resolution imaging problem at hand, it should become clearer which 3D method provides the best compromise. We contrast our opinion of the various strengths and weaknesses of the 3D methods in table II, but this is of course sample dependent. Therefore no single method will be best for all practical applications.

14.2.3 Analysis of Raw Data

The sections above have addressed the hardware aspects of generating high-quality data that should allow high-quality super-resolution information to be obtained. Much of the finesse of localization microscopy (in contrast to a method like STED where hardware and alignment are the ultimate determinant), lies in translation from the raw data into sets of single-molecule positions (e.g. [47]). It is desirable to use existing software packages to conduct this analysis—especially as more and more free or open source implementations are becoming available (see below). Nevertheless, a basic understanding of the various trade-offs in adopting different approaches is critical to choosing an analysis method and the ultimately determines the quality of the data coming out of a localization microscopy system. The major techniques fall into two broad categories—fitting methods in which an optimization method is used to fit a parametric model of the image of a single molecule (i.e., a lateral section of the specific PSF in question), and model independent methods such as centroid determination. Centroid-based methods are easier to implement and are significantly faster than fitting-based methods, but are less accurate, exhibiting both significant bias and a generally poorer precision (see Fig. 14.10 and [48]). In addition to having much lower bias, fitting methods allow a rigorous noise model to be considered, delivering both better precision and error estimates for the fitted parameters along with other indications of fit quality. Modern computer hardware is fast enough to conduct fitting based analysis in real time (at typical frame rates) as we explain in more detail below.

Extending the fitting approach to 3D requires a 3D model which describes how the PSF varies with defocus. While it is theoretically possible to use an analytical expression for the PSF, these are both computationally expensive and (in our experience) often a poor match for the actual microscope characteristics in the presence of aberrations and alignment limitations. The most common approach has been to measure a PSF using a fluorescent bead (or gold fiducial), and to base the model on this [29, 49]. A model which can represent sub-pixel PSF shifts can then be derived by on the fly interpolation of the measured PSF data. This approach is PSF-agnostic and works for all current 3D modalities.

To illustrate the potential problems associated with using analysis "shortcuts" for specific PSF models we show an example below. Initial 3D analysis of astigmatic data was accomplished by fitting a 2D Gaussian in which the x and y widths are independent parameters. The focus position was then determined by comparing the relative widths in the two dimensions to a pre-measured calibration curve that related the axes ratio to focus depth. Measuring the difference in x and y widths also forms the basis for the QuickPALM centroid-based 3D analysis method [50]. A downside of this method is that small amounts of coma (and of other higher order aberrations) that are typically present in high NA objectives can lead to coupling between the x and y centroid positions and estimated defocus. This results in a z-dependent systematic error in lateral position, as illustrated in Fig. 14.10.

The principal argument for centroid-based methods is their speed, but we can perform real-time analysis of both 2D and 3D data using fitting methods, on fairly modest computing hardware. 2D fitting can be performed in real time on a single workstation class machine (Xeon E5620, 2.4 GHz, 4 cores, HT) at 50 Hz acquisition frame rates—specifics depend on the typical number of events per frame, but we can typically analyse 70–100 frames per second. In connection with this point, we find that the bottleneck in our implementation of 2D fitting is often the initial event detection step, a step that also has to be performed prior to the application of centroid-based localization methods, rather than the fitting itself. Fitting 3D data is a bit slower and may require a small distributed cluster of 2–3 machines to allow analysis in real time. Our "cluster" is implemented by running a standard program on participating office PCs (which are present in the typical research laboratory setting) and we dynamically recruit additional PCs until the analysis keeps up with data acquisition. Given that the intrinsically more reliable fitting methods can be performed in real time, we feel that using a centroid estimator for its superior speed is an unnecessary compromise. Centroid-based estimation does, however, have a useful role in estimating start parameters for fitting-based methods, improving their speed and convergence. We use a bias-reduced derivative of the Quick-PALM algorithm [50] for this purpose when fitting astigmatic PSFs.

Positioned between the two categories are non-iterative methods based on concepts such as radial symmetry [51] and cross-correlation [52]. These typically have a reduced bias when compared to the centroid-based methods. Radial symmetry based localization is restricted to cases where the PSF is truly symmetric, and does not easily extend to 3D analysis. Cross-correlation-based localization extends well to 3D [52], although it is unclear if it is faster than a fitting based method. Neither provide estimates of parameter accuracy.

All the methods discussed so far rely on the single-molecule events being well separated both in space and time such that two or more fluorophores are not active simultaneously within the region of interest being fitted (typically on the order of 1 μm × 1 μm). In most cases it is possible to fulfil this requirement by using appropriate "switching" buffers, illumination power, and frame rates, but doing so constrains the acquisition speed. There is thus considerable interest in finding algorithms which permit scenarios where there are multiple simultaneous events within one ROI to be analyzed, and a number of methods for fitting multiple emitters have been demonstrated (e.g. [43–45]). To date, multi-emitter fitting approaches have been largely limited to 2D analysis (with the exception of [53]).

A number of fitting packages have been released into the public domain, examples of which are given below.

14.2.4 Visualization and Postprocessing

The result of any of the above analysis methods is a list of positions at which single molecules were observed. While containing all the information about the object, a list of fluorophore positions does not allow an easy interpretation of the data, and it has become common practice to reconstruct a visual density map of localizations, analogous to a conventional image. Initial attempts either binned the positions to create a 2D histogram, i.e., the "image" is a 2D pixel grid at a chosen spacing displaying how

Table 14.3 A selection of freely available localization microscopy packages

Package	URL
Quickpalm	http://code.google.com/p/quickpalm/
rapidSTORM	http://www.super-resolution.biozentrum.uni-wuerzburg.de/home/rapidstorm/
Localizer	https://bitbucket.org/pdedecker/localizer
rainSTORM	http://laser.cheng.cam.ac.uk/wiki/index.php/Resources
DaoSTORM	https://github.com/ZhuangLab/storm-analysis
3B	http://www.coxphysics.com/3b/
deconSTORM	https://github.com/ZhuangLab/storm-analysis
Palm3D	http://code.google.com/p/palm3d/
Python-microscopy	http://code.google.com/p/python-microscopy

many localizations fall within a given pixel (e.g. [8]). An alternative representation represents each event as a Gaussian distribution with a standard deviation equal to the localization error with all individual Gaussians being summed to create the final image [6]. While effective, both of these methods have limitations. The histogram method is sensitive to bin size and bin edge placement, with the optimum bin size depending on the local point density. The Gaussian rendering approach, while not subject to arbitrary bin sizes and positions, imparts excessive confidence to poorly sampled structures and effectively decreases resolution (by a factor of $\sqrt{2}$) in well-sampled areas (see [54]).

To address these shortcomings, we and others have proposed several improved methods. Adaptive histogramming based on quad or oct-trees allows the histogram bin size to vary so as to be appropriate to the local fluorophore density. Basing "bins" on a triangulation of the point data set itself is both locally adaptive, and ensures that at least three events are present in any given area before it is represented in the reconstruction, decreasing the chance that structures will appear by chance [54]. So-called "jittered histograms" [55] remove some of the arbitrary nature in histogram edge selection. Our preferred method is our jittered triangulation method which results in an image with good suppression of low-confidence structures, a decent visual indication of confidence, and minimal loss in resolution [54]. We show a qualitative comparison

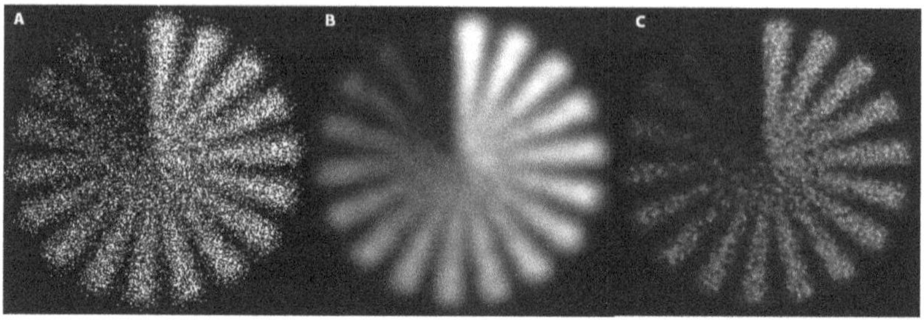

Figure 14.11 Comparison of event rendering methods. (A) Single points. (B) Each point is replaced with a Gaussian whose width matches the localization precision. The resolution is compromised by the fuzzy appearance, although it has an apparently high SNR. Note also that the low-density spokes (~11 o'clock) appear curved. (C) The data set rendered as a jittered triangulation with point density rather than localization precision determining the amount of blurring [54]. Note that the edges are crisper, and, due to the noisy appearance of the image, one is less likely to have excessive confidence in poorly sampled areas.

of two different methods in Fig. 14.11 showing the impact of reconstruction methods on resolution and perceived smoothness of the rendered images.

Once rendered and presented as an image, most conventional image processing tools can be used. There are, however, some areas where special consideration needs to be given to the method used to draw inferences about the data. An important example of this is colocalization analysis which attempts to quantify how much of one type of marker is associated with another. Colocalization is arguably problematic even when imaging at conventional resolutions, and ultimately, nothing is truly colocalized on a molecular scale.

Figure 14.12 illustrates this problem with a super-resolution image of two proteins in a cardiac muscle cell, ryanodine receptors (RyRs), and caveolin (Cav). Although both localize in close proximity to a small membrane network, the *t*-system, the colocalization when calculated using conventional colocalization measures such as Pearsons [.09] or Manders values [.08 (RyR vs Cav), .04 (Cav vs. RyR)] is very small. However, performing a distance transform shows that there is a significant association between the labels, with 50% of RyR labelling lying within 70 nm of the majority of caveolin

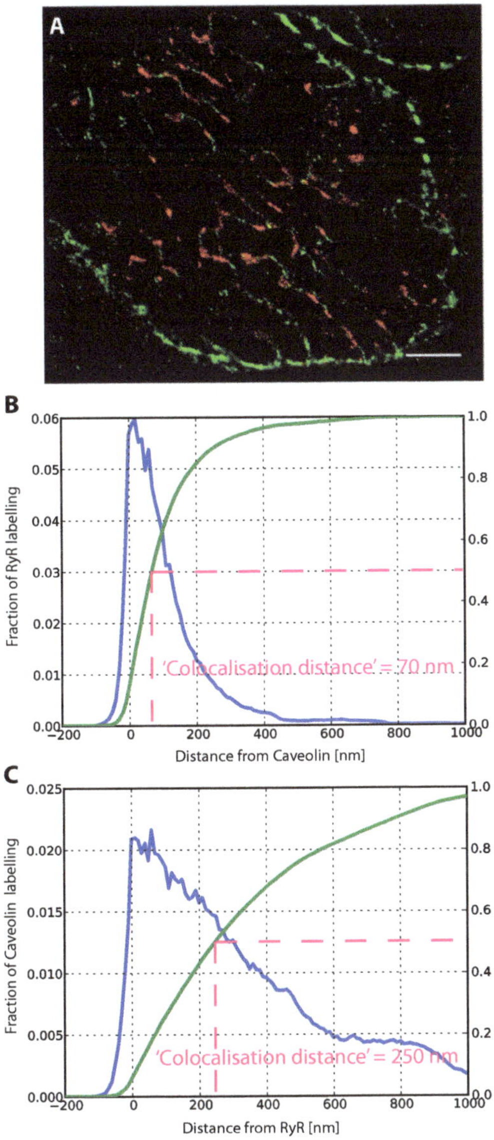

Figure 14.12 (A) dSTORM image of RyR (red) and Caveolin (green) in a transverse section through a cardiac myocyte. Scale bar 2 μm. (B) Distance transform showing the distribution of RyR with respect to a the Caveolin. While colocalization is low (~8%), the majority of RyR labeling is very close to some Caveolin, consistent with both being located at the t-tubule membrane. (C) Distribution of Caveolin with respect to RyR. In this case there is clearly still some association, but not all Caveolin is near RyR.

labelling (identified by a mask that captures most of the caveolin labelling). We suggest that the use of such distance-transform approaches is much more powerful than straight colocalization coefficients, allowing, for example, the detection of associated and non-associated fractions and should be the method of choice. If a single parameter is desired to quantify association on super-resolution data, we suggest using the colocalization distance, which is the average distance from the edge of one structure that incorporates 50% of the other structure.

In addition to the improved resolution (which is now approaching the molecular scale), there is a stochastic aspect to super-resolution data resulting from the potentially small numbers of molecules. Since super-resolution images provide a random sampling of underlying cellular structures (often with partially unknown completeness) colocalization may be underestimated, and this aspect needs to be considered in proper analysis.

14.3 Summary

Localization-based super-resolution microscopy is a surprisingly simple fluorescence microscopy technique which allows a dramatic improvement in resolution to be obtained with reasonably modest instrumental complexity. The method works by spreading the fluorescence from the structure of interest in time, so that in any given frame only a small number of the dye molecules attached to that structure are visible. The knowledge that one is only looking at a small number of molecules allows us to express the imaging problem in terms of a more highly constrained problem—namely only to find the locations of the molecules which are visible in each frame. This can be done with a precision limited only by the number of photons that can be extracted from a single molecule and the background noise. Current implementations generally achieve a lateral resolution of roughly 25 nm, a 10-fold improvement in over conventional fluorescence microscopy.

Although conceptually simple, localization microscopy relies on state of the art fluorescent dyes, labelling strategies, detection technologies, and computational power which have only recently reached a sufficient level of development to make the technique a viable proposition for most laboratories. As these technologies continue to be developed, the current limitations are likely to be reduced; for example, the development of sCMOS cameras has allowed STORM-type imaging at video frame rates [17]. There remain, however, a number of challenges. Correct interpretation of the data may remain problematic in some cases, especially when molecule density is low and physiological importance may not be simply reflected in protein amount.

References

1. Abbe, E. (1873). Beiträge zur theorie des mikroskops und der mikroskopischen wahrnehmung. *Arch. Mikrosk. Anatomie*, **9**(1), 413-418.
2. Rayleigh, L. (1886). LXIV. Notes, chiefly historical, on some fundamental propositions in optics. *Philos. Mag.*, 5th ser., **21**(133), 466–476.
3. Francia, G. (1955). Resolving power and information. *JOSA*, **45**(7), 497-499.
4. McCutchen, C. W. (1967). Superresolution in microscopy and the Abbe resolution limit. *J. Opt. Soc. Am.*, **57**(10), 1190–1192.
5. Hell, S. W., and Wichmann, J. (1994). Breaking the diffraction resolution limit by stimulated emission: stimulated-emission-depletion fluorescence microscopy. *Opt. Lett.*, **19**(11), 780–782.
6. Betzig, E., et al. (2006). Imaging intracellular fluorescent proteins at nanometer resolution. *Science*, **313**(5793), 1642–1645.
7. Rust, M. J., Bates, M., and Zhuang, X. (2006). Sub-diffraction-limit imaging by stochastic optical reconstruction microscopy (STORM). *Nat. Methods*, **3**(10), 793–795.
8. Hess, S. T., Girirajan, T. P., and Mason, M. D. (2006). Ultra-high resolution imaging by fluorescence photoactivation localization microscopy. *Biophys. J.*, **91**(11), 4258–4272.
9. Saxton, M. J., and Jacobson, K. (1997). Single-particle tracking: applications to membrane dynamics. *Annu. Rev. Biophys. Biomol. Struct.*, **26**, 373–399.
10. Folling, J., et al. (2008). Fluorescence nanoscopy by ground-state depletion and single-molecule return. *Nat. Methods*, **5**(11), 943–945.

11. Heilemann, M., et al. (2008). Subdiffraction-resolution fluorescence imaging with conventional fluorescent probes. *Angew. Chem.*, **47**(33), 6172–6176.
12. Baddeley, D., et al. (2009). Light-induced dark states of organic fluochromes enable 30 nm resolution imaging in standard media. *Biophys. J.*, **96**(2), L22– L24.
13. Schermelleh, L., Heintzmann, R., and Leonhardt, H. (2010). A guide to super-resolution fluorescence microscopy. *J. Cell Biol.*, **190**(2), 165–175.
14. Patterson, G., et al. (2010). Superresolution imaging using single-molecule localization. *Annu. Rev. Phys. Chem.*, **61**, 345–367.
15. Thompson, R. E., Larson, D. R., and Webb, W. W. (2002). Precise nanometer localization analysis for individual fluorescent probes. *Biophys. J.*, **82**(5), 2775–2783.
16. Deng, Y., and Shaevitz, J. W. (2009). Effect of aberration on height calibration in three-dimensional localization-based microscopy and particle tracking. *Appl. Opt.*, **48**(10), 1886–1890.
17. Huang, F., et al. (2013). Video-rate nanoscopy using sCMOS camera-specific single-molecule localization algorithms. *Nat. Methods*, **10**(7), 653–658.
18. Long, F., Zeng, S., and Huang, Z. L. (2012). Localization-based super-resolution microscopy with an sCMOS camera part II: experimental methodology for comparing sCMOS with EMCCD cameras. *Opt. Express*, **20**(16), 17741–17759.
19. Mlodzianoski, M. J., et al. (2011). Sample drift correction in 3D fluorescence photoactivation localization microscopy. *Opt. Express*, **19**(16), 15009–15019.
20. McGorty, R., Kamiyama, D., and Huang, B. (2013). Active microscope stabilization in three dimensions using image correlation. *Opt. Nanosc.*, **2**(1), 3.
21. Bates, M., Jones, S. A., and Zhuang, X. (2013). Stochastic optical reconstruction microscopy (STORM): a method for superresolution fluorescence imaging. *Cold Spring Harbor Protoc.*, **2013**(6), 498–520.
22. Shroff, H., et al. (2007). Dual-color superresolution imaging of genetically expressed probes within individual adhesion complexes. *Proc. Natl. Acad. Sci. U.S.A.*, **104**(51), 20308–20313.
23. Baddeley, D., et al. (2011). 4D super-resolution microscopy with conventional fluorophores and single wavelength excitation in optically thick cells and tissues. *PLoS One*, **6**(5), e20645.

24. Bates, M., et al. (2007). Multicolor super-resolution imaging with photo-switchable fluorescent probes. *Science*, **317**(5845), 1749–1753.

25. Testa, I., et al. (2010). Multicolor fluorescence nanoscopy in fixed and living cells by exciting conventional fluorophores with a single wavelength. *Biophys. J.*, **99**(8), 2686–2694.

26. Juette, M. F., et al. (2008). Three-dimensional sub-100 nm resolution fluorescence microscopy of thick samples. *Nat. Methods*, **5**(6), 527–529.

27. Huang, B., et al. (2008). Three-dimensional super-resolution imaging by stochastic optical reconstruction microscopy. *Science*, **319**(5864), 810–813.

28. Pavani, S. R., et al. (2009). Three-dimensional, single-molecule fluorescence imaging beyond the diffraction limit by using a double-helix point spread function. *Proc. Natl. Acad. Sci. U.S.A.*, **106**(9), 2995–2999.

29. Baddeley, D., Cannell, M. B., and Soeller, C. (2011). Three-dimensional sub-100 nm super-resolution imaging of biological samples using a phase ramp in the objective pupil. *Nano Res.*, **4**(6), 589–598.

30. Shtengel, G., et al. (2009). Interferometric fluorescent super-resolution microscopy resolves 3D cellular ultrastructure. *Proc. Natl. Acad. Sci. U.S.A.*, **106**(9), 3125–3130.

31. Aquino, D., et al. (2011). Two-color nanoscopy of three-dimensional volumes by 4Pi detection of stochastically switched fluorophores. *Nat. Methods*, **8**(4), 353–359.

32. Huang, B., Bates, M., and Zhuang, X. (2009). Super-resolution fluorescence microscopy. *Annu. Rev. Biochem.*, **78**, 993–1016.

33. Shannon, C. E. (1949). Communication in the presence of noise. *Proc. Inst. Radio Eng.*, **37**(1), 10–21.

34. Lippincott-Schwartz, J., and Patterson, G. H. (2009). Photoactivatable fluorescent proteins for diffraction-limited and super-resolution imaging. *Trends Cell Biol.*, **19**(11), 555–565.

35. Zhang, M. S., et al. (2012). Rational design of true monomeric and bright photoactivatable fluorescent proteins. *Nat. Methods*, **9**(7), 727–U297.

36. Dempsey, G. T., et al. (2009). Photoswitching mechanism of cyanine dyes. *J. Am. Chem. Soc.*, **131**(51), 18192–18193.

37. Olivier, N., et al. (2013). Resolution doubling in 3D-STORM imaging through improved buffers. *PLoS One*, **8**(7), e69004.

38. Uppoor, R., and Niebergall, P. J. (1996). β-D(+) glucose-glucose oxidase-catalase for use as an antioxidant system. *Pharm. Dev. Technol.*, **1**(2), 127–134.

39. Dempsey, G. T., et al. (2011). Evaluation of fluorophores for optimal performance in localization-based super-resolution imaging. *Nat. Methods,* **8**(12), 1027–+.
40. Axelrod, D. (2001). Total internal reflection fluorescence microscopy in cell biology. *Traffic,* **2**(11), 764–774.
41. Tokunaga, M., Imamoto, N., and Sakata-Sogawa, K. (2008). Highly inclined thin illumination enables clear single-molecule imaging in cells (vol 5, pg 159, 2008). *Nat. Methods,* **5**(5), 455–455.
42. Verveer, P. J., et al. (2007). High-resolution three-dimensional imaging of large specimens with light sheet-based microscopy. *Nat. Methods,* **4**(4), 311–313.
43. Huang, F., et al. (2011). Simultaneous multiple-emitter fitting for single molecule super-resolution imaging. *Biomed. Opt. Express,* **2**(5), 1377–1393.
44. Huang, F., et al. (2013). Video-rate nanoscopy using sCMOS camera-specific single-molecule localization algorithms. *Nat. Methods,* **10**, 653–658.
45. Wang, Y., et al. (2012). PALMER: a method capable of parallel localization of multiple emitters for high-density localization microscopy. *Opt. Express,* **20**(14), 16039–16049.
46. Pavani, S. R. P., and Piestun, R. (2008). Three dimensional tracking of fluorescent microparticles using a photon-limited double-helix response system. *Opt. Express,* **16**(26), 22048–22057.
47. Smith, C. S., et al. (2010). Fast, single-molecule localization that achieves theoretically minimum uncertainty. *Nat. Methods,* **7**(5), 373–375.
48. Cheezum, M. K., Walker, W. F., and Guilford, W. H. (2001). Quantitative comparison of algorithms for tracking single fluorescent particles. *Biophys. J.,* **81**(4), 2378–2388.
49. Mlodzianoski, M. J., et al. (2009). Experimental characterization of 3D localization techniques for particle-tracking and super-resolution microscopy. *Opt. Express,* **17**(10), 8264–8277.
50. Henriques, R., et al. (2010). QuickPALM: 3D real-time photoactivation nanoscopy image processing in ImageJ. *Nat. Methods,* **7**(5), 339–340.
51. Parthasarathy, R. (2012). Rapid, accurate particle tracking by calculation of radial symmetry centers. *Nat. Methods,* **9**(7), 724–726.
52. York, A. G., et al. (2011). Confined activation and subdiffractive localization enables whole-cell PALM with genetically expressed probes. *Nat. Methods,* **8**(4), 327–333.

53. Babcock, H., Sigal, Y., and Zhuang, X. (2012). A high-density 3D localization algorithm for stochastic optical reconstruction microscopy. *Opt. Nanosc.*, **1**(1), 6.

54. Baddeley, D., Cannell, M. B., and Soeller, C. (2010). Visualization of localization microscopy data. *Microsc. Microanal.*, **16**(1), 64–72.

55. Krizek, P., Raska, I., and Hagen, G. M. (2011). Minimizing detection errors in single molecule localization microscopy. *Opt. Express*, **19**(4), 3226–3235.

Chapter 15

Super-resolution Optical Microscopy with Structured Illumination

Liisa M. Hirvonen and Trevor A. Smith
School of Chemistry, ARC Centre of Excellence for Coherent X-Ray Science, University of Melbourne, Parkville, Victoria 3010, Australia
trevoras@unimelb.edu.au

In widefield fluorescence microscopy, resolution improvement can be achieved by illuminating the fluorescent sample with a spatially modulated excitation source. Structured illumination microscopy (SIM) is one such technique, where a spatially modulated illumination pattern is projected onto the sample, and the final image is reconstructed from a series of images in which the pattern is moved between the images in the set. With this technique, lateral resolution can be improved by a factor of 2 in comparison to widefield microscopy. Additional advances in this technique can lead to further improvements in optical resolution and wider applications.

15.1 Introduction

Since the seminal work of Ernst Abbe in 1873 [1], Lord Rayleigh in 1896 [2, 3], and others, the achievable resolution of optical

Fundamentals of Fluorescence Imaging
Edited by Guy Cox

ISBN 978-981-4774-85-7 (Hardcover), 978-1-351-12940-4 (eBook)
www.jennystanford.com

microscopy techniques has been limited by diffraction to approximately half the wavelength of the light used to illuminate the sample. The fundamental limiting factors to this resolution are the maximum half-angle of the emission collection cone of the objective lens (ϑ), the refractive index of the lens/immersion medium (n), and the wavelength of the light (λ). The wavelength of the excitation radiation can be shortened, for example to the ultraviolet [4–8], but this does not overcome the diffraction limit, and leads to potential sample photodamage. In the visible spectral range, the most sophisticated immersion lenses feature a maximum half-angle where the light is collected through a cone angle of about 73°, while the refractive index of the immersion medium is usually limited to less than about 1.5. For a typical widefield fluorescence microscope this results in a smallest resolvable distance between two features in the sample, $r \approx 0.61\lambda/(2n \sin\vartheta)$, of about 180 nm in the lateral, and about 600 nm in the axial direction.

In recent years a number of approaches have been developed in attempts to overcome the limitations in optical resolution, which has spawned the term "super-resolution" optical microscopy, defined here as providing optical resolution $<$100 nm. The fundamental laws of optical physics are not broken, but rather circumvented. As discussed by Hell et al. [9], many of these approaches have in common the exploitation and control of the populations of one or more electronic ground or excited states. The various approaches can be broadly classified into several categories including the following. The 4π [10–12] and I^5M [13, 14] approaches aim to increase the effective numerical aperture of the collection system by collecting emission using two opposing objective lenses, both of which focus to the same geometrical location. Near-field optical techniques, including evanescent wave-induced fluorescence microscopy (based on total internal reflection phenomena, e.g., standing evanescent wave) [15–19] and scanning near-field optical microscopy (SNOM), provide sub-100 nm resolution (axially in the former and laterally in the latter). A range of single emitter localization methods has also been developed, including stochastic optical reconstruction microscopy (STORM), photoactivated localization microscopy (PALM), and variants (dSTORM, FPALM, etc.). Stimulated emission depletion (STED) microscopy is the fourth broad approach to achieving super-resolution. As another variation on super-resolution optical microscopy, Raman microscopy has also achieved $\sim$10 nm resolution [20]. Finally, the use of structured illumination has also been developed in various forms to achieve this goal.

Each of these approaches has its inherent advantages and disadvantages, particularly in the context of considering the nature and characteristics of any given sample. Image acquisition time, required irradiation intensity, the photostability of the sample, suitability of fluorescent probes for the sample, expense, information to be gained, etc., all influence the choice of the most appropriate technique for the desired purpose. In this chapter we concentrate on the use of patterned illumination techniques and discuss the various advantages and disadvantages.

15.1.1 Spatially Modulated Illumination Microscopy

The spatial modulation of light has been used in microscopy, in one form or another, for decades and has been exploited in both emissive and non-emissive [21–24] modes. Techniques such as confocal microscopy and Hoffman contrast modulation microscopy [25] can be considered as relying on the spatial modulation of light. STED microscopy also uses spatially modulated illumination to achieve super-resolution fluorescence imaging, but this is discussed in detail elsewhere (Chapter 16). In this chapter, we restrict our discussion to widefield fluorescence microscopy techniques, where the whole sample is illuminated with a periodically modulated excitation source.

In 1997, it was shown that a coarse sinusoidal grating pattern projected onto the sample can be used to obtain optical sectioning in widefield fluorescence microscopy [26]. A year later, it was suggested that a fine sinusoidal grating pattern could be used to improve lateral resolution [27], and this was first demonstrated in 2000 [28]. The ability to gain improvements in lateral resolution using spatially modulated excitation and fluorescence detection comes about by exploiting the fact that the measured local intensity of the fluorescence emission is a convolution of the local concentration of the fluorophores and the local intensity of the excitation light. This convolution mixes information from different points in reciprocal space, and in particular moves some information from outside the observable region into the observable region, where it can now be detected [29]. The popularity of fluorescence-based spatially modulated illumination techniques has been growing ever since the demonstrated improvements in spatial resolution, to the point that several of the well-known microscopy companies now have commercial versions [30–32] of microscopes based on spatially modulated illumination. This approach exhibits various advantages: it requires only low to moderate irradiation flux, is applicable to any fluorophore over a range of concentration and can provide optical sectioning. Although several images are needed for image reconstruction, as a widefield technique the image formation is still acceptably fast for many samples and certain circumstances. No unusual properties are required from the sample or the fluorophore, and pulsed or high-intensity lasers are not needed.

15.1.2 The Principle of SIM

Resolution improvement by structured illumination is based on Moiré fringes. When two high-frequency patterns are multiplied together, Moiré fringes containing low-frequency information appear (Fig. 15.1). If an unknown sample pattern in a microscope is multiplied by a known high-frequency illumination pattern, and the illumination pattern can be removed from the recorded image containing low-frequency information, the unknown sample pattern

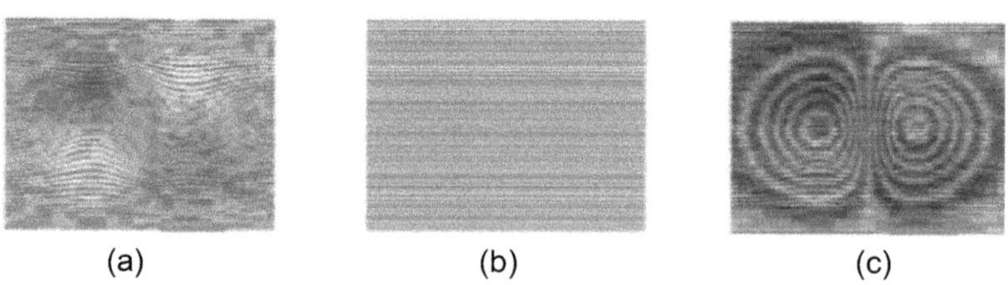

Figure 15.1 Moiré fringes: If an unknown sample pattern (a) is multiplied by a known illumination pattern (b), Moiré fringes appear (c). If the illumination pattern is removed computationally, previously unobservable details can be recovered.

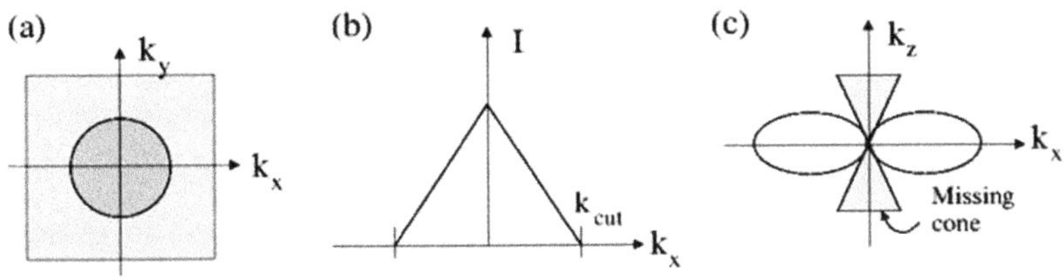

Figure 15.2 The OTF, a Fourier transform of the PSF, defines which frequencies are transferred through the microscope. In the lateral plane it defines a circular region with a radius k_{cut}, which defines the highest observable frequency (a, b). In the axial direction it is doughnut-shaped with a "missing cone" in the middle (c).

can be recovered. The resolvable frequency content of the sample pattern depends on the frequency of the illumination pattern.

Resolution improvement by SIM can be easily visualized in Fourier space. A microscope acts as a spatial filter in Fourier space; when an image is formed, the highest frequencies containing the smallest details of the object are lost. SIM improves resolution by extending the range of frequencies observable through a microscope.

The resolution of a microscope is often defined as the full width at half maximum of the point-spread function (PSF). Its Fourier transform, the optical Transfer function (OTF), defines the maximum transferable frequency, k_{cut}. In Fourier space it defines a circular region in the $k_x k_y$ plane with radius k_{cut} (Fig. 15.2a,b). In the axial direction the OTF is doughnut-shaped with a "missing cone" around the axis, resulting in poor optical sectioning (Fig. 15.2c). Any details about the object outside the support of the OTF are normally lost.

In SIM, a sinusoidal grating pattern is projected onto the sample, which creates two peaks in Fourier space on the lateral plane. The multiplication of the sample structure with the illumination pattern translates to convolution in Fourier space, and both these peaks now have the same information about the sample attached to them as the middle peak (Fig. 15.3a). Although all frequencies outside the support of the OTF are lost when the image is formed in the microscope, these peaks have transferred some frequencies to the observable region that otherwise would have been lost (Fig. 15.3b). The frequency components are unmixed and moved to coincide with

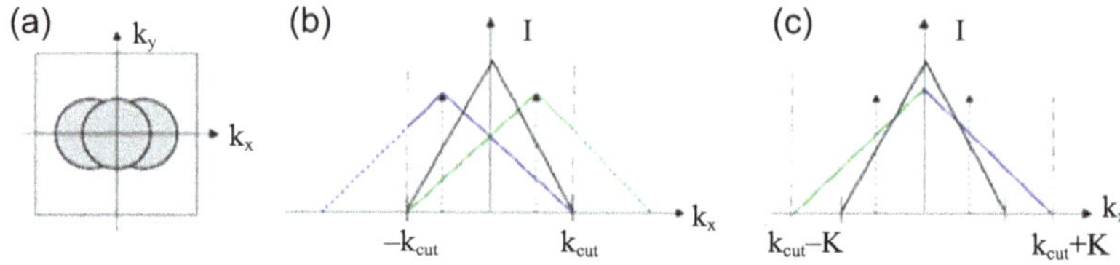

Figure 15.3 The sinusoidal grating pattern used in SIM creates two delta peaks in Fourier space that have the same information about the object attached to them as the middle peak (a). These peaks shift frequencies that would have otherwise been lost to the observable region (b). If the information attached to the peaks is unmixed and shifted to coincide with the middle peak, the new maximum observable frequency is $k_{\text{cut}} + \kappa$, where κ is inversely proportional to the grating period (c).

the middle peak, resulting in a new maximum observable frequency $k_{\text{cut}} + \kappa$, where κ is the distance of the peaks from the origin and is inversely proportional to the grating period (Fig. 15.3c).

The sectioning capability of SIM is achieved by filling the "missing cone" of the OTF (Fig. 15.4a). If only improved sectioning is required without lateral resolution improvement, a coarse grid can be used where the peaks lie relatively close to the origin (Fig. 15.4b). The missing cone is now filled, and the sectioned image can be obtained from a series of three images with different grating positions by calculating the degree of modulation for each pixel between the images.

When lateral resolution improvement is also desired, a coarse grid cannot be used. If the grating pattern on the sample is created by interfering only the first diffracted orders, no optical sectioning can be achieved with maximum lateral resolution improvement, as the peaks will lie at the edge of the OTF, resulting in a missing cone in the middle (Fig. 15.2c). If instead three illumination orders are used, two additional peaks are created in Fourier space on both sides of the focal plane at half the frequency of the grating, and the missing cone is now always filled (Fig. 15.2d). If the reconstruction of the SIM image is performed in 2D space, only the projection of these peaks to the focal plane is considered, and improved optical sectioning is achieved without actual resolution improvement in the axial direction. If the SIM image is reconstructed in 3D space, resolution can be improved also in the axial direction [33].

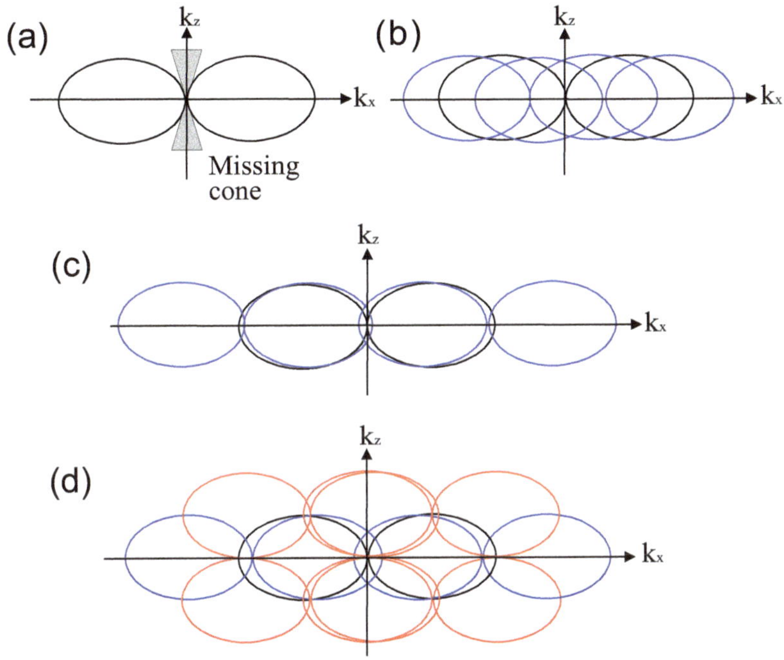

Figure 15.4 SIM in the axial direction. The widefield OTF has a "missing cone" in the middle, resulting in out-of focus blur and poor resolution in the axial direction (a). A coarse grid can be used to fill the "missing cone," resulting in improved optical sectioning (b). When maximum resolution improvement is required in the lateral direction, the peaks lie near the edge of the widefield OTF, and the missing cone appears (c). By using three interfering diffraction orders instead of two for the creation of the illumination structure, two extra peaks are created on both sides of the focal plane, and the missing cone is filled (d). If the reconstruction of the image is done in 3D space, resolution can be doubled in the axial as well as the lateral direction.

15.2 Optical Sectioning with Structured Illumination

In 1997 Neil et al. [26] demonstrated that it is possible to obtain optical sectioning by structured illumination by estimating the degree of modulation at each pixel of the image. A coarse, one-dimensional grid is used as the illumination pattern, and a minimum of three images Im_n, $n = 1, 2, 3$ with relative grating positions of

$2\pi/3$ have to be recorded. If the sample is illuminated incoherently, the light coming from in-focus regions of the sample is modulated between the images, whereas the light coming from the out-of-focus regions is blurred and the modulation depth is reduced. If coherent illumination is used, this technique only works with samples where the out-of-focus blur is on average the same between the images. Calculating the degree of modulation pixel by pixel over multiple images allows the discrimination between the non-modulated out-of-focus information and the modulated in-focus information. The final optically sectioned image, Im_s, can be obtained from

$$\mathrm{Im}_s = \sqrt{(\mathrm{Im}_1 - \mathrm{Im}_2)^2 + (\mathrm{Im}_2 - \mathrm{Im}_3)^2 + (\mathrm{Im}_3 - \mathrm{Im}_1)^2} \quad (15.1)$$

Whereas in the sum of the images, $\mathrm{Im}_{sum} = \mathrm{Im}_1 + \mathrm{Im}_2 + \mathrm{Im}_3$, gives a widefield image, the demodulation inferred by Eq. 15.1 removes the widefield component and calculates the final image from the degree of modulation for each pixel, thus removing out-of focus light.

This technique is relatively easy to implement and add to existing widefield microscopes, while providing optical sectioning capability comparable to confocal microscopy. Commercial versions employing variations on this approach are now readily available [34–36] and a cost-effective version based on a commercial data projector has also been reported [37].

15.3 Resolution Improvement with Structured Illumination

The use of patterned excitation for resolution improvement was conceived and demonstrated by Heintzmann [27] and Gustafsson [28] over a decade ago. They showed that by mixing the signal from the sample with a harmonic excitation pattern, extra information about the object can be brought to the transferable region of the OTF and used for reconstructing an image where the resolution is doubled compared to normal widefield microscopy. This technique is becoming increasingly popular in biological research, where low illumination intensity and fast image acquisition time are desirable. Its compatibility with any fluorophore, including autofluorescence,

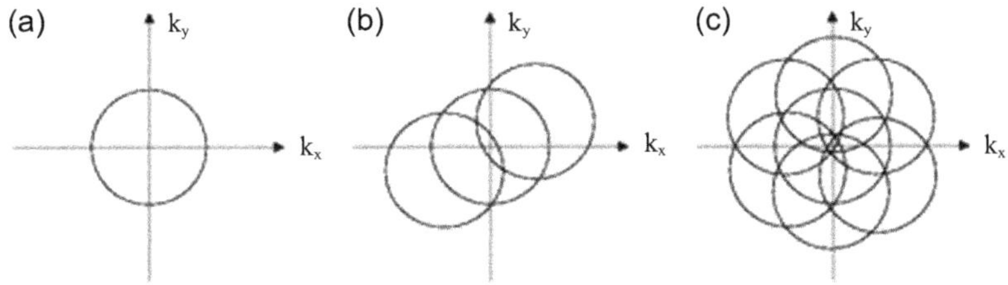

Figure 15.5 One grating direction improves the lateral resolution in only one dimension. A minimum of three directions are needed for isotropic resolution improvement on the lateral plane. Outlines of a widefield OTF (a), and reconstructed SIM OTFs for one (b) and three (c) grating directions.

is advantageous when it is difficult or undesirable to label the sample with fluorescent dyes or proteins.

Unlike optical sectioning SIM where the final image is calculated from three images, more images are required for isotropic resolution improvement with SIM. Using one grating orientation improves resolution only in that direction in 2D space. A minimum of three grating orientations are needed to fill an approximately circular area in Fourier space, yielding isotropic resolution improvement in the lateral plane (Fig. 15.5). If no optical sectioning is needed, a SIM image can be reconstructed from a minimum of nine images. If three orders are used for illumination such that two harmonics need to be separated, a minimum of five images are required for each grating direction. In commercial systems, a SIM optical slice is typically reconstructed from a total of 15 images with three grating orientations. The data acquisition rate and time for adequate image reconstruction make this technique slower than conventional widefield microscopy, but video-rate SIM has been demonstrated in living cells [38]. Two-dimensional patterns could potentially be used to speed up the data acquisition [39].

One disadvantage of structured illumination microscopy (SIM) over competitive super-resolution optical microscopy approaches is that when illuminating through the objective, the peaks have to lie within the support of the OTF, limiting the resolution improvement to a factor of 2 over conventional widefield fluorescence microscopy at the same wavelength. Further resolution improvement is possible

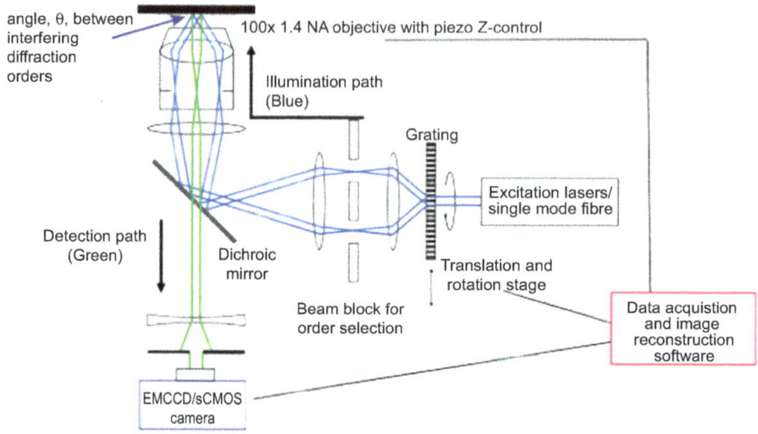

Figure 15.6 Schematic of SIM apparatus with transmission grating.

by saturated structured illumination microscopy (SSIM; see Section 15.2) [40, 41].

15.4 Practical Implementation of SIM

The practical implementation of a structured illumination microscope does not differ much from a conventional widefield microscope, and in principle SIM capabilities could be added to any widefield epi-illuminated microscope that provides access to the illumination path where the illumination structure is created. A diagram of a typical optical path, where a transmission grating is used for the creation of the illumination structure, is shown in Fig. 15.6.

The required illumination pattern can be generated by a number of approaches. Spatial light modulators are an attractive, active method of generating any desired pattern [42], but they can introduce some artefacts, e.g., if the unit is not particularly optically flat, and they can be expensive especially for high quality devices. A simpler approach is to use a transmissive grating that diffracts the beam into a number of orders. The ±1 diffracted orders are selected with or without the zero-order beam; a beam stop blocks the other orders. These orders produce the laser beams that overlap at the sample and produce an interference pattern that is a set of parallel lines with a sinusoidal intensity distribution in the excitation of the sample. The important factor is the contrast between the light and dark regions in the pattern. The greatest resolution enhancement is obtained with the highest frequency pattern that is still resolvable by the imaging system, which is usually limited by the NA of the objective. The grating period (i.e., the number of lines on the sample) can be varied by changing the angle, θ, between the interfering laser beams incident on the sample:

$$\Lambda = \frac{\lambda}{2\sin\theta} \tag{15.2}$$

where Λ is the period of the pattern and λ is the wavelength of the excitation light.

15.5 Mathematical Background and Image Reconstruction

The ability of SIM to achieve super-resolution is totally dependent on the computational reconstruction of the final image from the series of images acquired. The reconstruction is performed largely in Fourier space. As with most computational imaging techniques unwanted artefacts can be introduced, which need to be identified and can be minimized with inclusion of appropriate variables in the algorithms.

15.5.1 Image Formation

In a microscope, the measured image $\mathrm{Im}(\vec{r})$ can be described as the multiplication of the local excitation intensity $I_{\mathrm{ex}}(\vec{r})$ by the local fluorophore concentration $\rho(\vec{r})$ convolved with the point-spread function $\mathrm{PSF}(\vec{r})$:

$$\mathrm{Im}\,(\vec{r}) = [I_{\mathrm{ex}}\,(\vec{r})\,\rho(\vec{r})] \otimes \mathrm{PSF}(\vec{r}) \tag{15.3}$$

Using the convolution theorem, in Fourier space this translates to the convolution of the excitation intensity with the fluorophore concentration multiplied by the $\mathrm{OTF}(\vec{k})$:

$$\widetilde{\mathrm{Im}}(\vec{k}) = \left[\widetilde{I_{\mathrm{ex}}}(\vec{k}) \otimes \tilde{\rho}(\vec{k})\right] \mathrm{OTF}(\vec{k}) \qquad (15.4)$$

In case of uniform illumination, the excitation can be described as $I_{\mathrm{ex}}(\vec{r}) = \text{constant} \equiv I_0$, which in Fourier space translates to $\widetilde{I_{\mathrm{ex}}}(\vec{k}) = I_0\delta(\vec{k})$ where $\delta(\vec{k})$ is Dirac's delta distribution. In Fourier space, the resulting image can be described as

$$\widetilde{\mathrm{Im}}(\vec{k}) = I_0\tilde{\rho}(\vec{k})\mathrm{OTF}(\vec{k}) \qquad (15.5)$$

where $\mathrm{OTF}(\vec{k}) \equiv \widetilde{\mathrm{PSF}}(\vec{k})$ and has finite support with maximum transferable frequency k_0 which is limited by the wavelength of the light and the numerical aperture of the objective. Frequencies outside the support of the OTF are lost and cannot be recovered.

If instead the illumination pattern is a one-dimensional harmonic pattern with spatial frequency $\vec{\kappa}$ and initial phase ϕ_0 so that $I_{\mathrm{ex}}(\vec{r}) = I_0(1 + \cos(\vec{\kappa} \cdot \vec{r} + \phi_0))$, or, in Fourier space, $\widetilde{I_{\mathrm{ex}}}(\vec{k}) = I_0[\delta(\vec{k}) + \frac{1}{2}[e^{i\phi_0}\delta(\vec{k} + \vec{\kappa}) + e^{-i\phi_0}\delta(\vec{k} - \vec{\kappa})]]$, the resulting image as defined by Eq. 15.3 is transformed to

$$\widetilde{\mathrm{Im}}(\vec{k}) = I_0 \left[\tilde{\rho}(\vec{k}) + \frac{1}{2}\left[e^{i\phi_0}\tilde{\rho}(\vec{k} + \vec{\kappa}) + e^{-i\phi_0}\tilde{\rho}(\vec{k} - \vec{\kappa})\right]\right] \mathrm{OTF}(\vec{k}) \qquad (15.6)$$

i.e. in addition to the first zero-order term, the pattern creates two extra peaks that have the same information about the object attached to them as the zero-order peak. After multiplication with the OTF, there is now information inside the support of the OTF that otherwise would have been lost (Fig. 15.3b). If the information attached to each peak can be unmixed and shifted to coincide with the middle peak, the maximum transferable frequency is now extended to $k_0 + \kappa$ (Fig. 15.3c) [27]. However, when illuminating through the objective, the first-order peaks have to lie within the support of the OTF, which means that the maximum resolution improvement is limited to about a factor of 2[a] [27, 28].

[a]Because the excitation wavelength is shorter than the emission wavelength, it is possible to have the first-order peaks outside the emission OTF, resulting in resolution improvement of slightly over a factor of 2.

15.5.2 Image Reconstruction

The reconstruction of the data is performed mainly in Fourier space. The number of required images depends on the number of components to be separated. In general, to separate N harmonics (i.e., $2N + 1$ peaks), $M \geq 2N + 1$ images $\mathrm{Im_m}$ with different phases $\phi_m = 2\pi(m-1)/M$ images are needed.

A Fourier transform of an image vector containing the different phases of the illumination pattern has components

$$\widetilde{\mathrm{Im_m}}(\vec{k}) = \sum_{n=1}^{N} I_n \frac{1}{2}\left[e^{in\phi_m}\tilde{\rho}(\vec{k}+n\vec{\kappa}) + e^{-in\phi_m}\tilde{\rho}\left(\vec{k}-\vec{n\kappa}\right)\right]\mathrm{OTF}(\vec{k}) \tag{15.7}$$

where I_n are the strengths of the diffracted orders used for illuminating the sample. In matrix notation, this can be written as

$$\overrightarrow{\widetilde{\mathrm{Im}}}(\vec{k}) = A\vec{\tilde{o}}(\vec{k}) \tag{15.8}$$

where the vector $\vec{\tilde{o}}$ contains the contributions of the different harmonics:

$$\tilde{o}_n = \begin{cases} \frac{1}{2}I_{|N-n+1|}\tilde{\rho}\left(\vec{k}+(n-1-N)\vec{\kappa}\right)\mathrm{OTF}(\vec{k}), & n \neq N+1 \\ I_0\tilde{\rho}(\vec{k})\mathrm{OTF}(\vec{k}), & n = N+1 \end{cases} \tag{15.9}$$

and A is the mixing matrix:

$$A_{\mathrm{mn}} = \exp\left[\frac{2\pi i\,(n-1-N)\,(m-1)}{M}\right] \tag{15.10}$$

This is a linear set of equations, which can be solved for the $N+1$ different components if the M phases of the illumination pattern are known and $M \geq 2N+1$. The different components can be unmixed by inverting Eq. 15.8:

$$\vec{\tilde{o}}(\vec{k}) = A^{-1}\overrightarrow{\widetilde{\mathrm{Im}}}(\vec{k}) \tag{15.11}$$

If there are more images than separated components (i.e., $M > 2N+1$), the matrix is not square. In this case the pseudo-inverse approach can be used to extract information from the mixing matrix, M [43].

Once the components are separated, they are shifted to coincide with the zero-order peak using cross-correlation to find the amount of shift and combined with the other orders using weighted averaging. The final image is obtained by inverse Fourier transformation. This process is illustrated in Fig. 15.7.

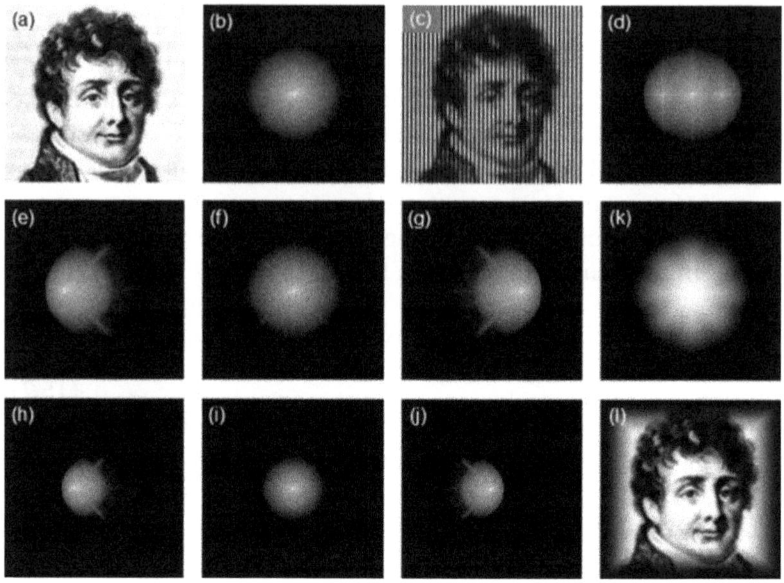

Figure 15.7 Schematic diagram of the image reconstruction process for structured illumination microscopy. A sample image (a) and its Fourier transform that has been multiplied by the OTF (b). The same image multiplied with a grid pattern (c), whose Fourier transform (d) now contains a delta function on both sides of the center created by the grid, convolved by the Fourier transform of the original image. The peaks and the information attached to them are separated (e–g). The peaks are then moved to the center, and the pixel size is decreased to accommodate increased resolution (h–j). All peaks from every grating orientation are combined with frequency-dependent weights to yield a goal function (k), and the final image is obtained by inverse Fourier transformation (l).

15.5.3 Artefacts

Successful reconstruction of a SIM image depends on the precise knowledge of the illumination structure and the imaging conditions. Although the image reconstruction in perfect systems is quite straightforward, no experimental system is perfect, and the reconstruction of experimental data requires several parameters to be optimized. The most typical artefact that can be seen in a reconstructed SIM image is a residual grating pattern [44], as seen

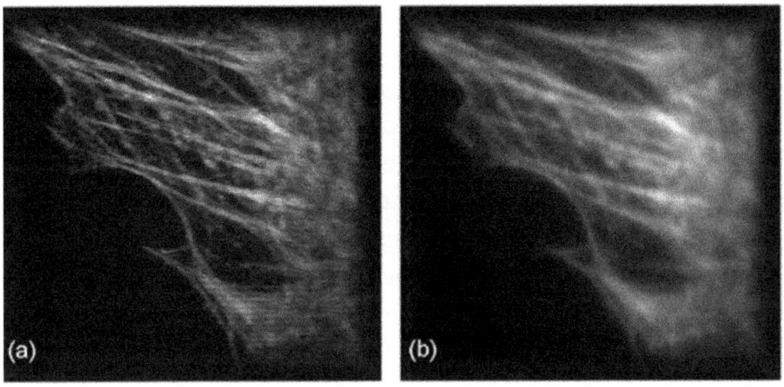

Figure 15.8 Residual grating pattern in a reconstructed SIM image (left) of actin fibers in a COS cell, stained with Alexa488 phalloidin. The horizontal residual grating pattern is clearly visible in the bottom right quarter. Wide-field image is shown on right. Image size: 30 μm.

in Fig. 15.8. These kinds of artefacts can be caused by many factors. Some of the most typical problems include the following [45]:

- Sample drift: Either between individual phases, different grating directions, or, with 3D imaging, between z-slices. This should be corrected for before the image reconstruction.
- Unknown grating period: Even though the grating period in the sample can be calculated from the optical parameters, it is never precise. The actual grating period in the sample has to be found by cross-correlation in Fourier space.
- Intensity fluctuation in the illumination and bleaching of the illumination pattern into the sample.
- Fluctuations in the phase of the grating, and unknown zero phase.
- Unknown grating order strengths, and fluctuations of the order strengths between images.

These problems can sometimes be corrected for in the reconstruction software [44], but it is essential to know good starting parameters. It is far preferable to remove these artefacts at the experimental source as much as possible wherever feasible.

Photo-toxicity, particularly relating to biological samples, can impose severe limitations on the application of any optical microscopy technique. SIM affords some advantages in this respect over certain other super-resolution techniques, being a widefield (nonpoint scanned) method. However, the excitation light flux should always be minimized as far as possible, often to the level that makes low signal-to-noise ratio in the images problematic. Modern electron multiplying charge-coupled device (EMCCD) or scientific complementary metal–oxide–semiconductor (sCMOS) cameras used in SIM apparatus exhibit remarkably low noise levels, but noise is still a major issue. Some success at overcoming noise issues has been demonstrated using denoising algorithms [46–48], prior to the

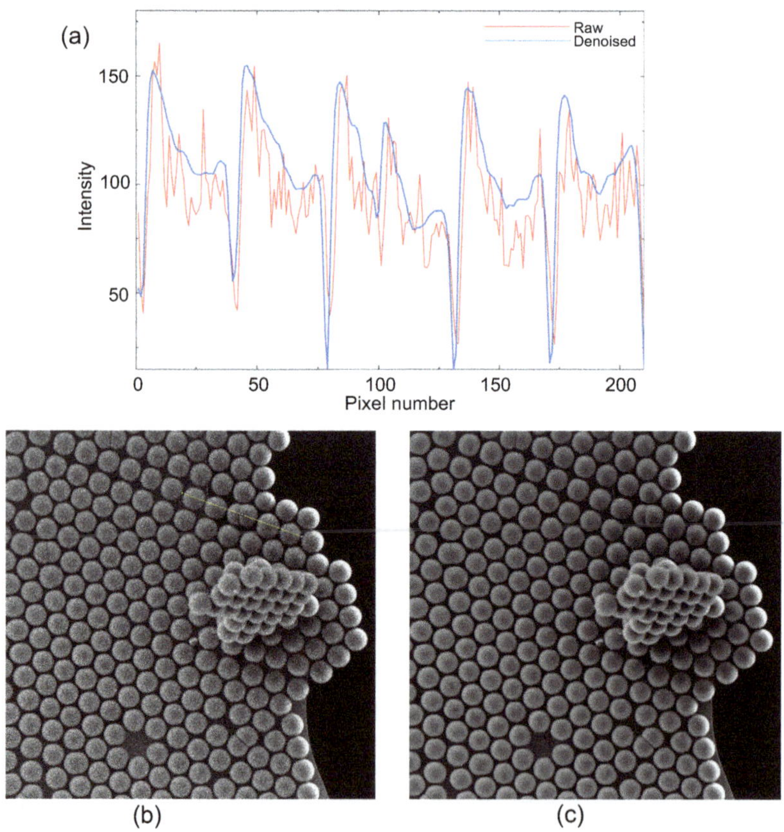

Figure 15.9 SEM image of polystyrene microspheres prior to (a) and after (b) use of the ND-SAFIR [46] denoising algorithm. (c) Profile across several spheres showing the enhancement of detail after denoising.

image reconstruction process. The effect of the ND-SAFIR denoising algorithm is illustrated in Fig. 15.9, for SEM an image of polystyrene microspheres. The ability to resolve two closely lying beads (middle of yellow profile line; Fig. 15.9b) is enhanced after denoising (Fig. 15.9b), as illustrated in the topology resulting from the profile line (Fig. 15.9c).

15.6 Applications

Although SIM can be used with any sample and fluorophore, including autofluorescence, it has been mostly adopted for applications in cell biology (Figs. 15.10–15.12). Due to the low illumination intensity required and the relatively fast data collection achievable, SIM can be used to image living cells in certain circumstances [38, 49].

In addition to the obvious applications in cell biology, super-resolution optical microscopy has largely untapped potential for application in other fields including the botanical and materials sciences [50, 51]. For example, the need for improved spatial resolution through optical techniques is also prevalent in aspects of nanotechnology [52] and SIM has the potential to become a widely used technique in this and related areas. In a botanical context, the structured illumination concept has been used for imaging pollen grains [37, 53], but it has far greater potential for

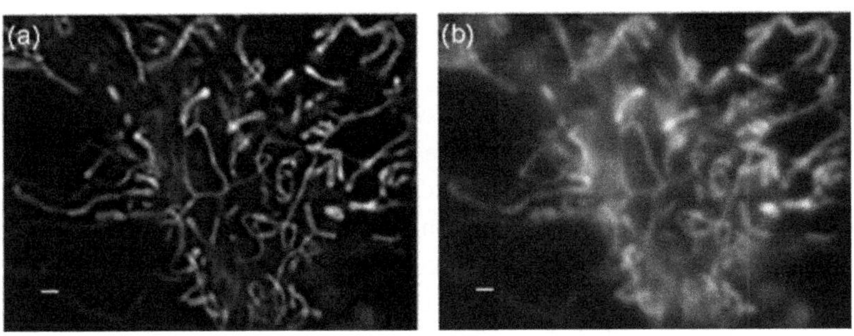

Figure 15.10 SIM (a) and widefield (b) images of mitochondria in a living COS cell, stained with MitoTracker. Scale bars: 1 μm.

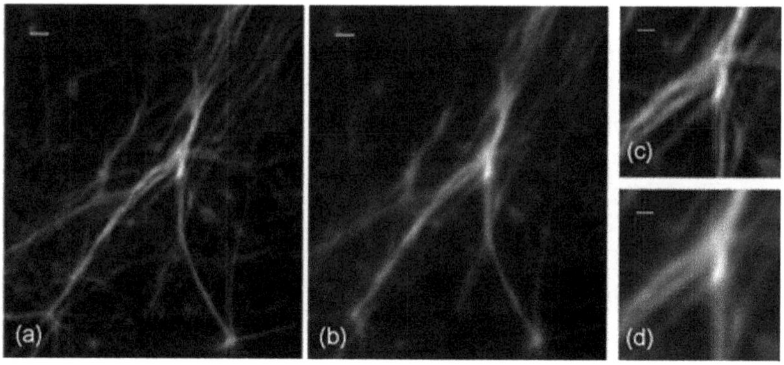

Figure 15.11 SIM (a, c) and widefield (b, d) images of actin fibers in a fixed COS cell, stained with Alexa488 phalloidin. Scale bars: 1 μm (a, b), 500 nm (c, d).

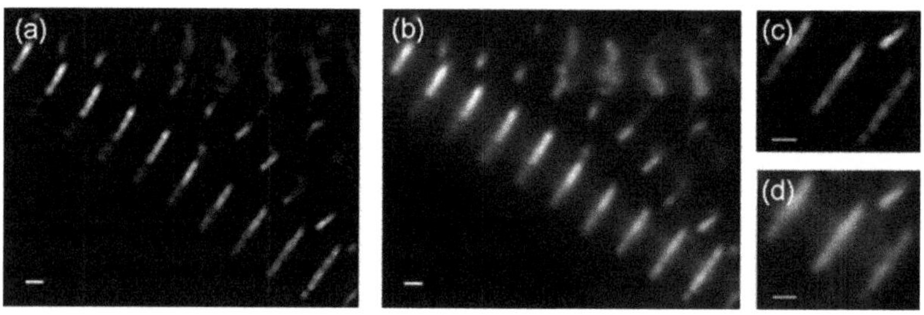

Figure 15.12 SIM (a, c) and widefield (c, d) images of isolated myofibrils from rat skeletal muscle stained by immunofluorescence. The doublets, measured to have an average separation of 177 ± 26 nm, are clearly resolved in the SIM images.

studying botanical samples. For example, we have applied SIM to the study of moss and moss spore samples (Figs. 15.13 and 15.14). The requirements imposed in such applications can be somewhat different to the cell biological context. Perhaps the sample cannot be readily labelled with an exogenous fluorescent probe, or the sample itself is inherently fluorescent and therefore does not require additional probes. This therefore requires a technique that can use any fluorophore, and SIM is ideally suited in this regard.

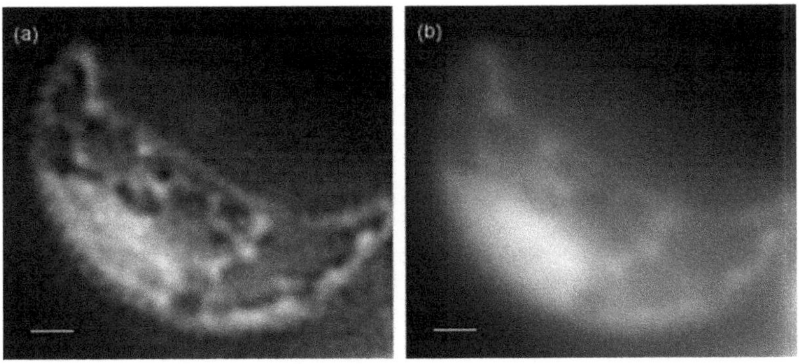

Figure 15.13 Comparison of (a) SIM and (b) widefield images of *P. commune* moss spore sample showing resolution enhancement through the SIM method. Scale bars: 1 μm. Figure prepared with the late Vassilios Sarafis.

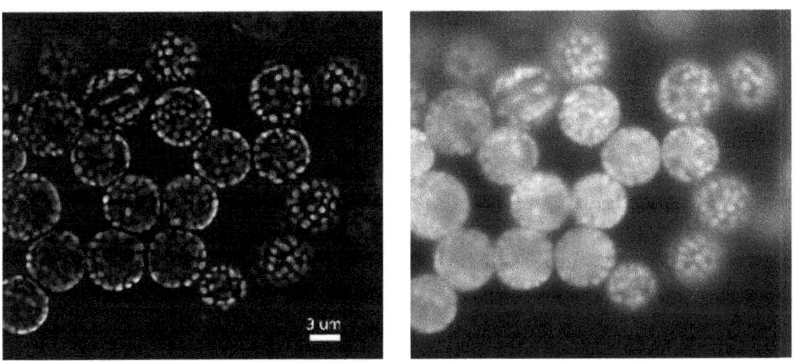

(a) (b)

Figure 15.14 Comparison of (a) SIM and (b) wide-field images of *Cyathophorum bulbosum* moss sample showing resolution enhancement through the SIM method. Figure prepared with the late Vassilios Sarafis.

One application of SIM in the materials science regime is in the study of the morphology of conjugated (light-emitting) polymers, and blends of these polymers with electron-accepting materials [51]. These polymers are becoming widely associated with thin-display technology and photovoltaic devices. SIM is an ideally suited super-resolution optical imaging method for the study of these films

(a) (b)

Figure 15.15 SIM images of conjugated polymer (P3HT)/electron acceptor (PCBM) annealed bulk heterojunction films. (a) Bright signals correspond to regions of high emission intensity from PCBM-depleted regions. (b) False relief map representation of a SIM image.

for a number of reasons: the inherent fluorescence (in the absence of electron–hole separation) requires methods that do not require special properties of the fluorophore; a widefield method requiring minimal illumination flux is desired to minimize photodegradation, nonlinear optical phenomena, or multi-exciton interactions; spatial resolution is required on the scale approaching the morphological features common within such films and the exciton diffusion length. Figure 15.15 shows SIM images of an annealed bulk heterojunction film cast from a blended solution of the polymer, poly (3-hexylthiophene) (P3HT), and an electron-accepting material, [6, 6]-phenyl C_{61}-butyric acid methyl ester (PCBM). The P3HT is weakly fluorescent in regions that are depleted of the PCBM (Fig. 15.15a), which is itself essentially nonfluorescent (rod- and dotlike regions). Enhanced emission is observed from some highly localized areas around the PCBM-rich regions, particularly at the tips of the rods. Figure 15.15b shows a false relief map representation of other regions within a similar film illustrating both the PCBM-rich crystalline regions and the large circular PCBM-depleted regions surrounding the PCBM crystalline areas.

15.7 Extensions of SIM

While commercial SIM systems are becoming widely available [30–32], new approaches have been developed that can overcome some of the limitations of SIM, and work in this direction continues. Polarization-resolved SIM can reduce the data acquisition time, while saturated SIM offers theoretically unlimited resolution. SIM can also be combined with other fluorescence techniques to achieve further resolution improvement in the axial direction, or to gain additional information about the sample.

15.7.1 Three-Dimensional SIM (3D SIM)

In this chapter, we have discussed the lateral resolution improvement achieved with SIM by creating a 2D illumination structure at the objective focal plane. The same concept can be expanded to provide enhanced axial resolution by forming a harmonic interference pattern that varies both laterally and axially and collecting SIM images from multiple focal planes [33]. This illumination structure can be created by interfering three diffracted orders at the sample (e.g., the 0th order in addition to the $\pm$1st orders). The image reconstruction is done in 3D space and provides doubling of the spatial resolution in all three dimensions, corresponding to an 8-fold decrease in 3D volume compared to conventional widefield microscopy.

15.7.2 Nonlinear Structured Illumination

Axial resolution improvement by "conventional" structured illumination is limited to a factor of 2 over the diffraction limit. Further resolution improvement can be achieved with SIM if the linear relationship between the excitation and emitted intensity is broken [40]. If the emission pattern is no longer sinusoidal but is replaced by a pattern of the same main frequency containing higher order harmonic terms, more peaks are created in Fourier space, all having identical information attached to them as the zero order peak [40, 41] (Fig. 15.16). In this case Eq. 15.6 can be generalized to include N

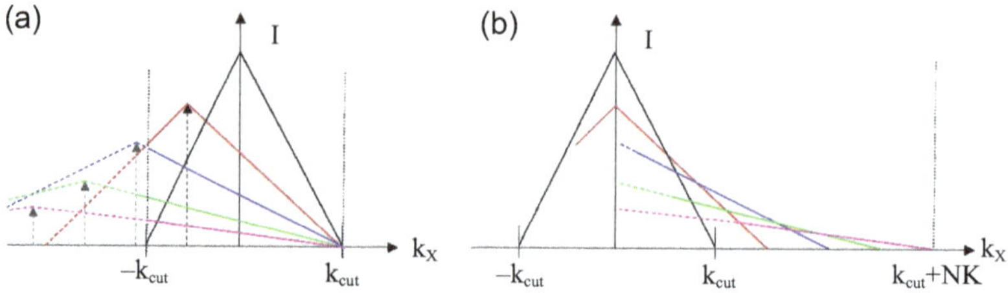

Figure 15.16 Nonlinear structured illumination microscopy is based on the creation of higher harmonic frequencies in Fourier space (a). These higher harmonics are then unmixed and moved to coincide the middle peak, yielding to theoretically unlimited resolution (b).

harmonics:

$$\widetilde{\mathrm{Im}}(\vec{k}) = I_0\left[\tilde{\rho}(\vec{k}) + \sum_{n=1}^{N}\frac{1}{2}\left[e^{in\phi_0}\tilde{\rho}(\vec{k}+n\vec{\kappa}) + e^{-in\phi_0}\tilde{\rho}(\vec{k}-\overrightarrow{n\kappa})\right]\right]\mathrm{OTF}(\vec{k}) \qquad (15.12)$$

Separating information from components lying outside the support of the OTF yields theoretically unlimited resolution. In practice, only a limited number of peaks are strong enough to rise above the noise level, and thus the resolution is limited by the signal-to-noise ratio, or other physical limitations.

There are several ways to create the required nonlinearity, many of which are based on saturating the transitions directly involved in the fluorescence emission process and require very high illumination intensity [40, 41, 54, 55]. 50 nm resolution has been demonstrated with fluorescent beads by saturating the excited state of the fluorophores [41], but the high intensity required to achieve this is not compatible with biological samples. With photoswitchable fluorophores the nonlinearity can be achieved at low illumination intensity by saturating the transitions between the bright and the dark state [56]. However, the higher the desired resolution, the larger the number of images required for reconstruction, e.g., ~50 images for ~50 nm resolution. This technique is currently limited by the photostability of the fluorophores, but has been demonstrated [57].

15.7.3 Rapid Data Acquisition

One of the advantages of SIM over other techniques is its relative speed of data acquisition; however, this is still a severe limitation in many applications, particularly live-cell imaging. A technique called polarization-illumination-coded structured illumination (picoSIM), which combines high data acquisition rate and high spatial resolution has been introduced by Heintzmann [58]. This technique encodes the three individual light patterns needed for conventional structured illumination microscopy in a polarization-coded light distribution. This allows the simultaneous acquisition of the three images required for the computational reconstruction of a sectioned image in one single exposure, allowing imaging with, in principle, very high data acquisition rate. The authors suggest that optical demodulation of the detected light would allow real-time observation of optically sectioned images. This technique has recently been realized experimentally [59], and shows great promise for some specific applications.

Another approach to achieving higher data acquisition in optical sectioning configuration is termed "continuously scanned" SIM (CSSIM) [60]. In conventional optical sectioning SIM the sinusoidal pattern is phase-shifted by 0, $2\pi/3$ and $4\pi/3$ and three images, are acquired at each axial position. An image is calculated, using Eq. 15.1, at every axial position, z, and then optically sectioned information determined at each pixel as a function of z. In contrast, in CSSIM, axial scanning is synchronized with the sinusoidal pattern shifting and only a single image is obtained at each axial position as opposed to three images in typical SIM.

15.7.4 Other SIM Approaches

One advantage of SIM is that is can be coupled with other modes of microscopy. Resolution below 100 nm in the axial direction has been achieved by total internal reflection fluorescence (TIRF) microscopy that creates an evanescent wave close to the coverslip to excite the sample, and 4π [11, 12] and I^5M [13, 14] techniques that use two opposing objective lenses to increase the angle from where the light is collected. While these techniques do not offer any lateral resolution improvement, TIRF and I^5M have been combined with SIM to achieve ~100 nm resolution in three dimensions [15, 38, 61, 62].

There are several advantages of resolving fluorescence intensity in time, both in bulk samples and at the microscopic regime, and time-resolved fluorescence imaging using both widefield and scanning confocal or multiphoton modes is now a mature technique [8, 63]. SIM has been combined with time-resolved detection by using a gated optical intensifier (GOI) before the CCD camera [64]. GOI devices can be operated with ~100 picosecond gate periods and synchronized to the tens of megahertz pulse repetition rates of conventional mode-locked lasers. A series of images can be collected, corresponding to various time delays after excitation and with a predefined gate width, in a "box-car" mode to provide temporal and spatial information.

While SIM has, to date, been largely limited to fluorescence microscopy, it has recently been shown that SIM can be extended to phase microscopies when coupled with spatial light interference microscopy (SLIM) [23]. The image formation process in the case of fluorescence microscopy is intensity-based and linear. Direct application of the structured illumination in the phase microscopy will result in artefacts and low-contrast phase. These authors propose a method to modify the SLIM system to accommodate the structured illumination and to modify the demodulation procedure of the structured illumination microscopy to accommodate the SLIM. The reconstructed phase images are high contrast with few artefacts. The SIM approach has also been applied in widefield coherent anti-Stokes Raman scattering (CARS) microscopy to provide sub-diffraction limited resolution using endogenous contrast generation [65].

"Correlative imaging"—the use of complementary imaging techniques with comparable spatial resolution in combination on the same sample—is becoming widely recognized as being highly beneficial in many cases to provide a greater level of detail than by any single technique alone. Conventional optical microscopy methods have been coupled with cryo- and scanning electron microscopy [66, 67] but other optical methods with higher spatial resolution are sought that can be coupled with other modes of microscopy, such as X-ray [68], scanning probe (AFM) [69]. SIM could contribute greatly to these developments.

15.8 Conclusion

Structured illumination microscopy is one of several young, but rapidly maturing, techniques of super-resolution fluorescence imaging. Each method has its advantages and disadvantages for a given imaging scenario. For many samples SIM affords a number of advantages over other methods, fuelling its rapidly growing uptake, particularly in the biological samples. Chiefly, the ability to image any fluorophore, in a low photon dosage, widefield, and reasonably fast mode, contrast it with some other techniques. While its obvious applicability to the biological sciences is already being exploited, there is also great potential for use in other areas, including the botanical and materials sciences.

Although the spatial resolution enhancement of SIM may appear a little modest when compared with the touted resolution of other techniques, it has been shown already to be capable of providing detail that has eluded previous studies, and there is the proven potential for this resolution to be increased further. Further developments in this technique are ongoing that will speed up the data acquisition rate, improve the image reconstruction process, and allow simultaneous coupling with other modes of microscopy.

References

1. Abbe, E. (1873). Beitrage zur theorie des mikroskops und der mikroskopischen wahrmehmung. *Arch. Mikrosk. Anatomie*, **9**, 413–480.
2. Rayleigh, L. (1896). On the theory of optical images, with special reference to the microscope. *Philos. Mag.*, **42**, 167–195.
3. Everett, J. D. (1904). Microscopic resolution: note on a point in lord rayleigh's paper of 1896. *J. R. Microsc. Soc.*, **24**, 26–27.
4. van der Oord, C. J. R. (1994). Confocal imaging and microvolume spectroscopy using synchrotron radiation: confocale microscopie en

microvolume spectroscopie met behulp van synchrotronstraling, PhD thesis. Universiteit Utrecht, Utrecht.

5. Li, Q., and Seeger, S. (2006). Label-free detection of single protein molecules using deep UV fluorescence lifetime microscopy. *Anal. Chem.*, **78**, 2732–2737.
6. Li, Q., Ruckstuhl, T., and Seeger, S. (2004). Deep-UV laser-based fluorescence lifetime imaging microscopy of single molecules. *J. Phys. Chem. B*, **108**, 8324–8329.
7. Schüttpelz, M., Müller, C., Neuweiler, H., and Sauer, M. (2006). UV fluorescence lifetime imaging microscopy: a label-free method for detection and quantification of protein interactions. *Anal. Chem.*, **78**, 663–669.
8. Smith, T. A., Lincoln, C. N., and Bird, D. K. (2009). Time-resolved fluorescence in microscopy. In: Goldys, E. M. (ed.), *Fluorescence Applications in Biotechnology and Life Sciences*. Wiley-Blackwell, New Jersey, pp. 195–221.
9. Hell, S. W. (2007). Far-field optical nanoscopy. *Science*, **316**, 1153–1158.
10. Cremer, C., and Cremer, T. (1978). Considerations on a laser-scanning-microscope with high resolution and depth of field. *Microsc. Acta*, **81**, 31–44.
11. Hell, S., and Stelzer, E. H. K. (1992). Properties of a 4π-confocal fluorescence microscope. *J. Opt. Soc. Am. A*, **9**, 2159–2166.
12. Hell, S. W., Stelzer, E. H. K., Lindek, S., and Cremer, C. (1994). Confocal microscopy with an increased detection aperture: type-b 4pi confocal microscopy. *Opt. Lett.*, **19**, 222–224.
13. Gustafsson, M. G. L., Agard, D. A., and Sedat, J. W. (1995). Sevenfold improvement of axial resolution in 3D widefield microscopy using two objective lenses. *Proc. SPIE*, **2412**, 147–156.
14. Gustafsson, M. G. L., Agard, D. A., and Sedat, J. W. (1999). I^5m: 3D widefield light microscopy with better than 100 nm axial resolution. *J. Microsc.*, **195**, 10–16.
15. Beck, M., Aschwanden, M., and Stemmer, A. (2008). Sub-100-nanometre resolution in total internal reflection fluorescence microscopy. *J. Microsc.*, **232**, 99–105.
16. Chang, B.-J., Chou, L.-J., Chang, Y.-C., and Chiang, S.-Y. (2009). Isotropic image in structured illumination microscopy patterned with a spatial light modulator. *Opt. Express*, **17**, 14710–14721.

17. Chung, E., Kim, D., Cui, Y., Kim, Y.-H., and So, P. T. C. (2007). Two-dimensional standing wave total internal reflection fluorescence microscopy: Superresolution imaging of single molecular and biological specimens. *Biophys. J.*, **93**, 1747–1757.
18. Cragg, G. E., and So, P. T. C. (2000). Lateral resolution enhancement with standing evanescent waves. *Opt. Lett.*, **25**, 46–48.
19. Frohn, J. T., Knapp, H. F., and Stemmer, A. (2000). True optical resolution beyond the rayleigh limit achieved by standing wave illumination. *Proc. Natl. Acad. Sci. U.S.A.*, **97**, 7232–7236.
20. NT-MDT NTEGRA Spectra (2012). http://www.ntmdt.com/device/ntegra-spectra.
21. Wang, C.-C., Lee, K.-L., and Lee, C.-H. (2009). Wide-field optical nanoprofilometry using structured illumination. *Opt. Lett.*, **34**, 3538–3540.
22. Lin, J.-Y., Huang, R.-P., Tsai, P.-S., and Lee, C.-H. (2009). Wide-field super-resolution optical sectioning microscopy using a single spatial light modulator. *J. Opt. A: Pure Appl. Opt.*, **11**, 015301/015301-015306.
23. Chu, K., Smith, Z. J., Wachsmann-Hogiu, S., and Lane, S. (2012). Super-resolved spatial light interference microscopy. *J. Opt. Soc. Am. A*, **29**, 344–351.
24. Vogel, M., Yang, Z., Kessel, A., Kranitzky, C., Faber, C., and Haüsler, G. (2011). Structured-illumination microscopy on technical surfaces: 3D metrology with nanometer sensitivity. In: Lehmann, P. H., Osten, W., and Gastinger, K. (eds.), *Optical Measurement Systems for Industrial Inspection VII*; *Proc. SPIE*, **8082**, pp. 80820S80821–80826.
25. Hoffman, R., and Gross, L. (1975). Modulation contrast microscope. *Appl. Opt.*, **14**, 1169–1176.
26. Neil, M. A. A., Juskaitis, R., and Wilson, T. (1997). Method of obtaining optical sectioning by using structured light in a conventional microscope. *Opt. Lett.*, **22**, 1095–1097.
27. Heintzmann, R., and Cremer, C. (1998). Laterally modulated excitation microscopy: improvement of resolution by using a diffraction grating. In: Bigio, I. J., Schneckenburger, H., Slavik, J., Svanberg, K., and Viallet, P. M. (eds.), *Optical Biopsies and Microscopic Techniques III*; *Proc. SPIE*, **3568**, pp. 185–195.
28. Gustafsson, M. G. L. (2000). Surpassing the lateral resolution limit by a factor of two using stuctured illumination microscopy. *J. Microsc.*, **198**, 82–87.

29. Gustafson, M. G. L., Agard, D. A., and Sedat, J. W. (2000). Doubling the lateral resolution of wide-field fluorescence microscopy using structured illumination. In: Conchelo, J.-A., Cogswell, C. J., and Wilson, T. (eds.), *Three-Dimensional and Multidimensional Microscopy: Image Acquisition Procesing VII*; *Proc. SPIE*, **3919**, pp. 141–150.
30. Zeiss ELYRA (2012). http://microscopy.zeiss.com/microscopy/en_de/products/superresolution-microscopy.html - inpagetabs-0.
31. Nikon N-SIM (2012). http://www.nikoninstruments.com/Products/Microscope-Systems/Inverted-Microscopes/N-SIM-Super-Resolution.
32. Applied Precision DeltaVision OMX (2012). http://www.api.com/deltavision-omx.asp.
33. Gustafsson, M. G. L., Shao, L., Carlton, P. M., Wang, C. J. R., Golubovskaya, I. N., Cande, W. Z., Agard, D. A., and Sedat, J. W. (2008). Three-dimensional resolution doubling in wide-field fluorescence microscopy by structured illumination. *Biophys. J.*, **94**, 4957–4970.
34. Perkin Elmer Improvision (2012). Volocity grid confocal; http://www.improvision.com.
35. Leica (2012). http://www.leica-microsystems.com/products/light-microscopes/life-science-research/upright-microscopes/details/product/leica-structured-illumination-1/.
36. Zeiss Apotome (2012). http://microscopy.zeiss.com/microscopy/en_au/products/imaging-systems/apotome-2-for-biology.html.
37. Delica, S., and Blanca, C. M. (2007). Wide-field depth-sectioning fluorescence microscopy using projector-generated patterned illumination. *Appl. Opt.*, **46**, 7237–7423.
38. Kner, P., Chhun, B. B., Griffis, E., Winoto, L., and Gustafsson, M. G. L. (2009). Super-resolution video microscopy of live cells by structured illumination. *Nat. Methods*, **6**, 339–342.
39. Heintzmann, R. (2003). Saturated patterned excitation microscopy with two-dimensional excitation patterns. *Micron*, **34**, 283–291.
40. Heintzmann, R., Jovin, T. M., and Cremer, C. (2002). Saturated patterned excitation microscopy - a concept for optical resolution improvement. *J. Opt. Soc. Am. A*, **19**, 1599–1609.
41. Gustafsson, M. G. L. (2005). Nonlinear structured-illumination microscopy: wide-field fluorescence imaging with theoretically unlimited resolution. *Proc. Natl. Acad. Sci. U.S.A.*, **102**, 13081–13086.
42. Hirvonen, L. (2008). Structured illumination microscopy using photo-switchable fluorescent proteins, PhD thesis. King's College, University of London, London.

43. Heintzmann, R. (2006). Structured illumination methods. In: Pawley, J. B. (ed.), *Handbook of Biological Confocal Microscopy*, 3rd ed. Springer Science+Business Media, LLC, New York, pp. 265–279.
44. Schaefer, L. H., Schuster, D., and Schaffer, J. (2004). Structured illumination microscopy: artefact analysis and reduction utilizing a parameter optimization approach. *J. Microsc.*, **216**, 165–174.
45. Wicker, K. (2010). Increasing resolution and light efficiency in fluorescence microscopy. PhD Thesis, Kings' College London, Randal Division of Cell and Molecular Biophysics.
46. ND-SAFIR software (2012). http://serpico.rennes.inria.fr/doku.php?id=software:nd-safir:index.
47. Boulanger, J., Kervrann, C., Bouthemy, P., Elbau, P., Sibarita, J., and Salamero, J. (2010). Patch-based nonlocal functional for denoising fluorescence microscopy image sequences. *IEEE Trans. Med. Imaging*, **29**, 442–454.
48. Carlton, P. M., Boulanger, J. r. m., Kervrann, C., Sibarita, J.-B., Salamero, J., Gordon-Messer, S., Bressan, D., Haber, J. E., Haase, S., Shao, L., Winoto, L., Matsuda, A., Knera, P., Uzawa, S., Gustafsson, M., Kam, Z., Agard, D. A., and Sedat, J. W. (2010). Fast live simultaneous multiwavelength four-dimensional optical microscopy. *Proc. Natl. Acad. Sci. U.S.A.*, **107**, 16016–16022.
49. Hirvonen, L. M., Wicker, K., Mandula, O., and Heintzmann, R. (2009). Structured illumination microscopy of a living cell. *Eur. Biophys. J.*, **38**, 807–812.
50. Stöhr, R. J., Kolesov, R., Xia, K., Reuter, R., Meijer, J., Logvenov, G., and Wrachtrup, J. (2012). Super-resolution fluorescence quenching microscopy of graphene. *ACS Nano*, **6**, 9175–9181.
51. Hao, X.-T., Hirvonen, L. M., and Smith, T. A. (2013). Nanomorphology of polythiophene-fullerene bulk-heterojunction films investigated by structured illumination optical imaging and time-resolved confocal microscopy. *Method. Appl. Fluoresc.*, **1**(1), 015004.
52. Lu, P. J. (2005). Confocal scanning optical microscopy and nanotechnology. In: Yao, N., and Wang, Z. L. (eds.), *Handbook of Microscopy for Nanotechnology*. Springer, pp. 3–24.
53. Wilson, T. (1995). The role of the pinhole in confocal imaging system. In: Pawley, J. (ed.), *Handbook of Biological Confocal Microscopy*, 2nd ed. Plenum Press, New York, pp. 167–182.

54. Hell, S. W., and Wichmann, J. (1994). Breaking the diffraction resolution limit by stimulated emission: stimulated emission depletion microscopy. *Opt. Lett.*, **19**, 780–782.

55. Hell, S. W., and Kroug, M. (1995). Ground-state depletion fluorescence microscopy, a concept for breaking the diffraction resolution limit. *Appl. Phys. B*, **60**, 495–497.

56. Hirvonen, L., Mandula, O., Wicker, K., and Heintzmann, R. (2008). Structured illumination microscopy using photoswitchable fluorescent proteins. *Proc. SPIE*, **6861**, 68610L.

57. Rego, E. H., Shaob, L., Macklin, J. J., Winoto, L., Johansson, G. A., Kamps-Hughes, N., Davidson, M. W., and Gustafsson, M. G. L. (2012). Nonlinear structured-illumination microscopy with a photoswitchable protein reveals cellular structures at 50-nm resolution. *Proc. Natl. Acad. Sci. U.S.A.*, **109**, E135–E143.

58. Wicker, K., and Heintzmann, R. (2010). Single-shot optical sectioning using polarization-coded structured illumination. *J. Opt.*, **12**, 084010.

59. Heintzmann, R., and Wicker, K. (2012). Picosim; https://sites.google.com/site/nanoimagingproject/projects/picosim.

60. Joo, K.-N. (2012). Fourier domain analysis on continuously scanned structured illumination microscopy. *Meas. Sci. Technol.*, **23**, 057002/057001-057005.

61. Fiolka, R., Beck, M., and Stemmer, A. (2008). Structured illumination in total internal reflection fluorescence microscopy using a spatial light modulator. *Opt. Lett.*, **33**, 1629–1631.

62. Shao, L., Isaac, B., Uzawa, S., Agard, D. A., Sedat, J. W., and Gustafsson, M. G. L. (2008). I^5s: wide-field light microscopy with 100-nm-scale resolution in three dimensions. *Biophys. J.*, **94**, 4971–4983.

63. Periasamy, A., and Clegg, R. M. (2010). *Flim Microscopy in Biology and Medicine*. Chapman & Hall/CRC, Boca Raton.

64. Webb, S. E. D., Gu, Y., Léveêque-Fort, S., Siegel, J., Cole, M. J., Dowling, K., Jones, R., and French, P. M. W. (2002). A wide-field time-domain fluorescence lifetime imaging microscope with optical sectioning. *Rev. Sci. Inst.*, **73**, 1898–1907.

65. Hajek, K. M., Littleton, B., Turk, D., McIntyre, T. J., and Rubinsztein-Dunlop, H. (2010). A method for achieving super-resolved widefield cars microscopy. *Opt. Express*, **18**, 19263–19272.

66. Sartori, A., Gatz, R., Beck, F., Kossel, A., Leis, A., Baumeister, W., and Plitzko, J. M. (2005). Correlation microscopy: bridging the gap between

light- and cryo-electron microscopy. *Microsc. Microanal.*, **11(Supp 2)**, 16–17.

67. Sims, P., Albrecht, R., Pawley, J. B., Centonze, V., Deerinck, T., and Hardin, J. (2006). When light microscope resolution is not enough: Correlational light microscopy and electron microscopy. In: Pawley, J. B. (ed.), *Handbook of Biological Confocal Microscopy*, 3rd ed. Springer Science, New York, pp. 846–860.
68. McDermott, G., Le Gros, M. A., Knoechel, C. G., Uchida, M., and Larabell, C. A. (2009). Soft x-ray tomography and cryogenic light microscopy: The cool combination in cellular imaging. *Trends Cell Biol.*, **19**, 587–595.
69. Micic, M., Hu, D., Suh, Y. D., Newton, G., Romine, M., and Lu, H. P. (2004). Correlated atomic force microscopy and fluorescence lifetime imaging of live bacterial cells. *Colloids Surf., B*, **34**, 205–212.

Chapter 16

STED Nanoscopy

Christian Wurm,[a] Andreas Schoenle,[a] and Christian Eggeling[b]
[a]*Abberior Instruments GmbH, Hans-Adolf-Krebs-Weg 1, 37077 Göttingen, Germany*
[b]*MRC Weatherall Institute of Molecular Medicine, University of Oxford, John Radcliffe Hospital, Headington, Oxford OX3 9DS, UK*
christian.eggeling@rdm.ox.ac.uk, a.schoenle@abberior-instruments.com

Optical super-resolution microscopy, also termed optical nanoscopy, is currently one of the most important developments in imaging for the life sciences. Based on far-field (lens-based) microscopy, multiple techniques have arisen that now allow imaging of (living) cells with a spatial resolution more than twice below the 200 nm dictated by the diffraction of light. The hallmark of true super-resolution is that diffraction no longer imposes a hard resolution limit and the maximum reachable resolution is thus dictated by the sample and the properties of the fluorescence marker. In this chapter we will explore the concept behind the first such technique, STED microscopy, and its development in the past two decades into a commercially available, quite universally applicable nanoscopy method. In the final chapter we will also shortly highlight other existing approaches and show that while the ever- increasing number of super-resolution acronyms might lead one to believe that there is more than one way to break the diffraction barrier, in fact

Fundamentals of Fluorescence Imaging
Edited by Guy Cox

ISBN 978-981-4774-85-7 (Hardcover), 978-1-351-12940-4 (eBook)
www.jennystanford.com

all these techniques are based on the same, universal principle and come in just two families with individual strengths and weaknesses.

16.1 Introduction: The Diffraction Limit

Optical microscopy using focused light is indispensable in life science technology, since it facilitates the investigation of the interior of living cells with minimal invasion. If combined with the fluorescence read-out, this far-field optical microscopy approach offers unique specificity and sensitivity when imaging molecular organizations in cells. However, the use of lenses for focusing the light comes at the cost of a limited spatial resolution. Due to diffraction at the lenses' aperture, light cannot be focused to an infinitely small spot, for visible light the minimum is roughly 200 nm in diameter. Objects of the same kind that are closer together than 200 nm can therefore not be discerned and details such as subcellular protein distributions can be visualized only down to about this scale [1, 2]. Unfortunately, the actual scale of relevance for protein distributions, the nanoscale, remains obscured.

Techniques such as near-field [3, 4] or total-internal-reflection fluorescence (TIRF) microscopy [5] introduce the possibility to image with a spatial resolution below the diffraction-limited 200 nm, but they rely on exciting with or collecting an evanescent wave, i.e., they require the placing of the sample into the very close proximity of an optical element—more specifically into its near-field; thus they are not capable of observing inner-cellular structures. Therefore, researchers mostly continued to rely on far-field techniques for their imaging needs. Over the years, multiple approaches were developed trying to surpass the diffraction limit, including special illumination patterns and mathematical approaches [6–8]. Due to a potential lack of accurate a priori information or due to the occurrence of other limitations such as strong side lobes, none of these approaches have been proven successful. Similarly, confocal microscopy (i.e., the detection through a pinhole, Chapter 3) [9–11] or multiphoton microscopy [12] (Chapter 4) do not give a practical improvement in spatial resolution, due to noise issues or the use of long wavelengths [13]. Several approaches have successfully suggested the use of structured illumination for increasing the spatial resolution of a far-field fluorescence microscope (e.g., [14–16], Chapter 15). For example, structured illumination microscopy (SIM) by a standing wave pattern allows the far-field imaging of living cells with a two-fold improvement in the lateral and axial resolution [17–20]. Similarly, SIM has also been realized based on a software-based confocal CCD detection approach using scanning single- [21] or multi-spot [22] illumination, or it has been combined with TIRF [23]. The use of two opposing objectives in either a spot-scanning 4Pi [24–26] or a widefield I^5M [27–29] microscope realized an improvement of the axial resolution of a far-field microscope from 400–800 nm down to 70–150 nm for far-field optics. The latter approach is often of special interest to the microscope users since the axial resolution of any standard far-field light microscope is at least three times poorer than the lateral resolution in the focal plane.

However, while improving the spatial resolution significantly, these microscopes still feature a limit (e.g., only two-fold increased spatial resolution in case of SIM), and they do not break the diffraction barrier; they rather push diffraction to its limits and until not very long ago, obtaining a spatial resolution on the nanometer scale with a far-field fluorescence microscope was considered impossible [30] until stimulated emission depletion (STED) microscopy was proposed in 1994 and in the following years proved to be a viable approach to obtaining resolutions that are no longer dictated by the limited focusing power of the lenses.

16.2 STED Nanoscopy

16.2.1 The Basics

A real breakthrough for the surpassing of the diffraction barrier was the insight that not solely the modification of the propagation of light but the properties of the labels can be used to attain in principle unlimited spatial resolution in the far field [31–33]. In all true optical nanoscopy techniques, individual features that are closer together than the diffraction barrier are discerned by ensuring that they do not contribute simultaneously to the measured signal. This is usually achieved by ensuring that the molecules providing the imaging contrast are transiently in discernible states and that only molecules inside a very small sub-region are in a state that contributes to the signal measured.

Thus, for fluorescent labels such states are, for example, states of different fluorescence emission characteristics, such as a dark OFF and a bright ON state. The transition between an ON and an OFF state ensures that the measured signal only stems from molecules within a region of the sample that is much smaller than 200 nm [2, 34–38] (Chapter 14); in the extreme case it is (the region covered by) a single molecule. It is important to stress, however, that in principle any contrast method and signal read-out can be chosen, as long as whatever marks the structure reveals two discernible states that can be controlled to a certain extent through irradiation by light.

As mentioned above, STED nanoscopy was the first such nanoscopy approach, proposed in 1994 [32] and first realized in 1999 and 2000 [39]. It uses stimulated emission to reversibly silence fluorophores in pre-defined regions in the sample to facilitate separation at sub-diffraction length scales. In a typical implementation of STED nanoscopy (Fig. 16.1), a laser is added to a scanning (confocal) far-field microscope, which forces excited fluorophores to their dark ground state, i.e., inhibits fluorescence emission. The wavelength of this second laser is tuned to the red edge of the fluorophore's emission spectrum so that it stimulates the de-excitation of the fluorophore, i.e., causes the transition from the fluorescent ON state (the electronically excited state) to the dark OFF state (the ground state). By detecting only the spontaneous (and not the stimulated) emission, the signal that can be registered from a fluorophore is decreased and effectively switched off when increasing the intensity of the STED laser above a threshold at which the optical de-excitation is nearly certain. The STED beam's wavefront can then be modified by an optical element (phase plate) such that it creates an intensity distribution which features one (or even several) local zero(es), such as a doughnut-shaped intensity distribution once focused into the image plane. While the size of this intensity pattern is still ruled by diffraction, large STED laser intensities force the area in which fluorophores are not effectively confined to the ground state to sub-diffraction scales.

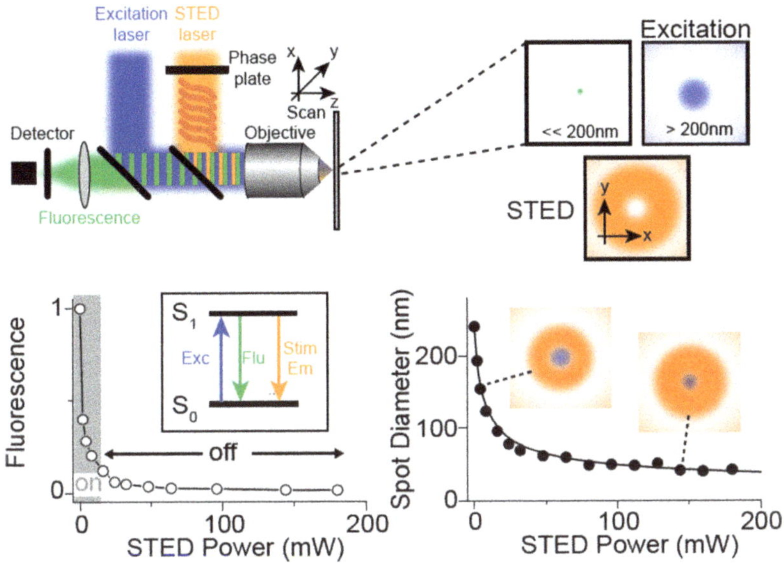

Figure 16.1 Principles of STED nanoscopy. (Upper panel) Schematic drawing of the setup of a STED nanoscope with the microscope objective which focuses the excitation laser (blue) and the STED laser (orange), and collects the fluorescence signal (green) for detection, and a three-dimensional (*x*, *y*, *z*) scanning device for moving the sample relative to the laser light. The excitation laser renders an ordinary diffraction limited focal spot, while the STED laser passes through a phase plate that realizes a focal intensity distribution with a local zero such as a doughnut-like pattern, creating an effective excitation volume much below the diffraction limit. (Lower left panel) When electrons are excited (Exc), the STED laser inhibits fluorescence (Flu) by de-exciting the excited state (S1) to the ground state (S0) via stimulated emission (Stim Em). Increasing the power of the STED laser over a certain threshold effectively precludes the occupation of the excited state S1. (Lower right panel) Sub-diffraction volumes to which the occupation of the excited state and hence fluorescence emission is restricted are created by driving the STED laser intensity up, and the volume's size decreases with increasing STED laser power. Adapted from Clausen et al. [40].

The observation region and thus the spatial resolution of the STED microscope is therefore tuned by the intensity of the STED laser. Scanning of the overlaid laser beams creates images with sub-diffraction resolution.

Concerns have first hampered the full implication of STED microscopy. For example, technical complexity has initially limited the spread of STED, due to the use of large and clumsy lasers, of complex optical arrangements (e.g., for creating the zero-intensity spot(s)), and of dedicated electronics. However, developments in laser and optical technology have surpassed these concerns [38, 40]. Also, due to the use of typical intensities of the STED laser in the range of 1–10 MW/cm^2, which are much higher than those of the excitation light, STED microscopy was and still is often regarded to be too phototoxic and thus not live-cell compatible. From theory, however, the STED light per se is not absorbed by the fluorescent label or sample (unlike the excitation laser light in two-photon microscopy, which is of even larger peak intensity) and thus, in this concept, should not produce any photo-reactive and thus -toxic species. Rather, compared to pure excitation, stimulated emission on top of excitation light should reduce photobleaching by shortening the time the fluorescent label spends in its excited (reactive) state (compare with [41]). Undesirably, the STED laser light is occasionally also absorbed by the excited states of the fluorophores, being—as for all types of fluorescence microscopy [42]—the main reason for photobleaching in STED microscopy [38, 43–46]. Yet, important insights have been gained how such photobleaching can be reduced by fast beam-scanning [43, 47–51], by a careful choice of the STED laser wavelength [45, 46] or by lowering the required STED intensities through the use of continuous-wave lasers and gated detection (gSTED) [52–54]. In addition, in recent years, a whole palette of fluorescent labels has been designed and tested for common cellular STED recordings [55–58], ranging from organic dyes for immuno- [55, 58] and related labelling approaches [59], as well as for single-molecule imaging [60] and specifically for gSTED microscopy [61], to fluorescent proteins [62, 63] and improved organic-dye based live-cell tagging methods such as Halo, Snap, or Clip and similar technologies [64–70]. All of these considerations minimized the concern of phototoxicity as well. On top, STED microscopy is nowadays available for multi-colour recordings of the relative organization of different cellular structures [40, 67, 71–82] (Fig. 16.2). As a consequence, STED nanoscopy has achieved imaging with a spatial resolution of down to 20 nm in (living) cells and down

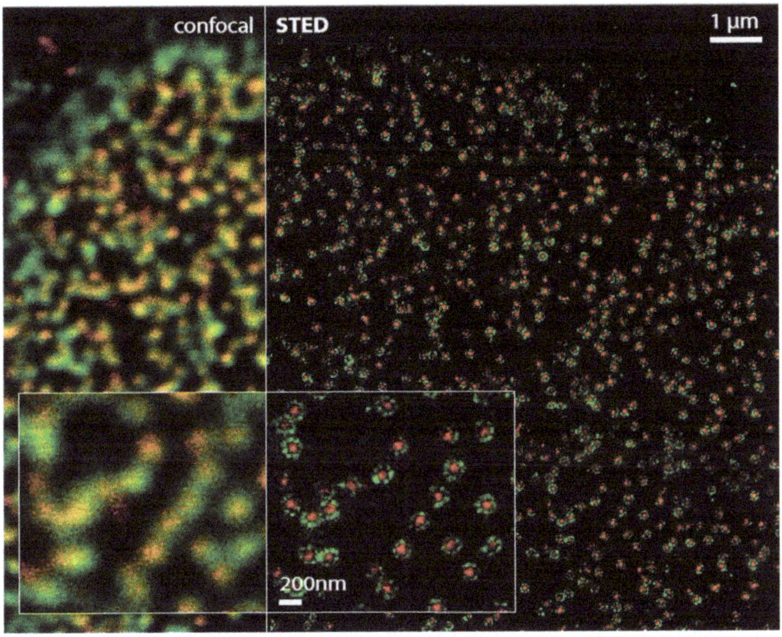

Figure 16.2 Nuclear-pore complexes imaged using conventional and two-color STED microscopy. Shown are a confocal (left) and a STED measurement (right) of nuclear pore complex subunits immuno-labelled with antibodies against gp210 (green) and a set of proteins in the central pore channel (red). The STED images reveal an octagonal symmetry of antibody-labelled gp210 homodimers within individual NPCs. Please note that only raw data is shown. As with all imaging methods, additional deconvolution or de-noising would further improve the perceived signal-to-noise ratio and resolution of the images albeit at the cost of possible artifacts.

to about 70 nm in vivo (upper brain layers of living mice), <5 nm in solid materials (diamond crystals), and of fast live-cell dynamics, even using commercial instrumentation [2, 37, 43, 50, 59, 67, 74, 80–87]. Nevertheless, coming along with the increased sensitivity of the nanoscopy approach, greater care should be exerted when labelling cellular samples, especially with respect to unspecific background staining [88]. In general, larger care has to be taken for accurate labelling in all optical nanoscopy techniques, since artefacts due to for example improper fixation in immune-labelling or unspecific binding may not be observed in confocal but may be visible in nanoscopy images, due to the improved spatial resolution in the latter [59, 89].

16.2.2 3D Resolution

Initial implementations of STED had focused on improving the lateral resolution because it offered a directly visible improvement over confocal images with features that remained invisible in the conventional image now being clearly visible. Nevertheless, it soon became clear that for many structures and imaging needs, improvement of both the lateral and the axial resolution was necessary. Conceptually, the path to 3D resolution for STED was clear: Instead of creating a doughnut-shaped distribution of inhibition light, the excitation spot had to be enclosed in all three directions by the STED intensity distribution. Arguably the most powerful approach to do this is to combine the "traditional" doughnut with a 4pi-like axial focus created by two opposing lens (forming a 4Pi configuration) and featuring a minimum in the focal plane. Building on initial success improving the z-resolution only in so called 4Pi-STED [90, 91], isoSTED implemented such a combination of a lateral doughnut with an axial 4pi-like focus [73, 92, 93] and isotropic resolutions of less than 30 nm have been demonstrated. However, the use of a second lens brings about changes to sample handling and a certain degree of complexity and vulnerability to stability issues. Also, in thick samples, controlling the relative phase of the opposing beams can become an issue.

Therefore, a second approach, based on the combination of a helical phase plate (producing the traditional doughnut) and a ring-like phase plate (producing a larger doughnut but also two strong central maxima on the optic axis above and below the focal plane [94]) has been put forward (Fig. 16.3a) [95, 96]. While this first implementation worked very well in the lab, it turned out to be vulnerable to relative drift of the two STED beams. The problem arises from the fact that the second beam, too, produces a doughnut shape in the focal plane and thus even a drift of <10 nm will illuminate the central zero of the doughnut with STED light and drastically reduce the SNR. Therefore, in its

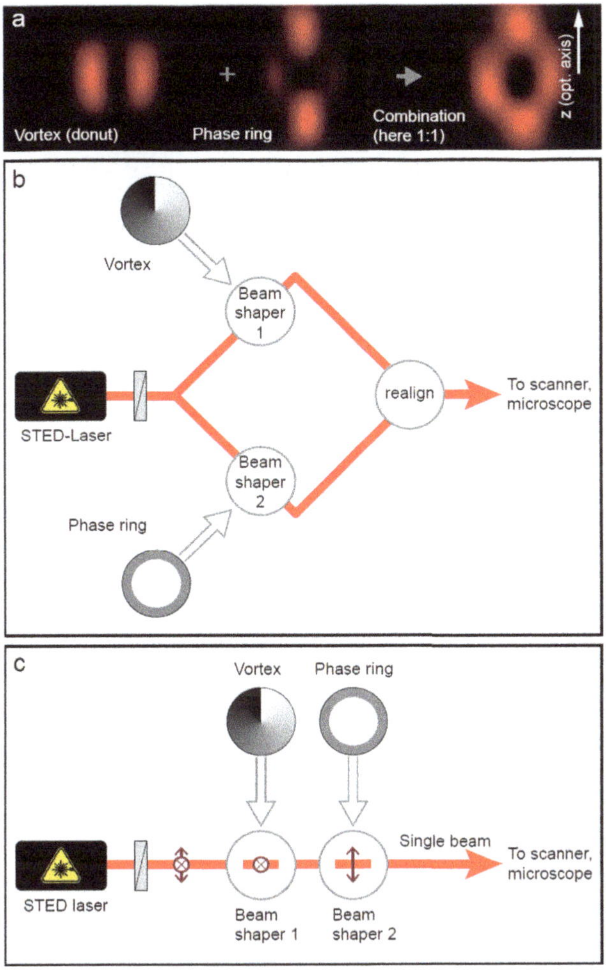

Figure 16.3 Technical realization of 3D STED microscopy. (a) instead of a single doughnut-shaped pattern restricting fluorescence to a thin needle along the optic axis, 3D STED requires the STED light to surround the focal spot in all directions. This is traditionally accomplished by combining two focal intensity distributions created by different phase patterns in the back aperture of the lens. (b) Traditionally this is achieved by imprinting the phase pattern using two beam shapers placed in different beam-paths and recombining the beams using polarizing beam splitters. However, this entails ultraprecise realignment of the beams that can lead to compromised long-term stability. (c) Using adaptive optics, the phase patterns can be imprinted on to orthogonal polarization components without splitting the beam, increasing flexibility and stability.

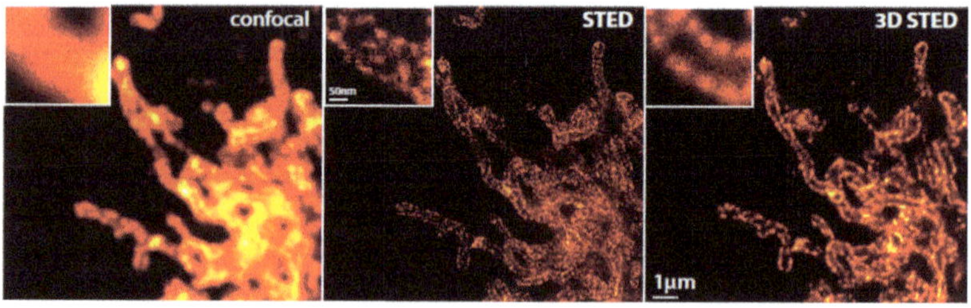

Figure 16.4 2D and 3D STED microscopy of immuno-labelled mitochondrial proteins in mammalian cells. Mitochondria within fixed mammalian cells were labelled using primary antibodies against Tom20 and secondary antibodies coupled to STAR RED. Then they were imaged using confocal microscopy, 2D STED microscopy and 3D STED microscopy. In the confocal image (left) the localization of the proteins cannot be determined within the 300 nm sized mitochondrial tubuli. Using 2D STED (middle) with resolutions down to 30 nm × 30 nm × 500 nm allows to determine whether a protein can be found in clusters or is distributed evenly within the mitochondria. However, it will not resolve whether proteins are localized at the mitochondrial surface or in its interior. This is achieved by 3D STED microscopy (right) with an isotropic resolution of <90 nm both laterally and along the *z* axis.

newest incarnation, a 3D STED microscope produces both intensity distributions from orthogonally polarized components of a single beam by modifying its wavefront with a spatial light modulator. This excludes the possibility of beam drift due to mechanical instabilities or temperature drift of mechanical components and greatly improves the long-term quality of the central intensity minimum (Fig. 16.3b).

By modifying the distribution of STED light between the two polarization components and thus the 2D (doughnut) and the 3D pattern, the shape of the PSF can be tuned between that of a 2D STED and an approximately isotropic resolution of down to 70 × 70× 70 nm. This ability allows access to information not available when the axial resolution remains that of a confocal microscope (Fig. 16.4).

16.2.3 Changing the Objective Lens

While, contrary to common belief, 2D STED with helical phase plates is relatively robust even in the presence of mild aberrations and works with most objective lenses, the 3D doughnut is much more easily compromised by spherical aberrations [97] and in addition, the wavefront of the STED beam needs to be exactly matched with the size of the back-aperture of the objective lens and its transmission characteristics. Therefore, when using fixed optical elements to do these modifications, the microscope will be limited to using a single objective lens—at least in 3D mode. This is not so much a problem when sticking to one immersion medium. After all, when using STED, one is after ultrahigh resolution and will typically want to employ the lens with the highest available numerical aperture. (While STED is not limited by diffraction, the size of the inhibition pattern and thus the resolution at a given STED intensity will still scale with the numerical aperture of the lens). However, especially when switching between fixed and live cells and different embedding media, it is important to match the immersion medium and lens to the diffractive index of the sample to avoid strong spherical aberrations when focusing more than a few microns deep into the specimen (Fig. 16.5a) [97, 98]. In fact, when imaging live cells in aqueous media, using a water immersion lens (or at least a glycerol objective with a refractive index of the immersion medium closer to that of water than of, e.g., oil) is paramount [98]. In such a setting, using adaptive optics to shape the STED wavefront proves especially useful as the pattern can be adjusted dynamically to both fit the lens in use and aberrations that might be introduced by the focusing into the sample and compromise resolution and signal strength (Fig. 16.5).

16.2.4 RESCue and New Dyes for Live Cell Imaging

As outlined above, photo-toxicity due to excessive exposure of the sample with STED light has been raised as an issue when applying the method to live cells. Additionally, as with every fluorescence method, bleaching of the dye can hamper long-term observations even in confocal microscopes. When using super-resolution this problem usually becomes more severe as the much higher resolution also commands a much denser sampling of the specimen und thus longer total exposure times and the need to collect more photons to reach acceptable signal-to-noise ratios in each image pixel.

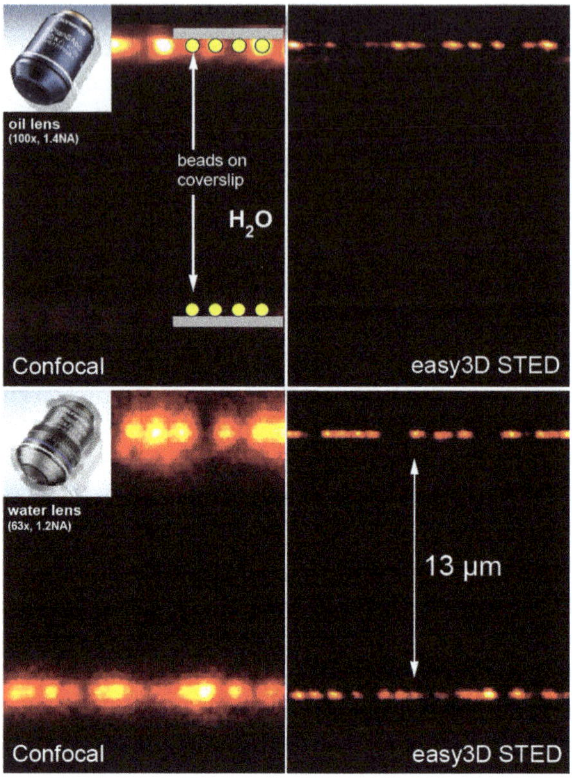

Figure 16.5 Importance of matching objective lens and immersion medium to the refractive index of the sample. When imaging aqueous samples with oil immersion lenses, the refractive index mismatch introduces strong spherical aberrations (top). Already at an imaging depth of 13 μm, a fatal loss of image intensity occurs even for confocal imaging, 3D STED is impossible under these conditions. The use of adaptive optics for shaping the STED patterns allows switching to a water immersion lens for such applications (bottom). Spherical aberrations are greatly reduced and no loss of image intensity occurs at the same imaging depth.

Bleaching is a complex spectroscopic process [41, 42, 99] (Chapter 1) and it is very difficult to give a general rule which imaging conditions (excitation wavelength, pulsed vs. CW excitation, timing, order in which different wavelengths are applied in multi-color measurements, etc.) are best. The most obvious statement is that less light of the same kind is always better. It is important to note that in this sense for STED, pulsed excitation combined with pulsed STED light of an optimal pulse length of >100 ps [100] will always outperform a CW STED laser [38]. The simple reason is that dye molecules remain in their excited state only a few ns after the excitation pulse and thus all STED light applied later (i.e., during the ~10 ns until the next excitation pulse arrives) does not contribute to the resolution gain and will potentially and unnecessarily harm the specimen or bleach dye molecules that are in a vulnerable long-lived state (triplet state or other electronic dark states). Thus, the benefit of gSTED (pulsed excitation, CW STED illumination combined with gated detection) [52, 53] is mainly that (i) STED can be applied at wavelengths where pulsed lasers with enough pulse energy are not available or too expensive and (ii) the resolution can be tuned a posteriori by adjusting a software gate [101], which is especially useful in combination with FCS [53] (see below) but may also be useful in imaging situations [101]. In applications where the required resolution is known a priori or can be determined by repeating the experiment multiple times and where a suitable pulsed laser is available, pulsed STED will typically reduce the STED light dose by more than a factor of 5 over gSTED implementations [102].

Nevertheless, further reduction of the light dose is desirable and might even be paramount especially in live cell applications. RESCue STED has been developed to address this issues without compromising the resolution. The approach is related to a technique first proposed for confocal microscopy, termed CLEM (controlled exposure microscopy) [103–105]. In each pixel the excitation light dose is reduced when either a certain upper threshold (UTh) is reached providing enough information about the density of labels at this position or if a lower threshold (LTh) is not reached after a certain illumination time indicating that no structure of interest is present. While considerable reduction in bleaching was demonstrated for CLEM in confocal applications—especially when imaging 3D volumes [103–106]—it has not found its way into routine application for two possible reasons: (i) Most importantly, the effect of CLEM in confocal microscopy is very limited when imaging specimen within a single plane. The reason is that in a confocal microscope, all dye molecules in or near the image plane

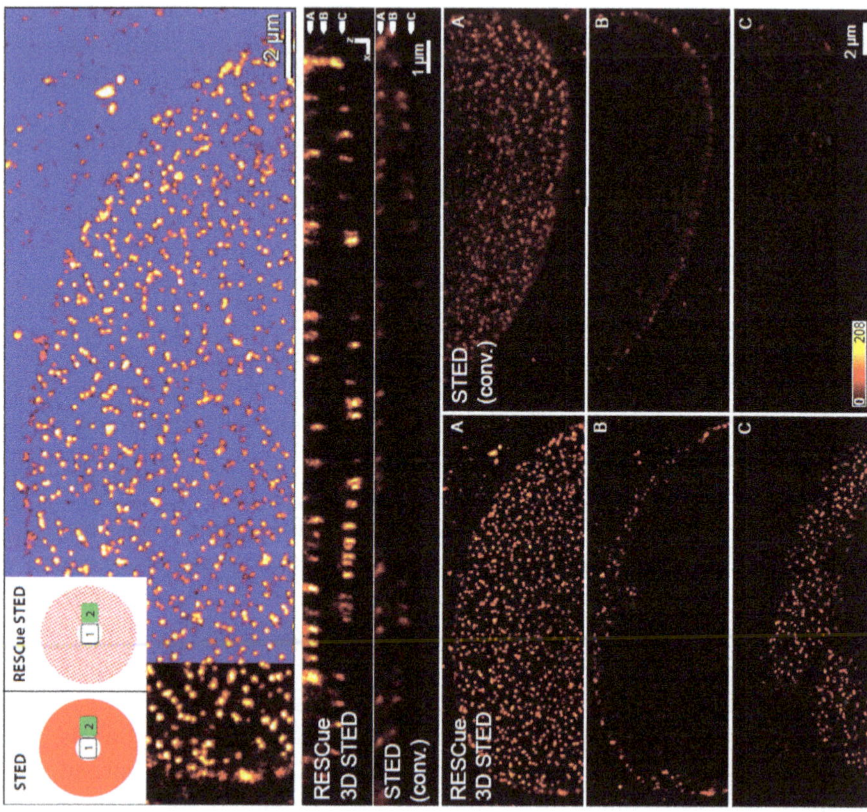
STED
RESCue STED
2 µm
RESCue 3D STED
STED (conv.)
A
B
C
1 µm
0
208
2 µm

Figure 16.6 Operating principle of RESCue STED (upper part). RESCue STED imaging is based on adaptation of the pixel dwell time during image generation. At positions where information is present (bright pixels) the dwell time is identical to conventional STED microscopy, whereas at pixels where no information is present (dark pixels, below a given threshold), the lasers are switched off. As illustrated in the inset, this means that while imaging a pixel without structural information (pixel 1), conventional STED will still illuminate neighboring pixels containing structure (pixel 2) with both STED and excitation light but without collecting any photons. In RESCue mode, such a situation is recognized after a fraction of the exposure time and illumination is stopped, thus reducing both light exposure of the sample and bleaching of neighboring structures. The image shows Vero cells labelled with antibodies against nuclear pore complex subunits in the central channel of the complex and secondary antibodies coupled to Abberior STAR RED. On the right side of the image, the blue color indicates the image pixels where the excitation and the STED laser were shut off after ~15% of the exposure time because the signal did not reach the lower threshold. Note that the blue positions thus correlate with the positions where no marker (i.e., target structure) is present. The displayed image is an imaging plane of a RESCue easy3D-STED image stack with 55 planes, 50 nm apart (lower part). In the given example, the total light exposure was reduced by a factor of 5 by RESCue. All images shown in the lower part have the same LUT demonstrating how RESCue enables imaging of 3D volumes where conventional STED cannot deliver high-SNR images of all planes due to bleaching of the marker.

that are illuminated also contribute to the signal. In other words, when the lower threshold is not reached and the excitation light dose is thus reduced in a pixel, it means that no or a very small amount of dye molecules are present and can benefit from this. Importantly, however, for out-of-focus planes, where fluorescence is generated and thus bleaching induced while the emitted photons do NOT contribute significantly to the measured signal this argument does no longer hold—hence CLEM mainly improves performance in 3D volumes. Still, routine application in 3D may have been hampered by (ii) the fact that the noise in resulting raw image stacks deviates from the expected Poissonian statistics, making deconvolution, which has become a standard tool in confocal 3D image analysis, and other quantitative approaches considerably more difficult to implement.

The picture entirely changes in STED. The hallmark of STED (and all other targeted read-out based super-resolution methods; see below) is that the *majority* of dyes that are inside the excitation volume do NOT contribute to the signal as they are not within the "allowed" area defined by the intensity zero of the inhibition light. In addition to that, when imaged at resolutions in the range of a few tens of nanometers, many structures start to appear sparse and in most samples the majority of sub-diffraction sized pixels do not contain structures. In such sample positions, RESCue test, whether a certain lower signal threshold has been reached after a short portion of the pixel dwell time, say 15%, signifying that imaging here is not necessary. If this is not the case, it will stop illumination and detection and neighboring structures that are well within the excitation volume but outside the STED probe volume will experience 85% less light exposure (Fig. 16.6). Experience shows that in most STED images the light dose to the sample can be reduced by a factor of at least 5 and bleaching—even within rather dense structures—is reduced significantly. In combination with pulsed STED lasers, the total light dose can thus be reduced by about a factor of 25 as compared to gSTED implementations common for GFP/YFP imaging, reaching levels that may well be lower than needed for the acquisition of a PALM/STORM image of the same image region [107]. Especially when imaging 3D volumes with 3D STED this approach is often the enabling factor (Fig. 16.6) [107, 108].

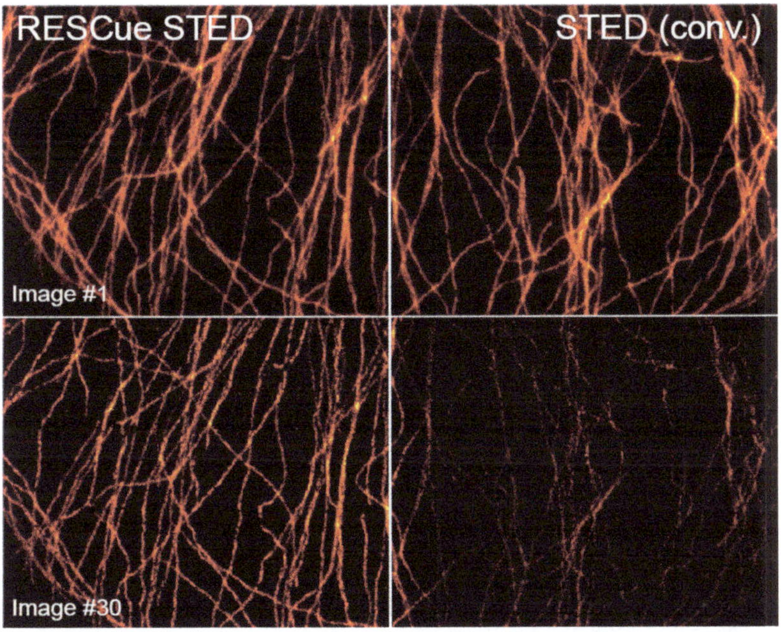

Figure 16.7 RESCue STED image series demonstrating the dramatic reduction of photobleaching under realistic imaging conditions even when imaging a single plane repeatedly (as opposed to 3D volumes). Shown are Vero cells labelled with antibodies against β-tubulin and secondary antibodies coupled to Abberior STAR 635P. The figure shows the first and last image of a series of 30 scans taken from neighboring areas in the same sample. While, especially for time-lapse imaging of a single plane, the density of labelled structures in the vicinity will influence how much a measurement benefits from RESCue at any given spot, in this example, the bleaching rate of STED (right) is cut by a factor of 2–3 on average when employing RESCue mode (left).

The full potential of RESCue is, however, unleashed in live-cell applications. In many cases the GFP/YFP dye combination is standard for live-cell STED, but real-live imaging with these dyes requires careful adjustment of all parameters to allow a long enough observation time [109] and even with RESCue bleaching can remain problematic. In addition to that, it should be noted that the typical STED wavelength of 595 nm for these applications is potentially more harmful to live specimen that the most common

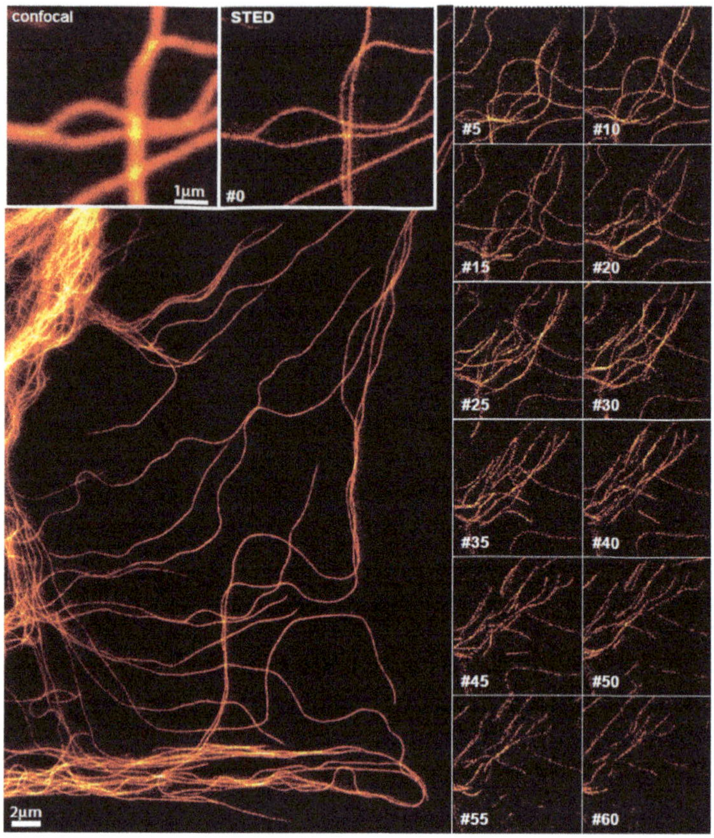

Figure 16.8 RESCue STED imaging of living cells. Shown are living human fibroblast cells labelled with SiR-tubulin. RESCue STED imaging at 775 nm was performed using a Nikon-based Abberior 2C STED 775 QUAD scan microscope in which the focal plane was continuously locked using a perfect focusing system. Note that the bleaching reduction in the RESCue STED mode enables imaging of more than 60 consecutive planes at the same imaging position. Total image size is 1230 × 1870 pixels with a pixel size of 20 × 20 nm, and the maximum pixel dwell time was 8 μs. Images were taken every 150 s.

red STED laser line at 775 nm. An important development is therefore the introduction of (fluorinated) silicone rhodamines (SiR, SiRF) and phosphorylated oxazines as organic dyes compatible with live cell applications [68–70, 110]. Besides enabling multicolor near IR imaging these dyes also prove to me considerably more photostable under STED conditions finally enabling routine long-term observation of live specimen with true super-resolution (Fig. 16.8).

16.2.5 Combination with Other Fluorescence Methods

Being essentially a confocal microscope with an additional inhibition beam, STED nanoscopy lends itself readily to the application of established fluorescence methods to nanoscale volumes. For example, time-domain fluorescence lifetime imaging (FLIM) can be used to, e.g., probe the micro-environment of a sensitive dye molecule [111], to detect and quantify FRET [112], or, quite simply, to separate the signal from two dye molecules with similar emission wavelengths but different lifetimes (Fig. 16.9) [113].

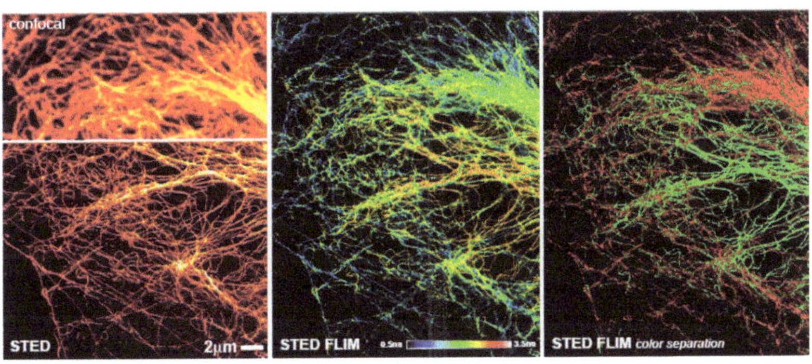

Figure 16.9 Dual-channel FLIM STED microscopy. Typically spectrally distinct dyes are used to label different structures within the cells. However to increase the number of channels that may be distinguished, the fluorescence lifetime of fluorophores may be used as further criterion for their separation. Thereby spectrally similar dyes may be distinguished. In the example shown above, the tubulin cytoskeleton of mammalian cells was immunolabelled using antibodies coupled to Atto647N and the vimentin cytoskeleton was immunolabelled using antibodies coupled to Abberior STAR635P. The fluorescence lifetime of Atto647N in antibody conjugates (1.8 ns) is significantly different from the fluorescence lifetime of Abberior STAR 635P (~3.6 ns). Imaging was performed using an Abberior Instruments 2C STED 775 QUAD Scan microscope equipped with a Becker & Hickl-based FLIM STED module.

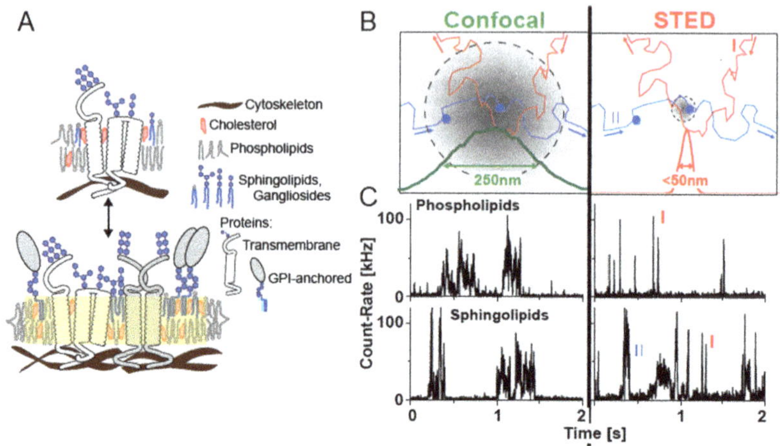

Figure 16.10 STED nanoscale spectroscopy of molecular dynamics in the plasma membrane of living cells. (A) Proteins and lipids may (transiently) interact in, potentially forming clusters assisted by cholesterol and the underlying cytoskeleton. (B) The diffusion of fluorescent phospholipid analogues in living mammalian cells has been found to be largely free (I, red trace), while that of sphingolipids was hindered (II, blue trace). (C) Detected fluorescence time-trace of single diffusing fluorescent phospholipid (upper panels) and sphingolipid analogues (lower panels): Only the STED (observation spot size down to below 40 nm) but not the confocal recordings (observation spot size 250 nm in diameter) can distinguish non-interacting (I) from interacting lipid transits. Adapted from Eggeling et al. [84], Eggeling [118], and Lingwood and Simons [149].

Likewise, STED nanoscopy also improves single-molecule-based fluorescence spectroscopy methods, improving the investigation of molecular interactions at the nanoscale. This has recently been shown by combining it with fluorescence correlation spectroscopy (FCS) [114, 115]. In such STED-FCS experiments the fluctuating intensity data was analyzed as labelled molecules moved in and out of the nonscanned observation volume, and heterogeneity in the observed diffusion characteristics revealed how these molecules interacted. While this approach had been routinely applied in confocal microscopes, the ability of STED to freely tune down the size of the probe to several tens of nanometers made the nanoscale accessible to the approach (Fig. 16.10) [84, 116–120]. Hindered diffusion modes could be highlighted by plotting the average transit times or apparent diffusion coefficients of the molecules through the observation spot against the spot size, which could be tuned by the power of the STED laser. As an example, it could be shown in the plasma membrane of living cells that, while certain fluorescently labelled phospholipid analogues such as a labelled phosphoethanolamine diffused almost freely, labelled sphingolipids such as a labelled sphingomyelin interacted with other molecules such as proteins on millisecond time and molecular spatial scales [84, 116], assisted by cholesterol and the underlying cortical actin cytoskeleton [117, 119]. This method and findings have given important insights into lipid membrane organization [118, 120–122]. While first experiments were recorded using a pulsed Titanium Sapphire STED laser, STED-FCS in combination with gated detection has now allowed using compact CW-STED lasers for these investigations of molecular membrane dynamics [53, 123–125]. An interesting feature of such gSTED-FCS measurements is that the position of the gate introduces an additional possibility for tuning the size of the observation volume [126, 127]. Further, scanning STED-FCS realizes exposing spatial heterogeneity in molecular diffusion modes as well as determination of diffusion characteristics with large statistics during measurements lasting several seconds [122, 128, 129]. In addition, STED-FCS also enables the investigation of anomalous diffusion modes of fluorescently tagged proteins, as well as the disclosure of hindered lipid and proteins diffusion due to compartmentalization by the cellular cortical cytoskeleton [84, 130].

16.3 Comparison with Other Nanoscopy Approaches

Based on the same principle for feature separation, namely keeping all but one feature dark through molecular OFF states, a second family of nanoscopy approaches arose in 2006. With both the underlying principle and the dyes needed at their disposal, several groups of researchers co-invented the method and a multitude of related techniques and accompanying acronyms was the result: stochastic optical reconstruction microscopy (STORM) [131] (Chapter 14),

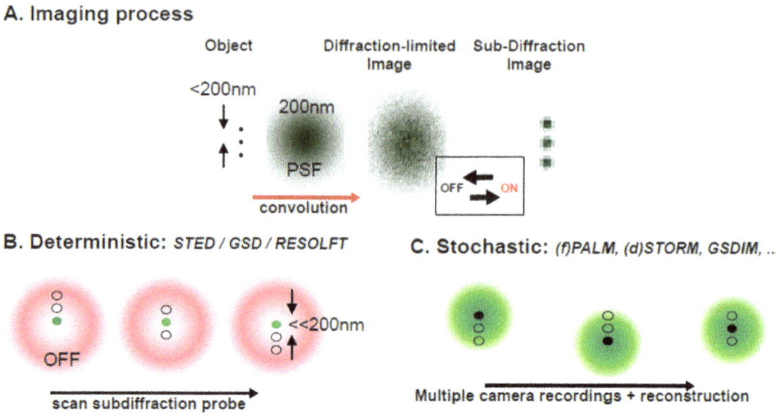

Figure 16.11 Principle of different far-field fluorescence nanoscopy approaches. (Upper panels) Due to the diffraction of light, object details that are closer together than about 200 nm cannot be discerned when imaged with focused light (diffraction spot or point-spread-function PSF ~200 nm in diameter), resulting in blurred images. Sub-diffraction images are realized by switching the fluorescence ability of the fluorophores on and off either in a deterministic or stochastic manner, resulting in the ability to resolve structures <200 nm in size. (Lower left panel) In the usual application of deterministic methods such as STED, GSD, or RESOLFT, a diffraction-limited spot of the fluorescence excitation (ON) laser beam is overlaid with a second laser beam which switches off fluorescence emission and which features at least one zero-intensity point. Consequently, the off-switching light selectively inhibits fluorescence everywhere but at the zero-intensity points, creating observation spots of nanoscopic size. Scanning fluorescence images therefore feature a considerable improvement in spatial resolution. (Lower right panel) Stochastic methods such as (f) PALM, (d) STORM, GSDIM, etc., are usually based on the recording of multiple widefield images using a sensitive camera. During these recordings only single molecules within the diffraction range are allowed to fluoresce in order to facilitate separation of the features. The position of the molecules is found while the molecules are in the ON state, emitting multiple fluorescent photons allowing the position to be determined using different localization approaches. By stochastically switching on and off all molecules during the recordings, the final image with sub-diffraction sized spatial resolution is reconstructed from the summation of all identified molecules. Adapted from Eggeling [37].

(fluorescence) photo-activated localization microscopy ((f)PALM) [132, 133] and then in quick succession ground-state depletion *microscopy* followed by individual molecule return (GSDIM) [134], direct STORM (dSTORM) [135], or single-molecule active control microscopy (SMACM) [136]. This family of nanoscopy methods relies on the stochastic on- and off-switching of the fluorescence emission of individual labels within the diffraction limit, combined with the determination of their exact spatial positions. In contrast, STED/RESOLFT/GSD use a deterministic switching approach where the fluorophore ON and OFF states are assumed at predefined positions.

Both spatially stochastic and deterministic fluorescence nanoscopy approaches have their specific advantages and disadvantages with respect to photobleaching, setup complexity, imaging speed, or image reconstruction. The fact that these fluorescence nanoscopy or super-resolution variants as well as techniques such a SIM or 4Pi oftentimes complement each other in their applications, promotes research environments with can easily access a whole range of microscopes and nanoscopes, depending on their suitability for the case in hand. Still one should not expect one microscope to be optimized in all aspects of microscopy: high spatial and temporal resolution, low phototoxicity (and thus large live-cell compatibility), and high image contrast. For example, (f) PALM/STORM-based nanoscopy approaches suffer from rather low temporal resolution and motion blurring and elevated phototoxicity, STED can reach speeds far beyond video rate [50, 137, 138] but even when combined with RESCue leads to exposure of the sample with rather high peak intensities which might be undesirable in or even incompatible with the application when the sample contains absorbers in the spectral range of the STED laser. RESOLFT [139–142] which relies on the same molecular switch as for example PALM (using reversibly photoswitchable fluorescent labels such as proteins) can reach acceptable speeds [143] and <70 nm resolution at low light levels in its parallelized version [144]; however, dye fatigue currently limits the obtainable resolution and number of exposures obtainable from the sample. Therefore, in some applications where true super-resolution is not needed to discern the features of interest, recent ultrafast, low-phototoxic light-sheet-based optical microscopes [145, 146], especially when combined with STED [147] or RESOLFT [148], may well fill a gap.

16.4 Conclusion

In this review we have highlighted the concepts of optical super-resolution microscopy. We have outlined the fast development that STED-based microscopy has taken over the last two decades towards a routine, commercially available imaging technique. Taking this, and recent advances in dye and labelling technology as an indicator, we expect that in the near future, imaging at resolutions well below 100 nm will be as common as confocal or widefield imaging. As the example of STED-FCS shows, the ability of restricting observation to sub-diffraction-sized spots also pushes related techniques and in fact this is true for all approaches outlined above. For example, the idea of stochastically imaging a small subset of fluorophores contained in a sample can be used to greatly widen the scope of both FCS, single-molecule-tracking applications and their combination with imaging. All taken together, the advent of optical nanoscopy has added a multitude of new instruments to the imaging toolbox of life scientists. Discoveries have already been made using super-resolution techniques that would not have been possible otherwise, but the full impact of this development is likely to become evident in the years to come.

References

1. Abbe, E. (1873). Beiträge zur theorie des mikroskops und der mikroskopischen wahrnehmung. *Arch. Mikrosk. Anatomie*, **9**, 413–468.
2. Hell, S. W. (2009). Microscopy and its focal switch. *Nat. Methods*, **6**, 24–32 (2009).
3. Pohl, D. W., Denk, W., and Lanz, M. (1984). Optical stethoscopy - image recording with resolution lambda/20. *Appl. Phys. Lett.*, **44**, 651–653.
4. Lewis, A., Isaacson, M., Harootunian, A., and Murray, A. (1984). Development of a 500 A resolution light microscope. *Ultramicroscopy*, **13**, 227–231.

5. Axelrod, D. (1981). Cell-substrate contacts illuminated by total internal reflection fluorescence. *J. Cell Biol.*, **89**, 141–145.
6. Toraldo di Francia, G. (1952). Super-Gain antennas and optical resolving power. *Nuovo Cimento B*, **9**, 426–435.
7. Bertero, M., Boccacci, P., Brakenhoff, G. J., Malfanti, F., and Van der Voort, H. T. M. (1990). Three-dimensional image restoration and super-resolution in fluorescence confocal microscopy. *J. Microsc.*, **157**, 3–20.
8. Cremer, C., and Cremer, T. (1978). Considerations on a laser-scanning-microscope with high-resolution and depth of field. *Microsc. Acta*, **81**, 31–44.
9. Minsky, M. (1961). Microscopy apparatus US Patent. 3,013,467.
10. Wilson, T. (1980). Imaging properties and applications of scanning optical microscopes. *Appl. Phys.*, **22**, 119–128.
11. Pawley, J. B. (2006). *Handbook of Biological Confocal Microscopy*, 2nd ed., Springer.
12. Denk, W., Strickler, J. H., and Webb, W. W. (1990). 2-photon laser scanning fluorescence microscopy. *Science*, **248**, 73–76.
13. Schoenle, A., and Hell, S. W. (1999). Far-field fluorescence microscopy with repetitive excitation. *Eur. Phys. J. D*, **6**, 283–290.
14. Lukosz, W. (1966). Optical systems with resolving powers exceeding the classical limit. *J. Opt. Soc. Am.*, **56**, 1463–1471.
15. Ash, E. A., and Nicholls, G. (1972). Super-resolution aperture scanning microscope. *Nature*, **237**, 510–512.
16. Bailey, B., Farkas, D. L., Taylor, D. L., and Lanni, F. (1993). Enhancement of axial resolution in fluorescence microscopy by standing-wave excitation. *Nature*, **366**, 44–48.
17. Gustafsson, M. G. L. (2000). Surpassing the lateral resolution limit by a factor of two using structured illumination microscopy. *J. Microsc.*, **198**, 82–87.
18. Frohn, J. T., Knapp, H. F., and Stemmer, A. (2000). True optical resolution beyond the Rayleigh limit achieved by standing wave illumination. *PNAS*, **97**, 7232–7236.
19. Gustafsson, M. G. L., et al. (2008). Three-dimensional resolution doubling in wide-field fluorescence microscopy by structured illumination. *Biophys. J.*, **94**, 4957–4970.
20. Schermelleh, L., et al. (2008). Subdiffraction multicolor imaging of the nuclear periphery with 3D structured illumination microscopy. *Science*, **320**, 1332–1336.

21. Muller, C. B., and Enderlein, J. (2010). Image scanning microscopy. *Phys. Rev. Lett.*, **104**, 198101.
22. York, A. G., et al. (2012). Resolution doubling in live, multicellular organisms via multifocal structured illumination microscopy. *Nat. Methods*, **9**(7), 749–754.
23. Fiolka, R., Beck, M., and Stemmer, A. (2008). Structured illumination in total internal reflection fluorescence microscopy using a spatial light modulator. *Opt. Lett.*, **33**, 1629–1631.
24. Hell, S. W. (1992). Double-scanning confocal microscope. European Patent 0491289.
25. Hell, S. W., and Stelzer, E. H. K. (1992). Properties of a 4pi confocal fluorescence microscope. *J. Opt. Soc. Am. A: Opt. Image Sci. Vis.*, **9**, 2159–2166.
26. Hell, S. W., Lindek, S., Cremer, C., and Stelzer, E. H. K. (1994). Measurement of the 4Pi-confocal point spread function proves 75 nm resolution. *Appl. Phys. Lett.*, **64**, 1335–1338.
27. Gustafsson, M. G. L., Agard, D. A., and Sedat, J. W. (1995). Sevenfold improvement of axial resolution in 3D wide-field microscopy using two objective lenses, in *Three-Dimensional Microscopy: Image Acquisition and Processing II* (eds. Wilson, T., and Cogswell, C.), *Proc. SPIE*, **2412**, 147–156.
28. Gustafsson, M. G., Agard, D. A., and Sedat, J. W. (1996). 3D widefield microscopy with two objective lenses: experimental verification of improved axial resolution, in *Three-Dimensional Microscopy: Image Acquisition and Processing III* (eds. Cogswell, C., Kino, G. S., and Wilson, T.), *Proc. SPIE*, **2655**, 62–66.
29. Gustafsson, M. G. L., Agard, D. A., and Sedat, J. W. (1999). (im)-m-5: 3d widefield light microscopy with better than 100 nm axial resolution. *J. Microsc.*, **195**, 10–16.
30. Alberts, B., et al. (2002). *Molecular Biology of the Cell*, 4 ed., Garland Science.
31. Hell, S. W. (1994). Improvement of lateral resolution in far-field light microscopy using two-photon excitation with offset beams. *Opt. Commun.*, **106**, 19–24.
32. Hell, S. W., and Wichmann, J. (1994). Breaking the diffraction resolution limit by stimulated-emission - stimulated-emission-depletion fluorescence microscopy. *Opt. Lett.*, **19**, 780–782.

33. Hell, S. W., and Kroug, M. (1995). Ground-state depletion fluorescence microscopy, a concept for breaking the diffraction resolution limit. *Appl. Phys. B*, **60**, 495–497.

34. Hell, S. W. (2003). Toward fluorescence nanoscopy. *Nat. Biotechnol.*, **21**, 1347–1355.

35. Hell, S. W., Dyba, M., and Jakobs, S. (2004). Concepts for nanoscale resolution in fluorescence microscopy. *Curr. Opin. Neurobiol.*, **14**, 599–609.

36. Hell, S. (2009). Far-field optical nanoscopy, in *Single Molecule Spectroscopy in Chemistry* (eds. Gräslund, A., Rigler, R., and Widengren, J.), Springer, pp. 365–398.

37. Eggeling, C., Willig, K. I., and Barrantes, F. J. (2013). STED microscopy of living cells - New frontiers in membrane and neurobiology. *J. Neurochem.*, **126**, 203–212.

38. Eggeling, C., Willig, K. I., Sahl, S. J., and Hell, S. W. (2015). Lens-based fluorescence nanoscopy. *Q. Rev. Biophys.*, **48**, 178–243.

39. Klar, T. A., and Hell, S. W. (1999). Subdiffraction resolution in far-field fluorescence microscopy. *Opt. Lett.*, **24**, 954–956.

40. Clausen, M. P., et al. (2013). Pathways to optical STED microscopy. *NanoBioImaging*, **1**, 1–12.

41. Eggeling, C., Volkmer, A., and Seidel, C. A. M. (2005). Molecular photobleaching kinetics of rhodamine 6g by one- and two-photon induced confocal fluorescence microscopy. *ChemPhysChem*, **6**, 791–804.

42. Eggeling, C., Widengren, J., Rigler, R., and Seidel, C. A. M. (1998). Photobleaching of fluorescent dyes under conditions used for single-molecule detection: evidence of two-step photolysis. *Anal. Chem.*, **70**, 2651–2659.

43. Donnert, G., et al. (2006). Macromolecular-scale resolution in biological fluorescence microscopy. *PNAS*, **103**, 11440–11445.

44. Ringemann, C., et al. (2008). Enhancing fluorescence brightness: effect of reverse intersystem crossing studied by fluorescence fluctuation spectroscopy. *ChemPhysChem*, **9**, 612–624.

45. Hotta, J., et al. (2010). Spectroscopic rationale for efficient stimulated-emission depletion microscopy fluorophores. *J. Am. Chem. Soc.*, **132**, 5021–5023.

46. Rankin, B. R., et al. (2011). Nanoscopy in a living multicellular organism expressing GFP. *Biophys. J.*, **100**, L63–L65.

47. Donnert, G., Eggeling, C., and Hell, S. W. (2007). Major signal increase in fluorescence microscopy through dark-state relaxation. *Nat. Methods*, **4**, 81–86.

48. Donnert, G., Eggeling, C., and Hell, S. W. (2009). Triplet-relaxation microscopy with bunched pulsed excitation. *Photochem. Photobiol.*, **8**, 481–485.

49. Moneron, G., et al. (2010). Fast STED microscopy with continuous wave fiber lasers. *Opt. Express*, **18**, 1302–1309.

50. Schneider, J., et al. (2015). Ultrafast, temporally stochastic STED nanoscopy of millisecond dynamics. *Nat. Methods*, **12**, 827–830.

51. Wu, Y., et al. (2015). Resonant scanning with large field of view reduces photobleaching and enhances fluorescence yield in STED microscopy. *Sci. Rep.*, **5**, 14766.

52. Moffitt, J. R., Osseforth, C., and Michaelis, J. (2011). Time-gating improves the spatial resolution of STED microscopy. *Opt. Express*, **19**, 4242–4254.

53. Vicidomini, G., et al. (2011). Sharper low-power STED nanoscopy by time gating. *Nat. Methods*, **8**, 571–573.

54. Vicidomini, G., et al. (2014). Gated CW-STED microscopy: a versatile tool for biological nanometer scale investigation. *Methods*, **66**, 124–130.

55. Kolmakov, K., et al. (2010). Red-emitting rhodamine dyes for fluorescence microscopy and nanoscopy. *Chem. Eur. J.*, **16**, 158–166.

56. Kolmakov, K., et al. (2010). A versatile route to red-emitting carbopyronine dyes for optical microscopy and nanoscopy. *Eur. J. Org. Chem.* **2010**, 3593–3610.

57. Mitronova, G. Y., et al. (2010). New fluorinated rhodamines for optical microscopy and nanoscopy. *Chem. Eur. J.*, **16**, 4477–4488.

58. Wurm, C. A., et al. (2012). Novel red fluorophores with superior performance in STED microscopy. *Opt. Nanosc.*, **1**, 1–7.

59. Opazo, F., et al. (2012). Aptamers as potential tools for super-resolution microscopy. *Nat. Methods*, **9**, 938–939.

60. Kasper, R., et al. (2010). Single-molecule STED microscopy with photostable organic fluorophores. *Small*, **6**, 1379–1384.

61. Beater, S., Holzmeister, P., Pibiri, E., Lalkens, B., and Tinnefeld, P. (2014). Choosing dfor cw-STED nanoscopy using self-assembled nanorulers. *Phys. Chem. Chem. Phys.*, **16**, 6990–6996.

62. Hein, B., Willig, K. I., and Hell, S. W. (2008). Stimulated emission depletion (STED) nanoscopy of a fluorescent protein-labeled organelle inside a living cell. *PNAS*, **105**, 14271–14276.

63. Morozova, K. S., et al. (2010). Far-red fluorescent protein excitable with red lasers for flow cytometry and superresolution STED nanoscopy. *Biophys. J.*, **99**, L13–L15.

64. Fitzpatrick, J. A., et al. (2009). STED nanoscopy in living cells using fluorogen activating proteins. *Bioconjug. Chem.*, **20**, 1843–1847.

65. Schröder, J., Benink, H., Dyba, M., and Los, G. V. (2008). In vivo labeling method using a genetic construct for nanoscale resolution microscopy. *Biophys. J.*, **96**, L1–L3.

66. Hein, B., et al. (2010). Stimulated emission depletion nanoscopy of living cells using SNAP-tag fusion proteins. *Biophys. J.*, **98**, 158–163.

67. Pellett, P. A., et al. (2011). Two-color STED microscopy in living cells. *Biomed. Opt. Express*, **2**, 2364–2371.

68. Lukinavicius, G., et al. (2013). A near-infrared fluorophore for live-cell superresolution microscopy of cellular proteins. *Nat. Chem.*, **5**, 132–139.

69. Lukinavicius, G., et al. (2014). Fluorogenic probes for live-cell imaging of the cytoskeleton. *Nat. Methods*, **11**, 731–733.

70. Butkevich, A. N., et al. (2016). Fluorescent rhodamines and fluorogenic carbopyronines for super-resolution STED microscopy in living cells. *Angew. Chem. Int. Ed. Engl.*, **55**, 3290–3294.

71. Donnert, G., et al. (2007). Two-color far-field fluorescence nanoscopy. *Biophys. J.*, **92**, L67–L69.

72. Meyer, L., et al. (2008). Dual-color STED microscopy at 30-nm focal-plane resolution. *Small*, **4**, 1095–1100.

73. Schmidt, R., et al. (2008). Spherical nanosized focal spot unravels the interior of cells. *Nat. Methods*, **5**, 539–544.

74. Friedemann, K., Turshatov, A., Landfester, K., and Crespy, D. (2011). Characterization via two-color STED microscopy of nanostructured materials synthesized by colloid electrospinning. *Langmuir*, **27**, 7132–7139.

75. Dean, C., et al. (2012). Distinct subsets of Syt-IV/BDNF vesicles are sorted to axons versus dendrites and recruited to synapses by activity. *J. Neurosci.*, **32**, 5398–5413.

76. Neumann, D., Bückers, J., Kastrup, L., Hell, S., and Jakobs, S. (2010). Two-color STED microscopy reveals different degrees of colocalization

between hexokinase-I and the three human VDAC isoforms. *PMC Biophys.*, **5**, 1–4.

77. Opaz, F., et al. (2010). Limited intermixing of synaptic vesicle components upon vesicle recycling. *Traffic*, **11**, 800–812.
78. Reisinger, E., et al. (2011). Probing the functional equivalence of otoferlin and synaptotagmin 1 in exocytosis. *J. Neurosci.*, **31**, 4886–4895.
79. Blom, H., et al. (2012). Nearest neighbor analysis of dopamine D1 receptors and Na1-K1-ATPases in dendritic spines dissected by STED microscopy. *Microsc. Res. Tech.*, **75**, 220–228.
80. Bethge, P., Chereau, R., Avignone, E., Marsicano, G., and Nagerl, U. V. (2013). Two-photon excitation STED microscopy in two colors in acute brain slices. *Biophys. J.*, **104**, 778–785.
81. Osseforth, C., Moffitt, J. R., Schermelleh, L., and Michaelis, J. (2013). Simultaneous dual-color 3D STED microscopy. *Opt. Express*, **22**, 7028–7039.
82. Bottanelli, F., et al. (2016). Two-colour live-cell nanoscale imaging of intracellular targets. *Nat. Commun.*, **7**, 10778.
83. Willig, K. I., Rizzoli, S. O., Westphal, V., Jahn, R., and Hell, S. W. (2006). Sted microscopy reveals that synaptotagmin remains clustered after synaptic vesicle exocytosis. *Nature*, **440**, 935–939.
84. Eggeling, C., et al. (2009). Direct observation of the nanoscale dynamics of membrane lipids in a living cell. *Nature*, **457**, 1159–1162.
85. Blom, H., et al. (2011). Spatial distribution of Na+-K+-ATPase in dendritic spines dissected by nanoscale superresolution STED microscopy. *BMC Neurosci.*, **12**, 16.
86. Lau, L., Lee, Y. L., Sahl, S. J., Stearns, T., and Moerner, W. E. (2012). STED microscopy with optimized labeling density reveals 9-fold arrangement of a centriole protein. *Biophys. J.*, **102**, 2926–2935.
87. Takasaki, K. T., Ding, J. B., and Sabatini, B. L. (2013). Live-cell super-resolution imaging by pulsed STED two-photon excitation microscopy. *Biophys. J.*, **104**, 770–777.
88. Wurm, C. A., Neumann, D., Schmidt, R., Egner, A., and Jakobs, S. (2010). Sample preparation for STED microscopy, in *Live Cell Imaging: Methods in Molecular Biology* (ed. Papkovsky, D. B.), Springer, pp. 185–199.
89. Tanaka, K. A. K., et al. (2010). Membrane molecules mobile even after chemical fixation. *Nat. Methods*, **7**, 865–866 (2010).

90. Dyba, M., and Hell, S. W. (2002). Focal spots of size lambda/23 open up far-field florescence microscopy at 33 nm axial resolution. *Phys. Rev. Lett.*, **88**, 163901.

91. Dyba, M., Jakobs, S., and Hell, S. W. (2003). Immunofluorescence stimulated emission depletion microscopy. *Nat. Biotechnol.*, **21**, 1303–1304.

92. Schmidt, R., et al. (2009). Mitochondrial cristae revealed with focused light. *Nano Lett.*, **9**, 2508–2510.

93. Ullal, C. K., Schmidt, R., Hell, S. W., and Egner, A. (2009). Block copolymer nanostructures mapped by far-field optics. *Nano Lett.*, **9**, 2497–2500.

94. Klar, T. A., Jakobs, S., Dyba, M., Egner, A., and Hell, S. W. (2000). Fluorescence microscopy with diffraction resolution barrier broken by stimulated emission. *PNAS*, **97**, 8206–8210.

95. Harke, B., et al. (2008). Resolution scaling in STED microscopy. *Opt. Express*, **16**, 4154–4162.

96. Harke, B., Ullal, C. K., Keller, J., and Hell, S. W. (2008). Three-dimensional nanoscopy of colloidal crystals. *Nano Lett.*, **8**, 1309–1313.

97. Gould, T. J., Burke, D., Bewersdorf, J., and Booth, M. J. (2012). Adaptive optics enables 3D STED microscopy in aberrating specimens. *Opt. Express*, **20**, 20998.

98. Urban, N. T., Willig, K. I., Hell, S. W., and Nagerl, U. V. (2011). STED nanoscopy of actin dynamics in synapses deep inside living brain slices. *Biophys. J.*, **101**, 1277–1284.

99. Eggeling, C., Widengren, J., Rigler, R., and Seidel, C. A. M. (1999). Photostability of fluorescent dyes for single-molecule spectroscopy: mechanisms and experimental methods for estimating photobleaching in aqueous solution, in *Applied Fluorescence in Chemistry, Biology and Medicine* (eds. Rettig, W., Strehmel, B., Schrader, M., and Seifert, H.), Springer, pp. 193–240.

100. Dyba, M., and Hell, S. W. (2003). Photostability of a fluorescent marker under pulsed excited-state depletion through stimulated emission. *Appl. Opt.*, **42**, 5123–5129.

101. Vicidomini, G., et al. (2013). STED nanoscopy with time-gated detection: theoretical and experimental aspects. *PLoS One*, **8**, e54421.

102. Leutenegger, M., Eggeling, C., and Hell, S. W. (2010). Analytical description of STED microscopy performance. *Opt. Express*, **18**, 26417.

103. Hoebe, R. A., et al. (2007). Controlled light-exposure microscopy reduces photobleaching and phototoxicity in fluorescence live-cell imaging. *Nat. Biotechnol.*, **25**, 249–253.

104. Hoebe, R. A., Van der Voort, H. T., Stap, J., Van Noorden, C. J., and Manders, E. M. (2008). Quantitative determination of the reduction of phototoxicity and photobleaching by controlled light exposure microscopy. *J. Microsc.*, **231**, 9–20.

105. De Vos, W. H., et al. (2009). Controlled light exposure microscopy reveals dynamic telomere microterritories throughout the cell cycle. *Cytometry A*, **75**, 428–439.

106. De Vos, W. H., et al. (2010). Four-dimensional telomere analysis in recordings of living human cells acquired with controlled light exposure microscopy. *J. Microsc.*, **238**, 254–264.

107. www.abberior-instruments.com.

108. Staudt, T., et al. (2011). Far-field optical nanoscopy with reduced number of state transition cycles. *Opt. Express*, **19**, 5644–5657.

109. Tonnesen, J., Nadrigny, F., Willig, K. I., Wedlich-Soldner, R., and Nagerl, U. V. (2011). Two-color STED microscopy of living synapses using a single laser-beam pair. *Biophys. J.*, **101**, 2545–2552.

110. Kolmakov, K., et al. (2015). Far-red emitting fluorescent dyes for optical nanoscopy: fluorinated silicon-rhodamines (SiRF dyes) and phosphorylated oxazines. *Chemistry*, **21**, 13344–13356.

111. Gadella, T. W. J., Jovin, T. M., and Clegg, R. M. (1993). Fluorescence lifetime imaging microscopy (flim) - spatial-resolution of microstructures on the nanosecond time-scale. *Biophys. Chem.*, **48**, 221–239.

112. Clegg, R. M. (1992). Fluorescence resonance energy transfer and nucleic acids. *Methods Enzymol.*, **211**, 353–388.

113. Bückers, J., Wildanger, D., Vicidomini, G., Kastrup, L., and Hell, S. W. (2011). Simultaneous multi-lifetime multi-color STED imaging for colocalization analyses. *Opt. Express*, **19**, 3130–3143.

114. Kastrup, L., Blom, H., Eggeling, C., and Hell, S. W. (2005). Fluorescence fluctuation spectroscopy in subdiffraction focal volumes. *Phys. Rev. Lett.*, **94**, 178104.

115. Blom, H., Kastrup, L., and Eggeling, C. (2006). Fluorescence fluctuation spectroscopy in reduced detection volumes. *Curr. Pharm. Biotechnol.*, **7**, 51–66.

116. Ringemann, C., et al. (2009). Exploring single-molecule dynamics with fluorescence nanoscopy. *New J. Phys.*, **11**, 103054.

117. Mueller, V., et al. (2011). STED nanoscopy reveals molecular details of cholesterol- and cytoskeleton-modulated lipid interactions in living cells. *Biophys. J.*, **101**, 1651–1660.

118. Eggeling, C. (2012). In *Fluorescent Methods to Study Biological Membranes*, Vol. 13 *Springer Series on Fluorescence* (eds. Mely, Y., and Duportail, G.), Springer-Verlag, pp. 291–309.

119. Mueller, V., et al. (2013). In *Methods Enzymol*, Vol. 591 (ed. Tetin, S. Y.), Elsevier, pp. 1–38.

120. Eggeling, C., and Honigmann, A. (2014). In *Cell Membrane Nanodomains: From Biochemistry to Nanoscopy* (eds. Lidke, D. S., and Cambi, A.), CRC Press, pp. 431–452.

121. Sezgin, E., et al. (2012). Partitioning, diffusion, and ligand binding of raft lipid analogs in model and cellular plasma membranes. *Biochim. Biophys. Acta*, **1818**, 1777–1784.

122. Honigmann, A., et al. (2014). Scanning STED-FCS reveals spatio-temporal heterogeneity of lipid interaction in the plasma membrane of living cells. *Nat. Commun.*, **5**, 5412.

123. Honigmann, A., Eggeling, C., Schulze, M., and Lepert, A. (2012). Super-resolution STED microscopy advances with yellow CW OPSL. *Laser Focus World*, **48**, 75–79.

124. Mueller, V., Eggeling, C., Karlsson, H., and von Gegerfelt, D. (2012). CW DPSS lasers make STED microscopy more practical. *Biophotonics*, **19**, 30–32.

125. Honigmann, A., Mueller, V., Fernando, U. P., Eggeling, C., and Sperling, J. (2013). Simplyfing STED microscopy of photostable red-emitting labels. *Laser + Potonik*, **5**, 40–42.

126. Clausen, M. P., et al. (2015). A straightforward approach for gated STED-FCS to investigate lipid membrane dynamics. *Methods*, **88**, 67–75.

127. Vicidomini, G., et al. (2015). STED-FLCS: an advanced tool to reveal spatiotemporal heterogeneity of molecular membrane dynamics. *Nano Lett.*, **15**, 5912–5918.

128. Mueller, V. (2012). Nanoscale studies of membrane dynamics via STED - Fluorescence Correlation Spectroscopy. PhD thesis, University Heidelberg.

129. Bianchini, P., Cardarelli, F., Di Luca, Diaspro, A., and Bizzarri, R. (2014). Nanoscale protein diffusion by STED-based pair correlation analysis. *PLoS One*, **9**, e99619.

130. Andrade, D. M., et al. (2015). Cortical actin networks induce spatio-temporal confinement of phospholipids in the plasma membrane – a minimally invasive investigation by STED-FCS. *Sci. Rep.*, **5**, 11454.

131. Rust, M. J., Bates, M., and Zhuang, X. W. (2006). Sub-diffraction-limit imaging by stochastic optical reconstruction microscopy (STORM). *Nat. Methods*, **3**, 793–795.

132. Betzig, E., et al. (2006). Imaging intracellular fluorescent proteins at nanometer resolution. *Science*, **313**, 1642–1645.

133. Hess, S. T., Girirajan, T. P. K., and Mason, M. D. (2006). Ultra-high resolution imaging by fluorescence photoactivation localization microscopy. *Biophys. J.*, **91**, 4258–4272.

134. Fölling, J., et al. (2008). Fluorescence nanoscopy by ground-state depletion and single-molecule return. *Nat. Methods*, **5**, 943–945.

135. Heilemann, M., et al. (2008). Subdiffraction-resolution fluorescence imaging with conventional fluorescent probes. *Angew. Chem. Int. Ed. Engl.*, **47**, 6172–6176.

136. Biteen, J. S., et al. (2008). Super-resolution imaging in live caulobacter crescentus cells using photoswitchable EYFP. *Nat. Methods*, **5**, 947–949.

137. Westphal, V., Lauterbach, M. A., Di Nicola, A., and Hell, S. W. (2007). Dynamic far-field fluorescence nanoscopy. *New J. Phys.*, **9**, 435.

138. Westphal, V., et al. (2008). Video-rate far-field optical nanoscopy dissects synaptic vesicle movement. *Science*, **320**, 246–249.

139. Hell, S. W. (2004). Strategy for far-field optical imaging and writing without diffraction limit. *Phys. Lett. A*, **326**, 140–145.

140. Hell, S. W., Jakobs, S., and Kastrup, L. (2003). Imaging and writing at the nanoscale with focused visible light through saturable optical transitions. *Appl. Phys. A*, **77**, 859–860.

141. Hofmann, M., Eggeling, C., Jakobs, S., and Hell, S. W. (2005). Breaking the diffraction barrier in fluorescence microscopy at low light intensities by using reversibly photoswitchable proteins. *PNAS*, **102**, 17565–17569.

142. Grotjohann, T., et al. (2011). Diffraction-unlimited all-optical imaging and writing with a photochromic GFP. *Nature*, **478**, 204–208.

143. Testa, I., et al. (2012). Nanoscopy of living brain slices with low light levels. *Neuron*, **75**, 992–1000.

144. Chmyrov, A., et al. (2013). Nanoscopy with more than 100,000 'doughnuts'. *Nat. Methods*, **10**, 737–740.

145. Wu, Y., et al. (2013). Spatially isotropic four-dimensional imaging with dual-view plane illumination microscopy. *Nat. Biotechnol.*, **31**, 1032–1038.

146. Chen, B.-C., et al. (2014). Lattice light-sheet microscopy: imaging molecules to embryos at high spatiotemporal resolution. *Science*, **346**, 439.

147. Friedrich, M., Gan, Q., Ermolayev, V., and Harms, G. S. (2011). STED-SPIM: stimulated emission depletion improves sheet illumination microscopy resolution. *Biophys. J.*, **100**, L43–L45.

148. Hoyer, P., et al. (2016). Breaking the diffraction limit of light-sheet fluorescence microscopy by RESOLFT. *Proc. Natl. Acad. Sci. U.S.A.*, **113**, 3442–3446.

149. Lingwood, D., and Simons, K. (2010). Lipid rafts as a membrane-organizing principle. *Science*, **327**, 46–50.

Index

absorption spectrum 11, 15, 17, 128, 281, 295
acceptor channels 295, 299
acceptor emission 294
acceptor fluorophores 281, 283–284, 293, 295
acceptor molecules 66, 73, 211, 281, 295–297
ADCs, *see* analog-to-digital converters
advanced fluorescence techniques 259–260, 262, 264, 266, 268, 270, 272, 274
analog-to-digital converters (ADCs) 338–339
anthozoans 123, 128–130
antibodies 81–89, 91–94, 96, 102, 104–105, 139, 145, 291, 350, 354, 421, 431
  polyclonal 82, 84, 87
  secondary 86, 424, 431
antigens 81–85, 88, 90–92, 99, 102–103, 105
APD, *see* avalanche photodiode
astigmatism 363–364, 366–369
autoantibodies 100–101
autofluorescence 25, 29, 31, 106–107, 116, 295, 390, 399
autoimmune diseases 100–101
avalanche photodiode (APD) 225, 314, 319–323, 335
avGFP 117, 119, 126–130

background photobleaching 271
bacteria 3, 118, 121, 172, 192–193, 289
BFP, *see* blue fluorescent protein
binding affinities 82–83, 87, 165
biosensor 209, 240, 242, 293
  cleaved 241–242
  intact 241–242
  linked 293–294
  MT1-MMP 240–241
bleaching 74, 77, 261, 263, 265–270, 283, 397, 425–427, 430
BLS, *see* bright light surgery
blue fluorescent protein (BFP) 127, 291–293
bright light surgery (BLS) 192

CAFs, *see* cancer-associated fibroblasts
calcium 74, 93, 164–167, 169, 293
calcium indicators 165, 169–170
cancer 100, 102, 169, 185, 192, 196–197
cancer-associated fibroblasts (CAFs) 193, 195
cancer cells 186, 188–190, 192, 196, 198
  dual-color 186, 188, 196
CCD, *see* charged coupled device
cell membranes 89, 103, 303
cells
  biological 234

dendritic 151–152
double-labeled 296
eye lens fiber 75
fluorescence-stained adenocarcinoma 2
fluorescent 274
healthy 242, 268
HeLa 23, 28, 200
HT-1080 190, 241
HT-1080-GFP-RFP 190
human fibrosarcoma 186, 190, 201
hybridoma 85
interphase 101
live 140–141, 143, 145, 147, 150, 153–154, 175, 234, 236–237, 239–240, 299, 303, 425
malignant 242
mammalian 118, 121, 127–128, 130, 139, 144, 149, 169, 288, 424, 433–434
myeloma 82, 85
peripheral 187
stromal 188, 196
tumor 193
Vero 125, 431
cellular imaging 127, 279
charge packet 328–330, 335–336
charged coupled device (CCD) 6, 227, 314, 327–329, 331–332, 335, 337–338, 341, 344, 364
chemotherapy 189, 199–200
chromophore 118, 120–121, 126–128, 287, 291
CLSM, *see* confocal laser scanning microscopy
confocal imaging 51–52, 57, 297, 426
confocal microscopes 49–52, 55–56, 63, 68, 72, 74, 262, 357, 424–425, 427, 433–434
confocal microscopy 28–29, 49–50, 52, 54, 56–58, 60, 62, 64, 76, 284, 348, 357, 385, 390, 427
confocal pinhole 53–55, 70–71, 78
corals 119, 122
cytoplasm 101, 103, 142, 146–147, 169, 186, 188, 193, 196, 263

depth information 356, 358, 365
detector phases 241
detectors 1, 3, 6, 28, 30, 50, 53, 76–77, 219–220, 222, 225, 285, 318, 320, 326–327
dichroic mirrors 36–37, 40–41, 52–53, 76, 226, 284, 341, 392
diffraction barrier 415, 417
diffraction limit 384, 403, 416, 419, 437
donor, unquenched 283, 299
donor emission 281, 294–295
donor excitation wavelengths 284, 295
donor fluorophore 281, 283
donor lifetime 236, 282
donor molecules 233, 281
donor–acceptor pairs 211, 283
dye-based small molecule indicators 166, 168
dye chemistries 353
dyes 36, 40, 42, 142–147, 150–152, 154, 164, 166, 173–174, 177, 355–356, 359, 430–431, 433, 435
ANEP 173
cationic 147–148
cyanine 145, 147–148, 349
fluorescent 14, 20, 49, 176–177, 359, 376, 391
neutral 152–154

EC, *see* extinction coefficient
eccrine glands 97–98

electrodes, charge-transfer 328, 331, 335
electron-accepting materials 401–402
electron multiplying charge-coupled device (EMCCDs) 343, 351, 398
electron spins 8–9
electronic states 8–10, 12–13, 17, 19–20, 29–30
  excited 8–9, 18, 22
  higher 11–12, 19
EMCCDs, *see* electron multiplying charge-coupled device
endocytosis 86, 143, 149, 151, 169
endoplasmic reticulum (ER) 29, 125, 150, 153–154, 168–169, 175, 265
ENF, *see* excess noise factor
epidermis 83, 97, 99
epifluorescence 187
epifluorescence confocal microscope 52–53
ER, *see* endoplasmic reticulum
excess noise factor (ENF) 316–319, 322, 324, 337
excitation
  donor 294
  fluorescence 22, 436
  long-wavelength 167
  multiphoton 22, 24, 66–69, 71
  photon 67–68, 215
  single-photon 64–66, 68–69, 72
  two-photon 23–24, 29, 64–69, 73, 77, 173, 221, 223, 238
  vibrational 11, 15–16
  wide-field 214, 221, 224–225, 229, 238, 362
excitation lasers 24, 392, 419
extinction coefficient (EC) 11–12, 140, 286
FCS, *see* fluorescence correlation spectroscopy
FFPE, *see* formalin fixed paraffin embedded
FGS, *see* fluorescence-guided surgery
FLAP, *see* fluorescence localization after photobleaching
FLIM, *see* fluorescence lifetime imaging
fluorescence
  background 176, 295
  basic principles of 2, 4, 6, 8, 10, 12, 14, 16, 18, 20, 22, 24, 26, 28, 30
  donor 282
  emitted 69, 355
  formalin-induced 108
  fundamental properties of 3, 32
  green 37, 121, 132
  in-focus 226, 362
  multiphoton 22–23
fluorescence correlation spectroscopy (FCS) 427, 434, 438
fluorescence decay 209–210, 213, 215, 224
fluorescence detection 191, 385
fluorescence emission 16, 106, 210, 212, 217, 221, 225, 385, 419, 436–437
  modulated 214, 222
fluorescence-guided surgery (FGS) 190–192
fluorescence-guided UV therapy 197
fluorescence illuminators 44–45
fluorescence imaging, time-resolved 210, 406
fluorescence lifetime imaging (FLIM) 5, 21, 30–31, 210, 234, 243–244, 279, 301, 303–304, 326, 433

fluorescence lifetime imaging microscopy 30, 279
fluorescence lifetimes 20, 30, 66, 209, 211–213, 219, 230–231, 237, 239, 433
fluorescence localization 259, 269
fluorescence localization after photobleaching (FLAP) 259, 262, 269
fluorescence loss in photobleaching 267
fluorescence markers 21, 25, 415
fluorescence microscope 35, 92, 100, 350
  far-field 417
  Ploem 35–36, 38, 40, 42, 44, 46
fluorescence microscopy
  biological 341
  confocal 53
  conventional 357, 376
  single molecule 304
  wide-field 284, 383, 385
fluorescence nanoscopy 437
fluorescence quantum yield 140, 172, 211
fluorescence quenching 21, 176
fluorescence recovery 27–28, 259, 262–265
fluorescence recovery after photobleaching (FRAP) 5, 27–28, 156, 259, 262–263, 265, 267–270, 272, 274
fluorescence resonance energy transfer (FRET) 5, 30, 127–128, 178, 211, 213, 233, 237, 240–241, 279–283, 285, 287, 289, 291, 293–304
fluorescence spectroscopy 281
fluorescent colors 192, 201
fluorescent conjugates 155–156
fluorescent dextrans 151–153
fluorescent lipid analogs 155
fluorescent microscopy 55, 347
fluorescent molecules 25, 54, 262–263, 271, 348, 350
fluorescent pigments 44
fluorescent probes 107, 139–140, 326, 348, 385, 400
  organic 239
fluorescent probes for FRET 285, 287, 289, 291, 293
fluorescent protein technologies 117
fluorescent proteins (FPs) 117–120, 122–132, 156, 164, 166, 175, 185–186, 197, 236–240, 274, 285–288, 291–293, 351–352, 357, 359
fluorescent proteins
  general properties of 125, 127, 129, 131
  photoswitchable 274
  visible 285, 291
fluorescent ubiquitination-based cell cycle indicator (FUCCI) 201
fluorophores 19, 21–23, 26–27, 29–31, 164–165, 170–171, 210–213, 262, 269–270, 348, 385–386, 399–400, 404, 418, 436–438
formaldehyde 86, 89–90, 92, 116, 169
formaldehyde-induced fluorescence technique 108
formalin 90, 98–99
formalin fixed paraffin embedded (FFPE) 90–91
Fourier space 387–388, 391, 393–395, 397, 403–404
Fourier transform 387, 395–396
FPs, *see* fluorescent proteins
FRAP, *see* fluorescence recovery after photobleaching
FRET, *see* fluorescence resonance energy transfer

FRET microscopy, application of 297, 299, 301, 303
FUCCI, *see* fluorescent ubiquitination-based cell cycle indicator
FUCCI-expressing MKN45 cells 198–199

gated optical intensifier (GOI) 406
GBM, *see* glomerular basement membrane
GECI, *see* genetically encoded calcium indicators
genetically encoded calcium indicators (GECI) 164, 166–167
GFP, *see* green fluorescent protein
GFP gene 185–186, 190, 197
GFP-like proteins 117, 121, 126, 288–289
GFP macrophage 189
GFP-tagged protein 267
glomerular basement membrane (GBM) 94–96
glomeruli 94, 96, 115
glycosphingolipids 154, 156
GOI, *see* gated optical intensifier
Golgi apparatus 154–156
green fluorescent protein (GFP) 27–28, 117–121, 124, 126, 131–132, 185–186, 195–196, 200, 236, 239, 260–261, 265, 271, 290, 292–293

high-speed confocal microscopy 59, 61
histochemical markers 81, 83, 85, 87

iFRAP, *see* inverse FRAP
IIF, *see* indirect immunofluorescence
indirect immunofluorescence (IIF) 99–101
inverse FRAP (iFRAP) 259, 265–267, 274

LEDs, *see* light-emitting diodes
Lewis lung carcinoma (LLC) 186, 197, 201
lifetime-resolved fluorescence imaging 236
light absorption 1, 14, 19, 22, 280
light-emitting diodes (LEDs) 44–45, 223
light microscopy techniques 282, 300
light scattering 69–70, 77, 127
lipofuscins 93, 97, 106
liver metastases 188, 196–197
LLC, *see* Lewis lung carcinoma
localization-based super-resolution microscopy 376
luminescence 1–2, 13, 64, 280

macrophages 94, 151, 190, 196–197
  tumor-educated 196–197
marker proteins 288
markers, fluorescent 313
melanomas 108, 197
metastases 102, 185, 187–188, 191, 200
micrometastasis 187
microtubules 124–125, 146, 350
mitochondria 70, 124–125, 142, 147–150, 172, 175, 268, 399, 424
mitochondrial membrane 141, 147–148, 268
MLCK, *see* myosin light chain kinase

molecular orbitals (MOs) 7, 10
MOs, *see* molecular orbitals
MPPCs, *see* multi-pixel photon counter
multi-pixel photon counter (MPPCs) 323, 325, 327
multiphoton microscopy 23, 25, 29–30, 60, 63–64, 66, 68, 70, 72, 74, 76, 78, 263, 416
mutations 118, 130, 240, 289
myosin light chain kinase (MLCK) 293

nanoscopy approaches 418, 421, 435, 437
NLS, *see* nuclear localization signal
non-invasive imageable tumorgraft model 186, 196
NPCs, *see* nuclear pore complex
nuclear localization signal (NLS) 290
nuclear pore complex (NPCs) 266, 421
nucleoli 101, 267–269
nucleoplasm 101, 267

optical microscopy 3, 6, 30–31, 384–385, 397, 399, 416
optical microscopy techniques 32
optical reconstruction microscopy 32, 156, 349, 384, 435
optical sectioning 28, 58, 221, 356, 385–389, 391
optical transfer function (OTF) 387–388, 390–391, 394–396, 404
OTF, *see* optical transfer function

pancreatic cancer 192, 196
PDE, *see* pulse-detection efficiency
photo multiplier tube (PMT) 53–54, 220, 222, 225, 285, 314–318, 321, 326–327, 336
photobleaching 19, 21–22, 25, 27–28, 75, 105, 107, 164, 223–224, 265, 267, 269, 271, 285, 420
photocathode 315–316, 319
photoconvertible probes 273, 275
photodamage 63, 68, 265
photodetectors 313–315, 317, 319, 326
photodetectors for fluorescence microscopy 313–314, 316, 318, 320, 322, 324, 326, 328, 330, 332, 334, 336, 338, 340, 342
photoelectrons 315–316, 319, 322, 328, 332
photolysis 73, 75
photomultipliers 1
photon emission 280, 300
photoswitchable probes 272–273
phototoxicity 140, 420
PIN photodiodes 320–321
PMT, *see* photo multiplier tube
point spread function (PSF) 54–56, 58, 72–73, 350, 357–358, 364, 366, 370, 372, 387, 393–394, 424
Poisson noise 316, 320, 334, 337, 342
PPIX, *see* protoporphyrin
protoporphyrin (PPIX) 242–243
PSF, *see* point spread function
pulse-detection efficiency (PDE) 324–325

QE, *see* quantum efficiency
quantum efficiency (QE) 285, 314, 317–318, 331, 342, 351

quantum yield (QY) 4, 20–21, 99, 127–129, 140, 145, 172–173, 211, 281–282, 286, 292, 296
quenching resistors 323–325
QY, *see* quantum yield

reactive oxygen species (ROS) 164, 174, 242, 272
red blood cells 3, 272
red fluorescent protein (RFP) 121, 126, 131–132, 185–186, 193, 195, 200–201, 292
RFP, *see* red fluorescent protein
rhodamine 11, 13, 36, 92, 174, 284, 353
ROS, *see* reactive oxygen species

scientific complementary metal–oxide silicon sensor (sCMOS) 314, 327, 338–339, 342–344, 351, 398
sCMOS, *see* scientific complementary metal–oxide silicon sensor
SIM, *see* structured illumination microcopy
structured illumination microcopy (SIM) 326, 383, 386–389, 391–393, 398–401, 403, 405–407, 417, 437
super-resolution fluorescence imaging 385, 407
super-resolution imaging 355, 357, 359
super-resolution microscopy 32, 125, 140
super-resolution optical microscopy 383–384, 386, 388, 390, 392, 394, 396, 398, 400, 402, 404, 406

TCSPC, *see* time-correlated single-photon counting
time-correlated single-photon counting (TCSPC) 215, 222, 225, 238, 327
TIRF, *see* total internal reflection fluorescence
total internal reflection fluorescence (TIRF) 155, 163, 361–363, 405–406, 416–417
tumors 102, 185–186, 192, 196–201
two-color STED microscopy 421
two-photon excitation microscopy 64
two-photon flash photolysis 73
two-photon microscopy 78, 420

UVC irradiation 185, 197, 200

vibrational energy 11–12, 14–19

yellow FPs (YFP) 118, 125, 129–130, 177–178, 232, 239, 261, 270, 287, 290, 292–293
YFP, *see* yellow FPs

www.ingramcontent.com/pod-product-compliance
Ingram Content Group UK Ltd.
Pitfield, Milton Keynes, MK11 3LW, UK
UKHW022200190926
475702UK00003B/120
9789814774857